PARASITES AND THE BEHAVIOR OF ANIMALS

Janice Moore

OXFORD

UNIVERSITY PRESS

2002

OXFORD
UNIVERSITY PRESS

Oxford   New York
Athens   Auckland   Bangkok   Bogotá   Buenos Aires   Cape Town
Chennai   Dar es Salaam   Delhi   Florence   Hong Kong   Istanbul   Karachi
Kolkata   Kuala Lumpur   Madrid   Melbourne   Mexico City   Mumbai   Nairobi
Paris   São Paulo   Shanghai   Singapore   Taipei   Tokyo   Toronto   Warsaw

and associated companies in
Berlin   Ibadan

Published by Oxford University Press, Inc.,
198 Madison Avenue, New York, New York, 10016

Oxford is a registered trademark of Oxford University Press.

Library of Congress Cataloging-in-Publication Data
Moore, Janice.
Parasites and the behavior of animals / by Janice Moore.
    p.   cm. — (Oxford series in ecology and evolution)
Includes bibliographical references
ISBN 0-19-508441-1
1. Host-parasite relationships.   2. Animal behavior.   I. Title.   II. Series
QL757.M559   2001
591.7'857—dc21   00-063690

9 8 7 6 5 4 3 2 1

Printed in the United States of America
on acid-free paper

For my parents

Doyle Liles Moore &
Tillie Spross Moore

Preface and Acknowledgments

My early interest and delight in learning (especially about everything that moved) was encouraged by my parents, who never left a doubt that doing what we love is one of life's greatest rewards. A grateful daughter dedicates this book to those two smart and loving people.

Robert May and Dale Clayton believed I should write this book before I believed it, and I appreciate that; the task has been a remarkable experience. John Raich, the dean of the College of Natural Sciences at Colorado State University, was instrumental in providing leave time with which I began this book; I'm thankful for his effort, and for the confidence he has in me.

Since that beginning, my research associate, Mike Freehling, has deftly handled all manner of things that have escaped my attention as I've explored the nooks and crannies of this book. I appreciate Mike's intelligence, ability, and friendship. Mike has worked with me throughout my career and has played a large role in everything I've done. He also assisted with this book, and I'm especially grateful for his help with the bibliography.

Paul Harvey, Eli Holmes, William Marquardt, Robert May, Randy Moore, and Robert Poulin read the manuscript in its various stages; their comments were invaluable. William Marquardt also produced the photograph of *Trichinella spiralis*. In addition, Ben Hart and Dennis Minchella and his students read various chapters and made useful suggestions. I also appreciate feedback from William Black, Dale Clayton, Larry Curtis, Douglas Futuyma, John Holmes, Mike Huffman, Hilary Hurd, Marc Klowden, Ed Levri, Manfred Milinski, Sarah Randolph, Heinz Richner, Dan Simberloff, Joe Travis, and Claus Wedekind, who read portions of the manuscript and commented on them. At Oxford University Press, Kirk Jensen and Lisa Stallings patiently answered questions and helped the book come to fruition.

Gary Raham of Biostration provided the cover illustration, depicting elevation-seeking behavior in ants. Conery Calhoon (C. Calhoon Illustration) created the other original illustrations and shared her characteristic good humor and encouragement. (Please note that we have taken liberty with scale in some drawings.) Photographs were provided by Greta Aeby, Richard Brusca, Norm Dronen, Terry Galloway, Ben Hart, Gary Hendrickson, Joseph Hinnebusch, Peter Hudson, Rick Karban, Paul Lewis, Richard Lucius, Barbara Maynard, C. Kirk Phares, Luc Plateaux, Laurent Péru, Larry Roberts, Marie Timmerman, and Claus Wedekind. Because the Colorado State library was devastated by a recent flood, some of the material I had to reprint was not in the best shape; nonetheless, the photographic services group at Colorado State University snatched visual victory from the jaws of some fairly grim photocopies. At Texas A&M University, Angus Martin of the Cushing Memorial Library and Pixey Mosley of the Sterling Evans Library went well beyond the call of duty to help me acquire and reproduce figure 5.9. The College of Natural Sciences and the Department of Biology at Colorado State University helped finance the publication of figures 3.4, 3.6, and 3.12. Susan Conway tracked down Justice Stewart's statement. Rob Kozusko, Melanie McCall, and Kristina Parkinson assisted a great deal with library work. Felicity Huntingford, Boris Kondratieff, Marc Mangel, Robert Paine, LeRoy Poff, and Daniel Simberloff provided useful information and discussion. I am especially grateful to Marijke de Jong-Brink, who was remarkably understanding when the completion of this book interfered with my participation in a collaborative project, and to Felicity Huntingford, in whose laboratory I proofread the final copy.

While writing about parasite transmission, I became enchanted (as you will see) by the legend of the rat king. Scholars on three continents, ranging from one of my invertebrate zoology students to a folklorist, helped me sort out this story and collect photographs. They are Tracy Carlson, Norman Dronen, Gary Hendrickson, Newton Kingston, Carol Mitchell, John Pearson, Nicole Rempel, and Claus Wedekind.

Shortly after beginning this book, I wrote to many colleagues, asking them for reprints, preprints, and insights they might have. The response was generous beyond my expectation, and I was the delighted recipient of many thoughtful letters. The community of scientists is one of the attractive things about the work we do, and the goodwill and help of my peers is a great personal and professional gift. Space prevents me from listing all of you, but I am indeed grateful.

Despite all this assistance, the book nonetheless has its limits, and some caveats are in order. The emphasis of the book is on eukaryotic endoparasites— helminths and protista. These animals have inspired much of the literature, with the exception of areas such as grooming and some avoidance behaviors. Ectoparasites are covered in more detail than parasitoids or noneukaryotes, and microherbivores are not included. This taxonomic variegation reflects both irregularity in the depth of the literature and in the extent of my own interests. In addition, some peripheral areas—such as evolution of virulence, sexual selection, parasitic castration/host reproduction, population consequences/models, and social parasitism—could each be the subject of a book; the discussion of these topics is necessarily truncated.

This book mentions a multitude of animal species, some briefly and others at length. For the sake of consistency, *I have used the Latin binomials that appeared in the original literature that I cite,* regardless of subsequent taxonomic revisions. In some cases, those original papers did not offer enough information for me to apply synonyms with confidence; rather than update some names, but not others, I chose to use the original designations in all cases.

To improve readability, all of the tables have been gathered in an appendix at the end of the book. Page references are provided where necessary.

As with all attempts to impose order in the form of categories, mine is unavoidably flawed. For instance, altered appearance is not, strictly speaking, behavior, and behaviors such as grooming can be seen as either avoidance or defensive maneuvers. It is my hope that these ambiguities and, at times, arbitrary designations, do not obscure the larger story I am telling.

That story is one that begins more than two decades ago with the application of theoretical rigor to ideas about the evolution of host–parasite interactions. This fusion showed that there were certain key elements (e.g., transmission, susceptibility, survival/recovery) that were likely to affect host–parasite interactions because they affected the fitness of the participants. After a brief excursion into the lives that parasites live, I explore the idea that transmission, susceptibility, survival, and recovery can have behavioral components as well as physiological ones. Like their physiological counterparts, host behaviors help define the multitude of evolutionary pathways taken by parasites and their hosts.

What you will see is that a grasshopper is not always truly a grasshopper, doing grasshopper things, nor is an ant always an ant. The questions that emerge from this realization can lead across all of biology, from neuroscience to ecology, into epidemiology and biological control. I imagine I haven't dreamed of most of these questions—I hope you do. And I hope you have a good time reading this book.

Janice Moore

Lafayette, Colorado
Summer 2000

Contents

Parasites and the Behavior of Animals

1

Introduction

> *A truly successful parasite is commensal, living in amity with its host,*
> *or even giving it positive advantages. . . . A parasite that regularly and*
> *inevitably kills its hosts cannot survive long, in the evolutionary sense,*
> *unless it multiplies with tremendous rapidity. . . . It is not pro-survival.*
>
> —Mr. Spock, *Star Trek 2*

Here we have the stuff of science fiction: Alien beings enter hosts and twist host behavior to sinister purposes. . . . This literary device has been used by many writers as a vehicle for exploring the nature of free will and the value of independent behavior. The thought of some slimy, abhorrent glob inhabiting one's body (the aliens rarely look like Greek gods) can also be trusted to evoke a mixture of loathing, disgust, and alarm in the reader, thereby making the story that much more engrossing.

Although most science fiction diverges from our perceptions of reality, indwelling aliens may find a parallel among parasites and their hosts (Moore 1984a). Consider *Plagiorhynchus cylindraceus*, a parasitic worm, that lives in the intestines of songbirds such as starlings (fig. 1.1). This worm is an acanthocephalan, a "thorny-headed worm," a member of a small but intriguing phylum. In the acanthocephalan life cycle, similar to that of some tapeworms (chapter 2), the gutless adult lives in the vertebrate intestine and produces eggs that exit with host feces. When the egg is eaten by a suitable arthropod, it hatches, and the liberated parasite burrows into the arthropod body cavity, where it develops to an infective stage called a *cystacanth*. This enters a new vertebrate host when the arthropod intermediate host is eaten.

In the case of *P. cylindraceus*, the starling feces and the eggs they contain are eaten by terrestrial isopods, which harbor the infective cystacanths after a

Figure 1.1. The starling-isopod lifecycle of the acanthocephalan *Plagiorhynchus cylindraceus* (Moore, 1984a)

few weeks of development. (The kind of isopod that rolls up—or "conglobates," in isopod-study parlance—is commonly called a "pillbug" in the southern United States) Normally, terrestrial isopods seek areas of high humidity; unlike insects, for instance, they do not possess a waxy epicuticle and are easily dehydrated. However, infected isopods are far more likely than uninfected ones to spend time in areas of relatively low humidity (75% vs. 98% relative humidity, or RH). Moreover, they spend more time on white surfaces, where they are highly visible, than they do on black surfaces, and they are unresponsive to overhanging shelter. This combination of behavioral shifts probably means that infected pillbugs are more conspicuous to visual predators than are uninfected ones, with the result that the infected animals are eaten in larger numbers than their prevalence in bird foraging areas would predict. It seems that the cystacanth is affecting pillbug behavior in a way that increases the likelihood of parasite transmission to the next host (Moore 1983a, 1984a). Perhaps the world of science fiction is not so distant after all.

There are many fascinating angles to this story of pillbugs, worms, and starlings. Does the cystacanth somehow "waterproof" its isopod, thus enabling it to withstand drier, more exposed environments? Perhaps, but my water-loss measurements did not support this hypothesis. Can the isopods, in which infection causes risk of increased predation as well as parasitic castration (for females, at least), avoid eating parasitized bird feces? Experiments with artificially "infected" (i.e., eggs added) feces did not support that hypothesis. Shouldn't starlings avoid eating infected isopods? Anecdotal accounts notwithstanding, the damage caused by *P. cylindraceus* to the starling is usually minor (Moore 1983a, 1984a; Moore and Bell 1983; Connors and Nickol 1991). Avoidance may not pay, given the minimal cost of infection and the ease of encountering infected isopods (chapter 4).

This system, however, illustrates at least two conditions that are shared by many host–parasite associations and that, far from being mysterious, illuminate the selective pressures and evolutionary constraints that can result in such a scenario. First, despite the maleficent reputation of parasites worldwide, the starling suffers little harm. There is evidence of some metabolic cost (Connors and Nickol 1991), but tissue damage is minimal, and infected birds in the wild exhibit weights comparable to those of uninfected birds (Moore and Bell 1983). Second, consequences for the pillbug are more severe. Both males and females begin to behave in dangerously abnormal ways, and females fail to develop ovaries.

Why is the parasite not equally benign (or pathogenic) in both hosts? Is it somehow less well adapted to the pillbug? What does a consideration of parasite fitness tell us about such different effects on two consecutive hosts?

One approach to answering these questions can be found in the models developed and elaborated by Anderson and May from the 1970s onward (e.g., Anderson and May 1978, 1979, 1982, 1986, 1992; May and Anderson 1978, 1979, 1983a,b, 1990). Beginning with first principles, Anderson and May focused on reproductive rate (R_0)—or "ratio," to be precise—as a measure of parasite fitness. R_0 is the number of reproductive offspring that a parasite can produce and is analogous to Fisher's intrinsic rate of increase (r in population dynamic models). For the purposes of their models, May and Anderson deliberately set about to dichotomize parasites into two groups: (1) microparasites (e.g., many protists, viruses, fungi, bacteria), which are small and have a short life span, and which reproduce directly within the host, usually inducing immunity, and (2) macroparasites (e.g., helminths, arthropods), which have larger bodies and longer life spans, and which disperse offspring from a host that will probably not remain immune and is therefore at risk of reinfection.

This bisection is an intentional oversimplification, but one that is nonetheless profitable. May and Anderson recognized that there are many variations on these themes, and that the dichotomy is, in reality, a continuum. The subdivision, although necessarily imprecise, generally reflects biological reality. In addition, it has utility in the discussion of the population biology and evolution of parasites. For instance, in the case of microparasites, the host population can be partitioned discretely—infected, immune, susceptible, and so on. The uneven distribution of macroparasites among hosts, on the other hand, leads to a more complicated situation, with hosts not so easily categorized.

Ultimately, these distinctions between parasites can influence our view of reproductive rate. Anderson and May noted that for macroparasites, the definition of R_0 is similar to that for free-living animals—the average number of female offspring that survive to reproduce. For microparasites, the definition must be modified, so that R_0 is the number of secondary infections that result from a primary one. In both cases, $R_0 = 1$ under equilibrial conditions; in the absence of other constraints, natural selection favors individual parasites that maximize R_0. In turn, R_0 can be influenced by a variety of host–parasite attributes.

What influences R_0? For microparasites, the potential number of secondary infections (R_0) is strongly influenced by the number of contacts with susceptible individuals. If previous exposure to the parasite confers immunity, then for any given population size, the effective value of R_0 is density dependent; the greater the proportion of infected or immune hosts in the population, the lower the likelihood of contacting a susceptible individual. For macroparasites, R_0 reflects the number of successful female offspring and can also be density dependent. Although immunity does not usually play as important a role in most cases of macroparasitic infection as it does in microparasitic infections, it is not without influence; in addition, macroparasites in heavily infected hosts may experience reduced fecundity or even host (and parasite) death. Thus, for macroparasites, R_0 is also density dependent.

The potential linkage of transmission or recovery rate to a parasite's impact on its host gives us insight into the ways that virulence (or lack of virulence) might evolve. For microparasites, R_0 can be expressed as the ratio of the production rate of successfully transmitted propagules to the summed death rate and recovery rate of the host, or

$$R_0 = y(N)/(a + b + v),$$

where y is transmission, N is host population density, a is host death rate from disease (virulence), b is host death rate from all but virulence, and v is recovery rate (May and Anderson 1990). Thus, R_0 increases as a decreases when virulence, transmission, and recovery rate are independent. What this basic model says is that if the virulence of a parasite is not related to either its transmission or to recovery rate, then that parasite should evolve toward a completely harmless state, thus maximizing its reproductive rate (May 1985). In the past, the last clause was part of medical/parasitological dogma: parasites evolve to live peacefully with hosts. If virulence *is* related to transmission or recovery rate, however (and this is not uncommon, although the precise details of the relationship may be elusive), then the nature of the relationship among those variables also influences the virulence of the parasite. In such a case, the formula for reproductive rate might be modified to

$$R_0 = y(a, N)/[a + b + v(a)]$$

(May and Anderson 1983a). For instance, if virulence enhances transmission, then virulence will enhance R_0 and be favored by natural selection. The important lesson that emerges here is one of "theoretical pluralism" (May and Anderson 1990, p. S91): Virulence in a host–parasite association depends to a great extent on the life history of the parasite—the mode of transmission, the way in which it uses host resources, the longevity, and reproductive schedules of host and parasite. To the extent that these vary among parasites, so will the evolution of virulence.

Macroparasites present a more complicated picture, but similar factors are influential. A few examples of complications are reinfection of hosts, location of mates (in the case of dioecious parasites), and great variation in intensities among hosts, with few hosts frequently harboring the majority of the parasites. (The tendency for most parasites to aggregate in a few hosts has spawned a rich literature examining the sometimes peculiar evolutionary and ecological consequences of such a skewed distribution for both parasites and hosts.) Anderson and May's (1992) macroparasite model, though more complex than the microparasite model shown above, nonetheless supports the idea that parasite reproductive rate, parasite life span, and transmission rate to the next host are critical influences on R_0.

The insights gained from May and Anderson's models can be applied to a wide variety of questions, ranging from vaccination efficacy to the evolution of virulence, from the role of introduced parasites in conservation biology to their role in biological control. The extent to which the models are applied successfully depends in large part on our understanding of the biology of the host–parasite association in question. The diversity of host–parasite interactions is the source of the vast array of "coevolutionary trajectories" (May and Anderson 1990, p. S91) that result in some associations being lethal, others being relatively benign, and others yielding a range of intermediate outcomes. The fact that such evolution can occur in a fraction of a human lifetime (e.g., the myxoma virus in European rabbits introduced into Australia; see Anderson and May 1982, May and Anderson, 1983a,b) is a powerful testament to the force of natural selection in a host–parasite system, when provided with variations in (and linkages among) virulence, transmission rate, and recovery rate.

In fact, the insights gained from Anderson and May's models can help explain the disparate impact of *P. cylindraceus* on its hosts. The outcomes of infection for the starling and the pillbug become more understandable as we consider the effect of those outcomes on parasite R_0. In the starling, the parasite mates and produces eggs that are distributed around the environment with feces, where they await ingestion by an isopod. Thus, successful transmission in the egg stage depends in large part upon a host that can be active during the parasite's reproductive life span. In that case, if virulence is low, R_0 will be increased. For the parasite in the pillbug, however, successful transmission depends on ingestion of the pillbug; compared to an infected pillbug, an uninfected pillbug is liable to escape notice. In such a case, transmission (and R_0) can increase as virulence increases. Therefore, if parasite R_0 is to be maximized, we expect the infected starling to remain relatively healthy and the infected pillbug to be in trouble.

The models discussed above have been developed and applied to situations where virulence and susceptibility are seen as reflections of host physiology. If one reviews the pillbug–starling example, however, it becomes obvious that the effects of parasitism extend beyond sickness and immunity. In fact, host behavioral alterations in response to parasitism can influence every parameter that in turn has an impact on R_0. The most obvious example is a behavioral change that leads to increased transmission (e.g., Dobson 1988). Some of these changes can be caused by gross pathology; invasions of the central nervous system are good examples. Intermediate hosts with nervous system lesions are probably easy prey. Other changes are more subtle. Although any deviation from normal might be said to be pathological, changes such as altered substrate preferences or hyperactivity are not routinely associated with an ordinary concept of illness. Pathology or not, such change will have the functional effect of increasing encounters between infected animals and susceptibles (y in the Basic Model). The actual number of infected and susceptible animals in the population need not change; if their distribution and subsequent interactions change, this will certainly affect the parasite's reproductive rate (chapter 3).

Finally, estimating the level of contacts between infected and susceptible animals becomes even more complicated when host avoidance behavior is taken into account (chapter 4). In the presence of parasites, many animals may shift habitats or engage in other avoidance behaviors, thus reducing contact between parasites and potential hosts. This is the opposite of the outcome seen in cases where altered behavior results in increased transmission—contacts between infected sources and susceptible animals are fewer than what might be predicted from their occurrence in the population. Nonetheless, in mediating the relative number of these contacts, such behavior influences parasite transmission and reproductive rates (chapter 4).

Although parasites may enhance their own life spans by reducing risky components of host behavior, other types of alterations in host behavior may diminish parasite fecundity or longevity. For ectotherms that behaviorally regulate their temperature, changes in activity or microhabitat choice can induce the equivalent of fever or chills, which may subsequently affect parasite development and even parasite survival. Other parasitized hosts may shift diet choices to medicinal items. In these situations, host behavior that maximizes host reproductive rate is being favored, affecting v in the Basic Model (chapter 5). The fact that such behavior can reduce parasite reproductive rate is a coincidental outcome of host behavior that seeks to ameliorate the damage associated with parasitism—damage that may be linked to parasite reproductive rate.

Less obvious are behavioral changes that may lead to increased longevity for the host and the parasite it contains and, in some cases, disperses. Many workers have suggested that parasitic castration and subsequent shifts in host resource allocation can lead to a more stable and rich environment for the castrating parasite. It is thought that a castrated host will not only shift energetic resources away from reproduction, but will also cease to engage in potentially risky courtship and mating behavior. Experimental tests of this hypothesis are rare; nonetheless, if parasites do enhance longevity of the host that disperses them by eliminating some risky behaviors, then such behavioral changes can directly affect parasite transmission and reproductive rate over time, as they affect a in the Basic Model.

Thus, host behavior can influence both parasite transmission rates and parasite/host survival. Some of the ways in which this can happen are the subject of this book. For the majority of host–parasite associations, we are not yet able to predict the nature of the influence, the way in which transmission and/or survival may be changed, without much investigation. To the extent that virulence is linked to parasite transmission, the evolutionary interests of parasite and host will diverge, and the current winner of the contest to maximize reproductive rates may not be clear or, for that matter, inevitable. In that respect, the "pluralistic" outcomes of the evolution of virulence are echoed in the possible outcomes of alterations in host behavior. As with the evolution of virulence, our ability to predict these alterations is severely constrained by our ignorance of trade-offs among virulence, reproduction, and transmission, among other things. Nonetheless, by affecting susceptibility, host–parasite life span, and transmission rate, host behavior affects parameters that are basic to our comprehension of how parasites invade host populations and, fundamentally, how parasites evolve.

2

Life Cycles: Blueprints for R_0

If parasite transmission, avoidance, or resistance are processes that greatly influence R_0, then parasite life cycles are the blueprints of those processes, the keys to understanding when and how transmission happens, when and how it can be avoided. Because the purpose of this chapter is to enhance appreciation of parasitological phenomena that subtend behavioral aspects of the Basic Model, it is organized according to modes of transmission.

In this chapter, I introduce some major routes of parasite–host encounters, especially those of metazoan animal parasites, along with a bit of terminology. For further details, please consult any of a number of parasitology texts and reviews (e.g., Prescott et al. 1990; Brooks and McLennan 1991, 1993; Marquardt et al. 2000; Roberts and Janovy, 2000). This introduction is not exhaustive; it is minimal and is meant to be auxiliary to this book. Where not specifically attributed, most of the information in this chapter was taken from Olsen (1974) and Roberts and Janovy (2000).

The first term I will dispatch immediately, if without much satisfaction—the word *parasite* itself. Although an abundance of fine scientists have had a try at this definition, some chagrin almost always flavors these attempts—the definition may be an improvement over previous exercises, but there remains a niggling feeling that something is still not quite right. This may be an occupational hazard of those who would define "parasite" because, although the word has come to be associated with a win–lose situation for the parasite and host, respectively, the parasite–host interaction itself is not static. Within a specific life cycle, the harm caused by the parasite may in many cases vary with the host or with the time since infection. Within a given species pair, the outcome may be density dependent or may hinge on a variety of other influences. Across host–parasite associations, the effect may range from inevitable mortality to one that is almost unmeasurable. In fact, in some situations, parasites may even be beneficial. This ambiguity mirrors Anderson and May's arguments (1992, and references therein) for the multiple evolutionary outcomes associated with their Basic Model.

Do we then define parasites by what they eat? By where they live? These attempts have also proven to be inexact. Moreover, in the evolutionary free-for-all that seems to have characterized the way many animals acquire resources, at least some representatives of almost every major taxon have seized upon parasitic methods to obtain those resources (Cheng 1986). As Brooks and McLennan (1993) point out, parasites are not a monophyletic group and thus lack the unifying traits enjoyed by groups that share a common ancestor. Parasites may therefore be condemned to perpetual ambiguity—in some ways, an appealing mystique.

Read (1970) argued for a different terminology. He suggested that *symbiosis* was the best word for these interactions—a word that simply means "living together," and carries no predictive baggage. Because the language we use can strongly and at times imperceptibly bias the way we think about things, this neutrality has value, and Read's arguments for resurrecting this term and an integrative approach were intelligent and forceful. Unfortunately, Read's attempts were no match for tradition, and this book is about "parasites." I will stop short of defining them, however. Like pornography, parasites elude definition, and to paraphrase Justice Stewart, may be known primarily when we see them.[1]

Other terms are more manageable. Parasites mature sexually in their *definitive*, or *final*, hosts; if sexual maturity is not so easily pinpointed (e.g., amoebae), the definitive host is, by convention, the one of greatest importance to humans. Parasites in *intermediate* hosts undergo development and perhaps multiplication, but they do not reach sexual maturity. They undergo no development in *paratenic* or *transport* hosts; these hosts may nonetheless be important if they provide ecological links between definitive and intermediate hosts that otherwise experience little overlap. In a *direct* life cycle, there is no intermediate host; the parasite goes from one definitive host to another. An *indirect* life cycle has at least one intermediate host. A *vector* is the vehicle of transmission from one host to the next; in a strict sense, it may itself be a host, such as a mosquito, or it may be an agent such as water or wind (Roberts and Janovy 2000).

I will keep terminology about developmental stages to a minimum and will concentrate on the helminths—specifically, the flatworms and acanthocephalans. Other parasitic groups have complex life cycles, but these either do not have terms associated with them that require introduction, or, if they do, those stages are rarely mentioned in this book.

Of the major groups, the parasitic flatworms (phylum Platyhelminthes) deserve special mention because of the developmental changes they undergo. Adult digenetic trematodes often have the flat, leaflike look of their free-living relatives. Their eggs either hatch into ciliated larvae called *miracidia* that penetrate the first intermediate host (almost always a benthic mollusc), or unhatched, await ingestion by a mollusc (fig. 2.1). In the mollusc, they typically proceed through one or more asexual generations of each of up to two forms: the *sporocyst* and the *redia*. These forms are, roughly speaking, sacs that contain asexually produced offspring, which may themselves become such sacs (more sporocysts or rediae) or which may become the next transmission stage (*cercariae*). Rediae are produced by sporocysts and have suckers and a small gut, thus laying claim to more recognizable structure than the sporocyst. [Rediae can also use these structures to harm and even ingest potential competitors (Lie et al. 1965; Lie 1966).] Not all life cycles include rediae, however, whereas the sporocyst is virtually ubiquitous.

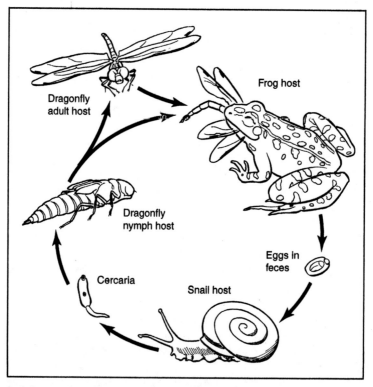

Figure 2.1. A sample trematode life cycle (*Haematoloechus medioplexus*). The adult lives in the lungs of the frog. Eggs are swallowed, then eliminated with feces. A planorbid snail eats the egg, which then hatches. The resulting miracidium migrates to the digestive gland of the snail, where it undergoes asexual reproduction as a sporocyst and produces cercariae. (In many other aquatic trematode cycles, the miracidium hatches in water and penetrates the molluscan host.) These cercariae leave the snail, enter the branchial basket of a dragonfly nymph and encyst as metacercariae in the tissues. When the nymph (or adult) is eaten by a frog, the life cycle is completed. (Drawing by Conery Calhoon.)

If asexual reproduction can be orgiastic, then the end result of this orgy of reproduction is a cercaria—or rather, up to a million or so cercariae from any one initial miracidium. The cercaria is a minuscule, juvenile digenean with a tail; it uses its limited energy reserves to find the second intermediate host, which it usually penetrates before encysting as a *metacercaria*. In this encysted stage, it awaits ingestion by the definitive host, usually a vertebrate. The variations on this entire pattern, from miracidium to adult worm, defy description (Shoop 1988; fig. 2.2).

Eucestodes (tapeworms) live as adults in the vertebrate intestine. They can rival the trematodes in seemingly extravagant reproduction, but they usually do so in the adult stage, which consists of a long, thin ribbon of many sets of reproductive organs, end to end, each set contained in its own segmentlike *proglottid* (fig. 2.3). Like most flatworms, eucestodes are usually hermaphrodites; each proglottid is protandrous, with the male reproductive system developing first, often persisting during the development of the female system. This ribbon of proglottids contains numerous testes, ovaries, and associated paraphernalia and is attached to a holdfast (scolex) that helps anchor the worm in the intestine.

The reproductive organs eventually degenerate, often leaving behind a uterus full of eggs that occupies most of the proglottid. At this point, the gravid proglottid releases the eggs and/or the proglottid itself sloughs off, and the tapeworm eggs confront the outside world the way their parent(s) did, from the dubious vantage point of a deposit of host feces. The proglottid may be capable of crawling away from the feces, thus distributing eggs more widely, and increasing the likelihood of exposure to more hosts, which need not be coprophagous (Mackiewicz 1988). The egg (or in some cases, hatched larva)

Trematode Transmission Patterns

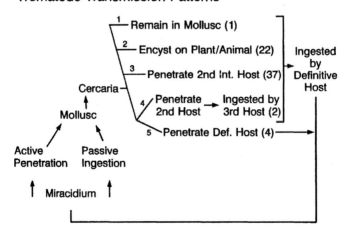

Figure 2.2. Variation in trematode life cycles. (N) = approximate number of families exhibiting the transmission pattern. (Shoop 1988; Reprinted with permission from *Journal of Parasitology*.)

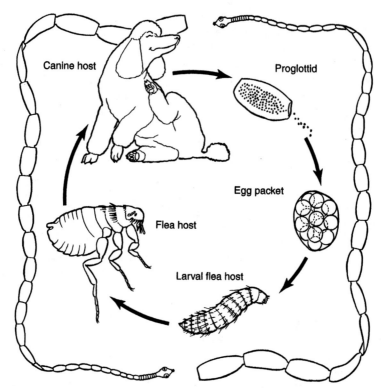

Figure 2.3.   A cestode life cycle. This cestode, *Dipylidium caninum*, lives in dogs (and some other mammals). Proglottids pass out with feces and liberate groups of eggs that may be eaten by larval fleas. When eaten, the egg hatches and the parasite invades the flea hemocoel. When the dog eats an infected flea, the life cycle is completed. (Drawing by Conery Calhoon.)

gets eaten by the intermediate host (there may be a succession of intermediate hosts), develops in any of a number of tissues, depending on species, and awaits ingestion by the next host. The names for the stages in intermediate hosts vary with taxa of cestodes, but include *procercoid, plerocercoid, cysticercoid,* and the more inclusive *metacestode.* Direct life cycles are extremely rare (Mackiewicz 1988).

The phylum Acanthocephala also deserves special mention because, although it is a minor group (~1000 species), it is a major player among parasites that induce behavioral alterations in hosts. Acanthocephalans were introduced in chapter 1; *P. cylindraceus,* the worm that alters isopod behavior and is thus transmitted to foraging starlings, is an acanthocephalan.

The behavioral changes that occur during acanthocephalan parasitism are largely characterized by altered responses to environmental stimuli on the part of the intermediate host. As Holmes and Bethel (1972) pointed out, this is a special category of behavioral alterations that are not easily attributable to gen-

eral illness or destruction of nervous tissue. Intermediate hosts of acanthocephalans often differ from uninfected counterparts in their responses to visual, chemical, or hygric stimuli. These responses are nonetheless vigorous
and frequently precise; they give no evidence of general malaise. The ability
to alter host behavior has been demonstrated for every acanthocephalan species
that has been investigated.

Parasites may either actively seek and enter their hosts or passively await
entry. Most arthropods, for instance, seek their hosts, whereas most protists,
viruses, and bacteria enter passively, often with the service of some vector,
such as water, food or a blood-feeding arthropod. Host seeking is a complex
activity, involving visual, chemical, and even auditory cues. Passive entry is
not without its charm, however, often because of the circuitous routes or host
manipulation involved.

Two flagellated protists,[2] *Giardia intestinalis* and *Histomonas meleagridis*,
offer examples of the life cycle diversity that can be subsumed under the head-
ing "passive." The former organism was discovered by Anton van Leeuwenhoek. It is the causative agent of "backpacker's diarrhea," a reference to the
fact that it can be acquired by drinking from cold, clear streams, where it is
deposited by any of numerous host species. The number and location of nu-
clei and flagella give the organism the appearance of a face (perhaps the only
protist that could encourage anthropomorphism), and we can only hope that
the pleasure of discovery mitigated whatever intestinal discomfort van
Leeuwenhoek might have been experiencing (fig. 2.4). *Giardia* has a direct
life cycle with passive host entry. Cysts leave the host in feces; when swallowed by the next host, they excyst in the small intestine and begin to divide.

Of course, passive host entry does not mean that there are no adaptive aspects to the transmission sequence. For instance, the diarrhea that often accompanies *Giardia* infection can be seen as a host attempt to expel the parasite, or a dissemination that benefits the protist. *Passive* as used here does not
mean nonadaptive; it means that the parasite is not actively seeking and entering the host.

For that matter, passive entry does not have to be associated with a straightforward life cycle. Unlike *Giardia*, *H. meleagridis* cannot survive in the external environment for long, in part because it does not form cysts. Instead,
Heterakis gallinarum, a nematode parasite of gallinaceous birds such as
turkeys that shares the ceca of the bird with the protist, eats *H. meleagridis*,
which proceeds to divide in the nematode and invade the nematode ovary. *Histomonas meleagridis* infects the juvenile nematode within the egg and exits
the turkey encased in the environmentally resistant nematode egg, which awaits
ingestion by another turkey—or an earthworm. Earthworms can serve as
paratenic hosts for *H. gallinarum*, and, if the nematodes are infected, for the
H. meleagridis they contain.

The life cycle of *H. meleagridis* is labyrinthine, even by parasitological
standards. It is included here as a contrast to that of *Giardia*, as a demonstration that passively transmitted parasites—in this case, both flagellated pro-

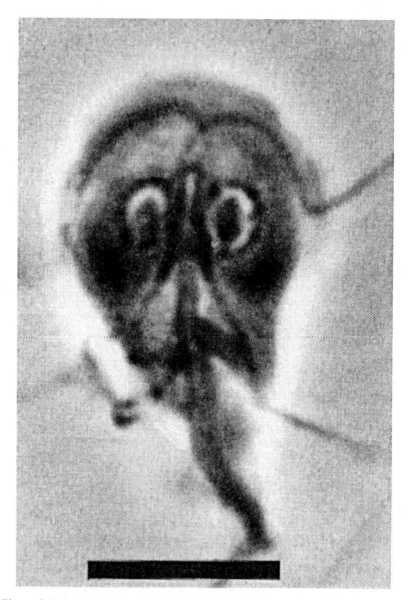

Figure 2.4.   Light micrograph of *Giardia* trophozoite. Bar = 5 μm. (Reprinted from D. E. Feely et al., "The biology of *Giardia*," in *Giardiasis*, E. A. Meyer, ed., with permission of D. E. Feely and Elsevier Press.)

tists, albeit in different orders—can differ greatly in life cycles and in the challenges they confront in transmission.

The remainder of the life cycles I describe here represent five major patterns: direct and indirect life cycles with passive transmission, direct and indirect life cycles with active transmission, and life cycles that incorporate living vectors. In choosing these parasites, I have indulged idiosyncrasy and chosen examples because they overlap other areas of biological interest, because they bear intriguing tales, or because they are downright bizarre. My purpose is not to teach a parasitology course, but to crack open a door leading to the mundane and the wildly unexpected variety of parasite transmission patterns that can influence the behavioral responses of hosts.

DIRECT LIFE CYCLE, PASSIVE TRANSMISSION

The simplicity of the *Giardia* life cycle—direct, with passive transmission— is shared by many other intestinal parasites. For instance, the large roundworm of humans, *Ascaris lumbricoides*, is almost identical to its congener in pigs, *A. suum*, and their divergence may have occurred after the time when humans and their livestock shared table and dwelling on a daily basis. These nematodes live in the small intestine, and eggs are released with host feces (fig. 2.5). After the embryos within the eggs develop, they hatch in the duodenum

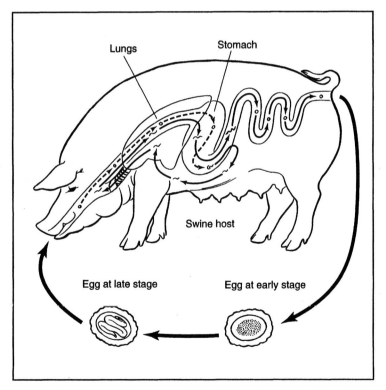

Figure 2.5. *Ascaris suum* lives as an adult in the intestine of pigs. Eggs that are highly resistant to environmental insult pass out with pig feces. After a developmental period, if the egg is eaten by a pig, it hatches in the small intestine, and the worm undergoes a journey through the tissues that eventually takes it to the lungs. After being coughed up and swallowed, the worm takes up residence in the small intestine and reproduces. (Drawing by Conery Calhoon.)

upon being eaten and begin a tissue migration that includes the liver, the heart, and the lungs. They are then swallowed (again!) and mature in the small intestine, where they began their journey.

Why such an indirect route? Many parasites undergo tissue migrations that begin and end in the intestine. The evolutionary impetus for this migration is not well understood. Several workers have suggested that in the case of ascarids, the tissue migration is a relic of a previous complex life cycle, something that occurred in the now-absent intermediate host (Sukhdeo et al. 1997, and references therein). Within another group of nematodes (order Strongylida), phylogeny appears to have influenced the nature of tissue migration. Sukhdeo and co-workers (1997) found that the order contained two clades, but in only one of these clades did juveniles migrate through the host tissues. They suggested that both clades originated as skin-penetrating forms, with subsequent tissue migration. Both adopted oral infection routes, and tissue migration was retained in only one of the two groups. In contrast, the early divergence of strongylids and ascarids (500 mya) increases the probability that tissue migration evolved independently in each of those lineages.

A comparative analysis of numerous parasitic nematode sister taxa differing in the presence/absence of tissue migration revealed that the tissue migrators were bigger than their nonmigrating relatives (Read and Skorping 1995). The tissue migrators did not take longer to reach the greater size, and because female nematode size is often linked to fecundity, this could be an advantage for migrators. Many questions remain about this phenomenon, but Read and Skorping's approach is a step toward applying evolutionary tools to an intriguing parasite life-history trait.

How parasites reach their preferred sites has puzzled parasitologists (Bailey 1982). How can a trematode, for instance, find its way to a certain section of the host's body? Sukhdeo and Sukhdeo (1994) have proposed a straightforward explanation based on the fact that once a parasite enters the host's body, it encounters a fairly predictable landscape. The common liver fluke, *Fasciola hepatica*, is a parasite with an indirect life cycle, but illustrates the explanation well. The encysted metacercaria, once ingested by a definitive host, performs a specific set of activities in response to intestinal chemicals that results in penetration of the small intestine and entry into the abdominal cavity. At this point, migrating along the body wall invariably results in contact with the liver, which *F. hepatica* can recognize. The body wall is unavoidable once the worm leaves the intestine, and the liver is unavoidable thereafter. Certain fixed motor behaviors occur at each stage that increase the probability of arriving at the next location. Such an explanation, based on a kind of minimalist response to a predictable environment, is appealing in its parsimony (see also Haseeb and Fried 1988; Bansemir and Sukhdeo 1996; Sukhdeo and Bansemir 1996).

Returning to direct life cycles, pinworms also exhibit passive entry. The eggs of the human pinworm, *Enterobius vermicularis*, are deposited on the perianal skin. Whereas *Ascaris* spp. eggs are remarkably hardy, embryonating successfully in a variety of wretched environments including most of the reagents in a student chemistry laboratory, *E. vermicularis* eggs are more delicate and desiccate quickly. They are easily airborne, however, and can be inhaled, and then swallowed. When I first learned of this phenomenon as an undergraduate, I found it surprisingly disturbing. I had grown up with the knowledge that I might acquire all manner of viruses and bacteria simply by breathing; I was, however, unprepared for the idea of aerial attack by worms. [As a last blow to complacency, it happens that *Ascaris* spp. eggs have been found on circulating German banknotes (Schmidt and Roberts 1989, citing Dold and Themme 1949).]

In contrast to nematodes that are transmitted when the egg is eaten, some nematode eggs hatch before ingestion, and the hatched larvae are eaten as part of a direct life cycle. *Trichostrongylus tenuis* is in this group and is of special interest because of its role in red grouse population cycles. The infection was present on moors where grouse cycled, and absent where they did not. Such population effects could be caused by host mortality or by reduced host fecundity (Hudson and Dobson 1991; Dobson and Hudson 1992). Eggs exit the grouse with feces and the larvae live in the soil; after undergoing two molts, they crawl up on vegetation, where they become dormant, awaiting ingestion (fig. 2.6).

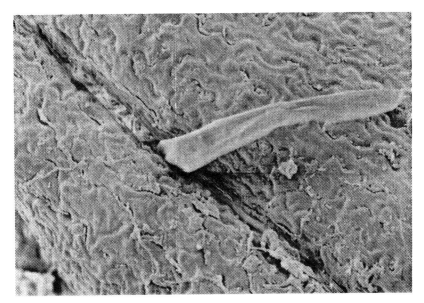

Figure 2.6.   An electron micrograph of a third-stage *Trichostrongylus tenuis* crawling into a curled heather leaflet. The curled leaflet protects the nematode and provides a vehicle for its entry into heather-eating red grouse. (Photo courtesy of Peter Hudson.)

Strongyloides spp. have a similar direct life cycle, but add a twist: parasitic females are parthenogenetic (there are no parasitic males) and produce eggs that develop into larvae that may infect a new host when the larvae are eaten or when they penetrate the host. Yet other larvae develop into free-living males and females that reproduce in the soil. Autoinfection (infection of the same host without leaving) is also a possibility for some species in this versatile genus.

This parthenogenesis and sexuality in *Strongyloides* spp. do not occur haphazardly; a variety of factors are influential. For instance, hosts that have mounted an immune response to *S. ratti* yield nematode larvae that are more likely to become sexual adults than are larvae from immunosuppressed hosts—intriguing experimental support for the Red Queen hypothesis, and the adaptive significance of genetically diverse offspring in changing or adverse environments (Gemmill et al. 1997).

Many bacteria and viruses depend on direct or aerial contact to reach new hosts. The causative agents of leprosy, chickenpox, measles, mumps, tuberculosis, tetanus, diphtheria, and anthrax are among these, as is the fungus that causes histoplasmosis. Others, such as *Salmonella, Shigella, Clostridium botulinum*, and *Vibrio cholerae* are more likely to be ingested (Murray et al. 1994).

INDIRECT LIFE CYCLE, PASSIVE TRANSMISSION

The life cycles I have described so far have been direct life cycles in which transmission could be passive. Passive transmission is also a hallmark of many indirect life cycles. One common example involves the ingestion of an intermediate host by the next host. Typically, an intermediate host becomes infected by an immature parasite, which then develops in that host to a stage that is infective for a subsequent host when consumed. For instance, members of the phylum Acanthocephala (thorny-headed worms) take up residence in vertebrate intestines when their arthropod intermediate host is eaten (chapter 1). Likewise, virtually all tapeworms reach their final hosts when those hosts eat infected intermediate hosts.

Dipylidium caninum, the common tapeworm of dogs and cats, depends on another parasite as its intermediate host (fig. 2.3). Larval fleas eat the eggs; after metamorphosis, the parasitized flea, now feeding on the mammalian host, may itself be eaten in the process of grooming. The host loses a flea, but gains a tapeworm. I have often wondered if infected fleas are as agile as uninfected ones, if they are slower at escaping the rough tongue of a cat or the snapping teeth of a dog. (By the way, *D. caninum* is among those tapeworms that has detachable, mobile proglottids.)

Many tapeworms of humans reach us when we eat meat that has not been thoroughly cooked. Nelson (1990) credits the German parasitologist Kuchenmeister with this discovery when Kuchenmeister fed cysts from a pig (prepared, we are told, in most delicious fashion in soup and black pudding) to a condemned convict. After the guillotine, autopsy revealed immature *Taenia solium*, the pork tapeworm of humans.

Diphyllobothrium latum found fame (by cestode standards) in Desowitz's

(1981) book, *New Guinea Tapeworms and Jewish Grandmothers: Tales of Parasites and People*. Eggs exit with final host feces and, once in water, hatch into ciliated larve that are ingested by copepods. The immature worm makes its way to the hemocoel of the copepod, where it grows and awaits ingestion by a fish. In the fish, it penetrates the muscles, growing some more before taking up residence in the intestines of a variety of fish-eating mammals, including humans. Infectivity depends on insufficient cooking. Jewish grandmothers were particularly at risk of acquiring the infection because they tasted gefilte fish before it was thoroughly cooked. The worm has a strong affinity for vitamin B_{12} and can cause a pernicious anemia similar to that which results from failure to absorb the vitamin.

An indirectly, passively transmitted nematode, *Dracunculus medinensis*, may have provided the medical profession with a symbol and Moses with a miracle of sorts. The nematode is acquired when humans ingest infected copepods along with drinking water. Simply filtering drinking water can eliminate much of the risk of this disease, which has been known since antiquity.

The intriguing aspect of *D. medinensis* is not how it gets into humans, but how it gets out. For once, this has nothing to do with feces, and little to do with the intestine. After release from the ingested copepod, the juvenile nematodes penetrate the small intestine and undergo a complex journey through the tissues. They mate, and approximately 8–10 months after ingestion, the gravid female migrates to subcutaneous areas, often those of the lower extremities, which are most likely to be submerged in water during bathing, water collection, or wading. The ovoviviparous worm's uterus and body wall begin to degenerate, and the host response to the liberated juveniles causes a blister to form, through which the juveniles can exit. Cool water can speed the process, causing muscular contractions of the worm's body wall that expel hundreds of thousands of juveniles, ready to be eaten by copepods. Ironically, the arid areas of the Middle East and Africa favor the transmission of this waterborne disease, and have done so for thousands of years, with people repeatedly visiting the few predictable sources of water and wading into the pools.

Over the centuries, a common form of treatment for this ailment has been to pull the worm out by winding it on a stick, a process that takes many days (fig. 2.7). Moses and his people may have encountered *D. medinensis* on their travels: "And the Lord sent fiery serpents among the people, and they bit the people; and much people of Israel died. . . . And the Lord said unto Moses, Make thee a fiery serpent, and set it upon a pole: and it shall come to pass, that every one that is bitten, when he looketh upon it, shall live" (Numbers 21: 6,8; King James Version). The Israelites were in an endemic area and had probably been exposed to the parasite for some time. The "fiery serpent" may reflect the burning sensation accompanying the blister (making cool water all the more desirable), and the pole is consistent with the stick cure that is used in many cases even today. In addition, Apollo, the god of medicine, is often shown carrying a caduceus with serpent (until he traded it to Hermes for a lyre); that staff, probably another representation of *D. medinensis*, remains the symbol of medicine.

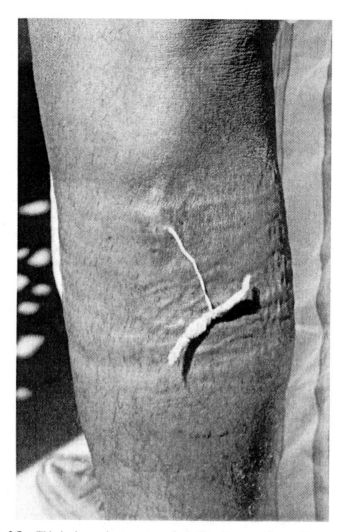

Figure 2.7. This is the ancient process of winding *Dracunculus medinensis* onto a stick, slowly removing it from its subcutaneous location. (Reprinted from Ralph Müller, "*Dracunculus* and Dracunculiasis," *Advances in Parasitology*, vol. 9, p. 123, 1971, with permission of Academic Press.)

Other nematodes, those of the genus *Trichinella*, never leave the body of the host, which serves in both definitive and intermediate capacities. This worm was found by a London medical student dissecting a cadaver in 1835, and the life cycle discovered in 1860, when a German housemaid had the bad luck to consume an apparently undercooked Christmas dish (Nelson 1990). When juveniles are ingested, they rapidly mature to adults, enter the intestinal mucosal epithelium, and mate. The resulting juveniles migrate through the circulatory system and eventually penetrate skeletal muscle fibers, where they await ingestion by the next host (fig. 2.8). A variety of mammals can be infected—bears, foxes, cats, walruses, pigs, rats, dogs, badgers, humans—(the only prerequisite is a bite of poorly cooked muscle), and the worms have been known to survive freezing for months. Dinners of polar bear have proven fatal to more than one Arctic explorer.

Depending on the degree of host resistance or the likelihood of subsequent exposure, such *Trichinella* populations may be fairly genetically isolated, transmitted in related groups from host to host (La Rosa et al. 1992; Wakelin 1993). The implications of this for sex ratios, for the influence of kin selection, and for speciation are tantalizing and mostly unexplored. In general, the peculiar population structure of many parasites offers a wide range of intriguing genetic questions that scientists have only begun to address (Price 1980; Nadler 1995).

Several nematodes have found yet another mode of passive transmission that does not involve leaving the host—transmammary transmission—in which they move from mothers to offspring during lactation (Olsen and Lyons 1965; Stone and Smith 1973; Miller 1981). *Uncinaria lucasi* lives in the lower small intestine of northern fur seal pups (*Callorhinus ursinus*). These parasites are acquired with milk when the pup nurses and they then go directly to the intestine. There they reproduce, and eggs are shed with the feces, but the intestinal infection does not persist in pups more than 5 months old. These eggs take most of the summer to hatch and develop in the soil of the rookery, and the larvae penetrate the returning seals the following year. The larvae undergo a tissue migration that distributes them throughout the adult seal's body, especially in blubber and mammary glands, where they await the next pup.

Some trematodes and even cestodes have also added the transmammary route to their repertoire (Conn and Etges 1983; Shoop and Corkum 1984, 1987; Shoop 1994). Transmammary transmission may be suspected of many para-

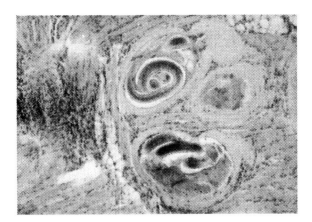

Figure 2.8.   *Trichinella spiralis* encysted in muscle. (Photograph by W. C. Marquardt.)

sites that undergo a tissue migration. A variety of parasites, ranging from nematodes to viruses, also engage in transplacental and transovarial transmission (see *H. meleagridis*, above; Ewald 1987, 1994; Dunn et al. 1993).

Paratenic hosts may bridge ecological gaps required for passive tranmission (e.g., Poinar et al. 1976). For instance, *Alaria americana*, typical of digenetic trematodes, requires a molluscan first intermediate host. The resulting cercariae seek tadpoles. Amphibians may not be available or palatable to the canid definitive host, but *A. americana* does quite well in paratenic hosts such as water snakes, which may eat the infected amphibian and in turn be eaten by the definitive host. (Because *A. americana* undergoes a migratory phase before settling in the intestine, a fondness for undercooked froglegs has had lethal consequences for humans.)

Paratenic hosts are probably underreported in the literature because there is often little to say about their discovery other than for the record. There is evidence that in some cases even adult worms, if consumed by an appropriate predator, may continue their residence in that predator (e.g., Lassiere and Crompton 1988).

Unusual and unexpected transmission pathways are not the exclusive domain of paratenic hosts. For instance, trematode metacercariae do not have to encyst in prey animals for passive transmission to occur. The fasciolid trematodes (e.g., *Fasciola hepatica*) are successful parasites of mammalian herbivores because, instead of entering a second intermediate host, their cercariae encyst on vegetation, in fact, they encyst on many objects in the water. In a similar manner, herbivores may acquire parasites from insects. *Dicrocoelium dendriticum* uses ant second intermediate hosts to enter grazing herbivores, which accidentally eat the ant along with vegetation (see chapter 3). Other parasites can broaden or shift their range of potential hosts when the intermediate host changes its habitat. For instance, *Haematoloechus medioplexus* moves from an aquatic environment to a terrestrial (or aerial) one when its dragonfly nymph host metamorphoses to an adult (fig. 2.1). This opens a new world of potential predators and hosts.

DIRECT LIFE CYCLE, ACTIVE TRANSMISSION

Active transmission involves host seeking and can be found in both direct and indirect life cycles. The eggs of many intestinal nematodes hatch in the environment and, unlike *T. tenuis*, do not ascend vegetation and await ingestion, but ascend vegetation and await a warm footpad. They move toward sources of warmth, actively burrow into the new host, and migrate through the tissues, ultimately reaching the small intestine. Hookworms are a good example of this sort of host seeking.

Some monogenean flatworms also actively seek their hosts. They are ectoparasitic and often live on the gills and skin of aquatic animals. A ciliated larva (*oncomiracidium*) hatches from the egg and swims until it finds a host, where it sheds its ciliated cells and develops into an adult (fig. 2.9). The search is not random. Oncomiracidia of some species may delay hatching until

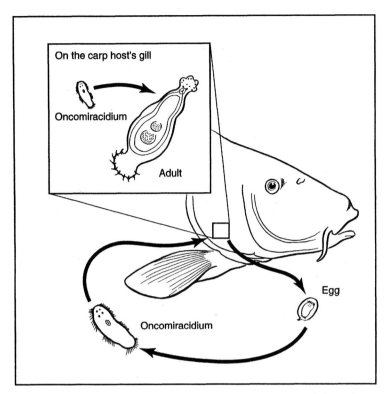

On the carp host's gill

Oncomiracidium

Adult

Egg

Oncomiracidium

Figure 2.9. The direct life cycle of a typical monogenean. The adult produces eggs that hatch into ciliated larvae (oncomiracidia). These actively seek new hosts, where they become adults. (Drawing by Conery Calhoon.)

prompted by environmental cues. Hatched oncomiracidia can also respond to physical and chemical signals (Kearn 1980, 1986; Gannicott and Tinsley 1997, Yoshinaga et al. 2000). In some monogenean species, reproduction is tied to host hormones and occurs when host reproduction does, thus assuring a supply of new hosts (e.g., Tinsley and Jackson 1988, Tinsley 1989; see chapter 3).

The majority of parasitic arthropods actively seek hosts. This is a large group that includes the parasitic crustaceans, such as copepods, branchiurans, barnacles, amphipods, isopods, and decapods. In most cases the adult crustaceans are parasitic, and the larval stages disperse. These crustaceans also use environmental cues to increase probability of encounters with hosts (e.g., Mikheev et al. 1998). The morphological adaptations to parasitism can be dramatic in the adults, ranging from animals that only resemble crustaceans in their larval stages to animals that are worthy of the tabloids. Among the most intriguing is the parasitic isopod *Cymothoa exigua*, which essentially replaces the tongue in snappers (*Lutjanus guttatus*); the tongue is reduced to a stub, and the isopod takes up the space formerly occupied by the tongue (fig. 2.10). It may even function as a tongue (Brusca and Gilligan 1983).

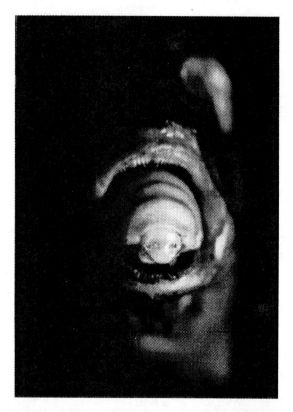

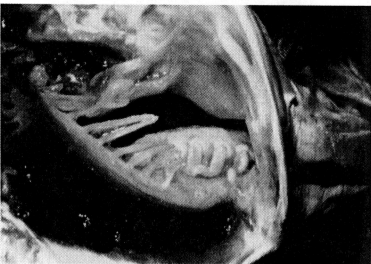

Figure 2.10.   Lateral and (top) frontal views of a *Cymothoa exigua* "tongue" in a snapper. (Photographs by M. Gilligan and O. Feverbacher; provided by Richard Brusca.)

The insects are another important group of arthropods that has adopted parasitism and active host seeking. These include lice, bugs, fleas, flies, and even a few Lepidoptera. The literature on parasitoid host location (e.g., flies, wasps, and strepsipterans) is voluminous (e.g., van Alphen and Vet 1986; Godfray 1994; Feener et al. 1996). The third extant arthropod subphylum, the chelicerates, contains the vast armies of ticks and mites.

There are two basic ways for an arthropod to locate a host: it may search for one or it may live on or near a host. Within these two strategies lies a world of variation. In the case of diurnal feeders, which includes many of the flies, visual cues such as color, movement, and shape may be important, whereas nocturnal insects may be attracted to warm temperatures (Moore 1993). Chemical cues such as carbon dioxide and water vapor are used by a wide range of ectoparasites (Ribeiro 1996).

Hematophagous flies have been most thoroughly studied (Klowden 1990, 1996; Bowen 1991; Takken 1991; Moore 1993). For female mosquitoes, blood feeding is hardly a random snack, but is part of a carefully regulated sequence of behaviors essential for reproduction. There are chemical stimuli that function in long-distance attraction, and in decisions to alight, begin feeding, and continue feeding (Klowden 1996). A successful blood meal distends the abdomen, which in turn inhibits host seeking; in mated females, the inhibition continues under hormonal intervention, also in response to abdominal distention (Klowden 1990, 1995). Hormonal changes ultimately affect olfactory sensitivity to some host cues such as lactic acid (Bowen 1991).

Perhaps the most notable attractants identified to date are Limburger cheese volatiles, fatty acids that are apparently considered delectable by the malaria mosquito, *Anopheles gambiae*. It turns out this is not a laboratory artifact. The bacterium *Brevibacterium linens* is responsible for the odor of Limberger cheese. Human foot odor (in Dutch, *tenenkaas*, or "toes-cheese") is also caused by skin bacteria, among them, a related bacterium, *B. epidermidis*. In fact, some bacteria involved in cheese-making may have originated from human skin. Short-chain fatty acids common to human sweat can also be found in Limburger cheese (Knols and De Jong 1996; Knols et al. 1997).

The attraction of tsetse flies to their hosts has also been scrutinized. Tsetses are attracted to dark colors (an attractant shared with many biting flies), certain patterns and shapes, motion, and a variety of chemical stimuli. Tsetse flies have a fairly slow reproductive schedule; the female nourishes one maggot at a time until it is almost ready to pupate, and then she larviposits. Knowledge of attractants and their use in traps has proven useful in tsetse population control.

Parasitoids are not to be outdone in the arena of host seeking. Female *Ormia ochracea* seek crickets as hosts for their young. Conveniently, *Gryllus* males call over relatively long distances at frequencies >3 kHz to attract female crickets. The acoustic organ of most dipterans is an antennal structure that responds to local air movements. However, the auditory organ of the acoustically orienting parasitoid flies is tympanal in nature, like those of crickets and grasshoppers; it can sense sounds at a distance. Moreover, the auditory organ of female *O. ochracea* is most sensitive in the range of cricket song, much more sensitive in this range than same structure in male *O. ochracea*, which do not have to locate hosts; "for a fly to act like a cricket, it must hear like one" (Robert et al. 1992, p. 1137).

INDIRECT LIFE CYCLE, ACTIVE TRANSMISSION

Many parasites with indirect life cycles also actively seek hosts during at least one phase of the life cycle. Digenetic trematode miracidia and cercariae do this, and it is all the more apparent in the life cycle of *Schistosoma* spp., which possibly has eliminated its second intermediate host, the host that figures most prominently in passive transmission of most other digeneans. The schistosome cercariae leave the molluscan intermediate host and seek not a second intermediate host, but the final host, a dispersal that is coordinated with final host circadian activity (Therón 1984). [This does not remove the risk of passive exposure. Goats have been shown to acquire *S. bovis* from drinking water (Boulanger et al. 1999).]

Whereas miracidia of almost all digenean species face a similar challenge—locating a molluscan first intermediate host, usually a snail or clam, often benthic and slow—digenean cercariae invade a wide range of second intermediate hosts, and their behavior and morphologies reflect this diversity (reviewed by Combes et al. 1994). Cercariae respond to environmental stimuli such as light, gravity, and turbulence to move to spaces where host encounter is likely or, in fewer instances, to hosts themselves. In the case of *Maritrema misenensis*, this requires assistance from the waves; cercariae on the surface of the water are washed onto the shoreline seaweed habitat of the amphipod intermediate host (Bartoli and Combes 1986).

The tails of many cercariae are essential for dispersal, but other species have virtually no tails (fig. 2.11). One such cercaria, that of *Paragonimus kellicotti*, creeps to the next benthic intermediate host, the crayfish. In the case of *Bucephalus elegans,* the forks of the tail are so long that passing fish may become entangled in them, thus acquiring metacercariae. In some cases, the tails can be extremely long and thin, up to 20 times the length of the cercaria's body, bifurcating and forming a net of cercariae. This net may facilitate host contact (Wardle 1988; fig. 2.12).

Some cercariae do enter hosts passively. A few cercariae are large enough to be seen by the next host and presumably mistaken for prey. The production of such a large cercaria is probably more costly than the production of many tiny ones and may involve some life-history trade-offs (Lewis et al. 1989). Dronen (1973) described three cercariae with large tails, which use those tails to enter the minnow second intermediate host in different ways (fig. 2.13). One uses the tail to take advantage of the respiratory current of the fish; in this species, the tail is approximately three times the length of the body of the cercaria, and nearly twice as wide. Another species with a gargantuan tail (seven times the length of the cercarial body) uses that tail to attract preda-

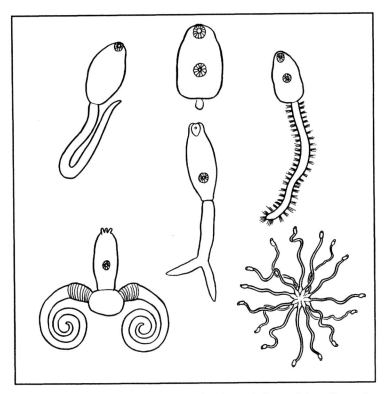

Figure 2.11.   All cercariae are not alike. The size and shape of the tail reveal much about the second intermediate host and what the cercaria needs to do (swim? crawl? get eaten?) to find that host. (Drawing by Conery Calhoon.)

tion. A third species may use the respiratory current, or may form a fascinating aggregation called a rat king.

Rattenkönigcercariae, or rat king cercariae, are cercarial aggregations that occur in some species; this is a large, noticeable mass that can be mistaken for a food item (fig. 2.14). The cercariae that engage in this behavior move slowly when solitary and may be poorly adapted to active host seeking (Hendrickson and Kingston 1974). Beuret and Pearson (1994) summarized the characteristics of 11 known *Rattenkönigcercariae*, which occur across five trematode superfamilies. Depending on the species, the aggregations may contain anywhere from a few to hundreds of individual cercariae. Beuret and Pearson described the most organized rat king to date from an intertidal snail (*Clypeomorus batillariaeformis*) on Heron Island (Great Barrier Reef). (The adult form was unknown to them, so they resisted the impulse to describe a new species.) The cercariae emerge from the mollusc's ctenidia into the mantle cavity, where they assemble and then leave through the exhalant siphon. This assembly is no mean trick: each cercaria is oriented in the same direction, and the cluster spirals around a central lumen. Individual cercariae swim poorly; the mem-

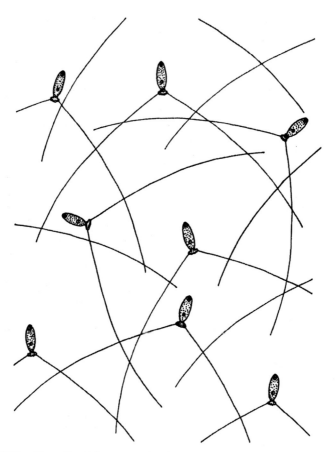

Figure 2.12. The tails of these cercariae form a net, which may increase the probability of contacting the next host. (Wardle 1988; Reprinted with permission from *Journal of Parasitology*.)

bers of the aggregation rotate in the same direction, generating a current that moves the batch of up to 700 cercariae efficiently. They are positively phototactic and, using mechanisms that have yet to be understood, can reverse direction immediately if the light source moves. Beuret and Pearson suspect that many types of *Rattenkönigcercariae* evolved from large-tailed, prey-mimicking types described above; the large tails facilitate the rat king clusters.

Officially, *Rattenkönigcercariae* are described as zygocercous, referring to the "yoked" tails. The term *Rattenkönig*, however, certainly captures the imagination (fig. 2.15). The word may have originally been associated with the Mouse King, the seven-headed rodent of an early Nutcracker tale who is the Nutcracker's nemesis. Other German fairy tales refer to rat royalty seated in carriages with knotted tails. In the Middle Ages, folks had the idea that there was a rat leader, and eventually there arose the notion of the Rat King as a group of powerful rats joined at their tails and fed by other, subservient rats.

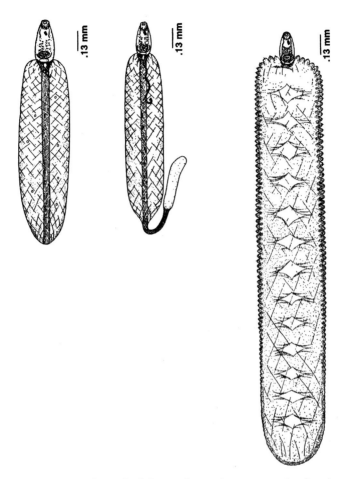

Figure 2.13. The cercaria on the left uses the respiratory current to invade a minnow; the one on the right must be ingested and may mimic a prey item. The middle cercaria may use the respiratory currents of larger minnows, but also depends on ingestion. (Reprinted from Dronen 1973, with permission.)

When Martin Luther inveighed against the pope, he called him (among other things) a *Rattenkönig*. (Luther had a fairly large suite of rodent monikers for Vatican habitués.) Presaging ecumenicism, he did not exclusively reserve these names for Catholics, but decided that some Protestant groups as well were kingdoms of rats.

Rat catchers do, on occasion, tie rats together by their tails. However, the natural occurrence of rat kings, not mediated by rat catchers, has been described all over the world. There are even published records, complete with X-rays, showing rats firmly attached (accidentally entangled?) to one another at the ends of their tails. One empiricist published his account of an attempt to produce a rat king by gluing rat tails together. He carefully recorded weight

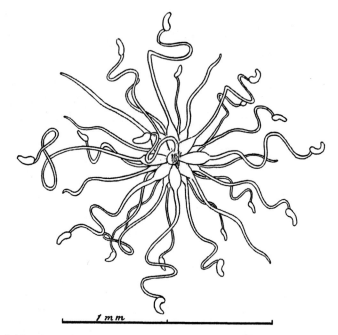

Figure 2.14. *Rattenkönigcercariae*, or Rat King cercariae. (Ward 1916, Plate B; Reprinted with permission from *Journal of Parasitology*.)

and gender, only to come to the disappointing conclusion that most rats do not want to become rat kings and are eager to abdicate (see Becker and Kemper 1964).

An eminent parasitologist argued for *zygocercous* over *Rattenkönig* on the basis of classical versus barbaric origins, respectively. As I consider the legend of the Rat King, in a carriage with his queen (tails entwined, of course) drawn by other loyal rats, my vote is for the barbarians. What wonderful trematodes!

INDIRECT LIFE CYCLES, LIVING VECTORS

Many living vectors are ectoparasites, which may also transmit other parasites to their hosts. These parasites include the rickettsiae that we share with ticks and lice, the bacteria brought to us by ticks and fleas, and the viruses, protists, and worms we acquire from biting flies.

Because all these parasites depend on the blood-feeding habits of their ectoparasitic hosts for transmission, their life cycles may appear to be superficially similar. For instance, both *Dirofilaria immitis* and *Brugia pahangi* are transmitted to mammals by the bite of mosquitoes. The larval nematodes (microfilariae) enter the bloodstream and mature to adults, which produce another generation of microfilariae that can be picked up by other mosquitoes. However, the developmental site of these nematodes in the mosquitoes differs, with

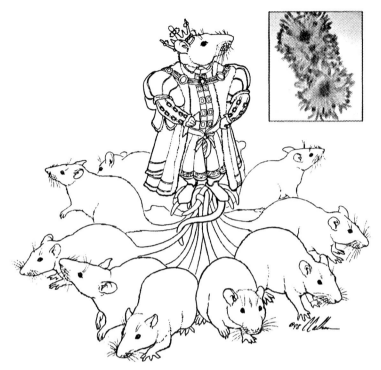

Figure 2.15. The legendary antecedent of the *Rattenkönigcercariae*. (Drawing by Conery Calhoon; inset photo of a real *Rattenkönigcercariae* courtesy of Gary Hendrickson.)

D. immitis entering the Malpighian tubules and *B. pahangi* the flight muscles. These locations can have different physiological and behavioral consequences for the mosquito (Berry et al. 1986, 1987a). (See chapter 3 for ways that some nematodes affect vector foraging.)

Another nematode, *Wuchereria bancrofti*, causes a notorious condition in a minority of human hosts that is known as elephantiasis. The ancients noticed the thick, rough skin of the sufferer and thought it similar to the skin of an elephant. [Schmidt and Roberts (1989) wryly noted that if taken literally, elephantiasis is a condition caused by elephants.] Elephantiasis is an infrequent outcome of the infection and can be the result of a buildup of scar tissue following repeated lymphatic inflammation. Adult worms live in the lymphatic ducts, and females produce juvenile nematodes (microfilariae) that enter the lymph and then the bloodstream. Mosquitoes ingest them from the peripheral blood; the worms develop in the mosquito to a later larval stage and are then ready to enter a new host. These microfilariae exhibit a circadian periodicity in the peripheral blood; their peak occurrence there corresponds with the feeding time of the local vector, and may therefore be nocturnal in some areas and diurnal in others (Pichon 1981; chapter 3).

Members of the *Trypanosoma brucei* group that are infective for humans cause African sleeping sickness. These trypanosomes are also ingested with a blood meal and develop over a 15- to 35-day period in tsetse flies, first in the midgut, then in the foregut, and finally in the salivary glands. In the vertebrate host, their population levels fluctuate. The decline signals an effective immunological response on the part of the host; this is followed by a resurgence that occurs because a subset of that trypanosome population exhibits a novel antigenic coat that allows them to escape temporarily the immune response. These "games parasites play," to quote Bloom's (1979) review, can be deadly.

CONCLUSION

Beyond this breathless sprint through parasitology, much remains, from the varied behavioral strategies of the parasitic wasps, to the wormlike gastropod parasites of sea cucumbers (lacking eyes, tentacles, hearts, and shells), to the scolices of tapeworms in rays, which mirror the distinct intestinal morphologies of the hosts (H. H. Williams 1960). With such variety, there is no one-size-fits-all parasite, no universal host. Behavioral alterations that may enhance transmission in one association can easily fail to do so in another, and the same is true for host defensive behaviors. When superimposed upon the environmental heterogeneity that usually exists over the range of a host–parasite association, the potential for variation is even more impressive. This simply underscores Anderson and May's argument that from virulence to near benevolence, there may be as many outcomes of parasite–host associations as there are associations, with no class of outcomes claiming a priori primacy.

Time alone will not inevitably produce a finely coevolved, "good" parasite that limits harm to its host, nor will new arrivals always bungle the situation. Natural selection favors parasites with enhanced transmission and reproduction—that is, those with increased fitness. It favors hosts that limit or avoid any fitness losses associated with those parasites. That selection occurs within a great diversity of life cycles. The ways in which host behavior influences parasite transmission and host survival, parasite avoidance and host defense—and in so doing, influences the evolutionary trajectory of the host-parasite relationship—are the subjects of the following chapters.

ENDNOTE

1. In a 1964 U.S. Supreme Court case, Justice Potter Stewart concurred with the majority opinion of the Court, which overturned the conviction of Nico Jacobellis for possessing and exhibiting an obscene film. Justice Stewart wrote, "I have reached the conclusion . . . that under the First and Fourteenth Amendments criminal laws in this area are constitutionally limited to hard-core pornography. I shall not today attempt further to define the kinds of material I understand to be embraced within that shorthand description, and perhaps I could never succeed in intelligibly doing so. But I know it when I see it, and the motion picture involved in this case is not that." *Jacobellis v Ohio,* 378 U.S. 184, 197; 84 S. Ct. 1676, 1683 (1964) (Stewart, J., concurring).

2. The higher taxonomy of the protista is in a state of flux. I use descriptive terms (e.g., *flagellated protist*) that are not meant to have strict taxonomic implications.

3

Behavioral Alterations
and Parasite Transmission

As we have seen, parasites are transmitted in a variety of ways, and almost all major transmission modes offer examples of transmission-enhancing behavioral alterations that are associated with parasitism. In May and Anderson's (1992, and references therein) Basic Model, an increase in transmission rate results in an increase in reproductive ratio if everything else is equal. Thus, when parasites induce such behavioral alterations, not only may they be favored by natural selection, but several aspects of parasite epidemiology, including host population levels necessary for their establishment, also diverge from those predicted by models that do not assume such enhancements.

Altered host behavior that enhances parasite transmission can be found in both direct and indirect life cycles. There are countless transmission modes, but most involve predation upon intermediate hosts, vector transmission, or direct transmission (host seeking or consumption of actual parasite). Of these, host behavioral alterations that result in increased predation upon intermediate hosts are the most dramatic; after all, when predators eat parasites and their intermediate hosts, the cost to the intermediate host is conspicuously high. Although demonstrating the effect of altered behavior on transmission can be daunting, representatives of every major taxon of predation-transmitted parasite have been shown to enhance the risk of predation experienced by their intermediate hosts. Hematophagous vectors also exhibit altered behaviors, many of which involve feeding anomalies that probably increase the transmission of blood parasites. Even gastropod intermediate hosts of trematodes may behave in ways that favor contact between the recently emerged trematode larva and its next host, and sit-and-wait parasites have developed characteristics that enhance contact with hosts. Among other things, this means that R_0 for these parasites will be higher than if their contact with hosts were a result of random encounters.

The suggestion that behavioral changes associated with parasitism might be important in parasite transmission is not new (Cram 1931; Rothschild 1962). In fact, there are scores of examples of altered behavior that suggest the possibility of enhanced transmission; of these, the case of transmission by intermediate hosts is perhaps most amenable to investigation, for the fate of prey is often easier to determine than the fate of vector-borne microorganisms or the fate of parasite propagules. Below, I summarize some examples in which predation-mediated enhanced transmission has been demonstrated. Increased transmission by vectors or on the part of parasite propagules is harder to document. I discuss behavioral aspects of these transmission modes that likely result in more efficient host-to-host transfer, but for which evidence admittedly is not as robust as that for predation-transmitted parasites. Finally, I review some types of behavioral alterations for which transmission has not been assessed.

TRANSMISSION BY INTERMEDIATE HOSTS

In the case of parasites transmitted with prey items, immature stages occur in intermediate hosts that must be eaten by predators (definitive hosts) before the parasite can reach adulthood and reproduce in the definitive host. This is a striking phenomenon because the transmission event, predation, is by and large hostile to the intermediate host, and the last thing we normally expect from an animal is a change in behavior that might be seen as suicidal. In contrast, the favored outcome for the parasite (transmission) is, with few exceptions, in direct and unambiguous opposition to what we expect to be favored for its host (predator avoidance). [Smith Trail (1980) has suggested that such "suicide," in fact, may have been beneficial to host kin at one time (possible examples in chapter 5); for some major groups of parasites, this seems unlikely (Moore 1984b).]

The examples of behavioral alterations that I cite immediately below share one thing in common: They all have been shown to enhance predation on the intermediate host. In so doing, they increase transmission opportunities above what would be predicted based on the occurrence of intermediate hosts in the prey population.

I have chosen case studies that can be allocated to "laboratory" and "field" categories, although in reality there is much overlap. Crustaceans star in the laboratory, where they have been shown to sustain increased predation when infected by any of several different parasites, and where there are hints of some mechanisms that might be involved. In the field, the unlikely combination of isopod and corallian hosts lead the way. In addition, one of the first field studies ever, that of fish and the large tapeworm *Ligula*, directs us toward a treasure trove of information from a closely related tapeworm back in the laboratory.

Beyond these examples, there are abundant studies of altered behavior that could reasonably be argued to favor predation but that have not been tested; I discuss a few of these later in this section, concluding with a discussion of pathology and adaptation. Helminth parasites dominate this area of study.

In the Laboratory (Mostly)

The best known examples from the laboratory are those involving amphipod intermediate hosts. In 1973, Bethel and Holmes reported that uninfected amphipods avoided light and, if disturbed, dived and burrowed; this eminently appropriate antipredator behavior also occurred in amphipods infected with *Polymorphus paradoxus* acanthellae (young acanthocephalans that are not infective to the next host) and with metacestodes (*Lateriporus* spp.). On the other hand, amphipods containing cystacanths infective for ducks [i.e., *Polymorphus paradoxus, P. marilis,* and *Corynosoma constrictum,* in their amphipod intermediate hosts, *Gammarus lacustris* (for *Polymorphus*) and *Hyalella azteca* (fig. 3.1)] behaved in more conspicuous ways. These amphipods preferred being in the lighted zone of a light/dark tank, and most of those with *P. paradoxus* and *C. constrictum* were especially photophilic, concentrating in the portions of the light zone nearest the light source. Amphipods infected with *P. paradoxus* were especially sensitive to disturbance, either direct (touch) or indirect (disturbance of surroundings). They skimmed on the surface and/or clung to floating objects; both behaviors increased the occurrence of infected amphipods at the surface of the water. Amphipods infected with *P. marilis* moved to lighted areas, but did not go to the surface (fig. 3.2).

The three acanthocephalan species induced different combinations of behavioral changes in their hosts—attraction to a lighted area, attraction to light, skimming and clinging—and these varied combinations meant that the parasite–host associations were not equally distributed. The alterations appeared to predispose the host–parasite combinations to occur in different microhabitats, thus increasing their chances of encountering the definitive host for that parasite (Bethel and Holmes 1973). Accordingly, when given infected

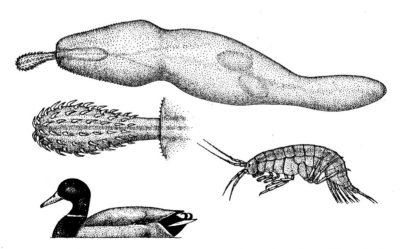

Figure 3.1. *Polymorphus paradoxus* has a typical acanthocephalan life cycle, living as an adult in waterfowl and using gammarid intermediate hosts (Moore 1984a).

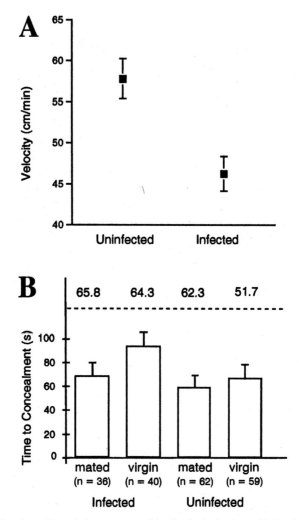

Figure 3.2. Acanthocephalans are not alone in the induction of behavioral alterations in intermediate hosts. Mature metacestodes of *Hymenolepis diminuta* cause beetles (*Tenebrio molitor*) to slow down and reduce both photophobia and responses to aggregation pheromone (Hurd and Fogo 1991). In the flour beetle *Tribolium confusum,* both (A) mean velocity and (B) time required for concealment were negatively affected by the parasite (Figure reprinted from Robb and Reid 1996). (Figures reprinted with permission from *Canadian Journal of Zoology*.)

and uninfected amphipods in equal numbers in a tank in the laboratory, naive mallards (a definitive host) ate almost four times as many *P. paradoxus*-infected amphipods; the skimming and clinging behaviors seemed to make the infected amphipods particularly vulnerable, as the ducks were especially interested in the floating vegetation, and were quick to notice the skimming amphipods. Both mallards and diving ducks are hosts for *C. constrictum,* and in uninfected/infected amphipod comparisons, ducks ate more of the latter, but in much lower numbers than they had eaten *P. paradoxus*-infected amphipods. When provided with equal numbers of uninfected, *P. paradoxus*-infected, and *P.marilis*-infected amphipods, mallards ate only *P. paradoxus*-infected prey; *P. marilis* is a parasite of diving ducks (e.g., scaup), but not of mallards. Wild-caught muskrats, also a host for *P. paradoxus,* accidentally ingested one-fourth of the available *P. paradoxus*-infected amphipods, along with the vegetation to which the amphipods clung, but ate none of the uninfected ones. Upon being offered uninfected and *P. marilis*-infected amphipods in a tank, the muskrats ate neither. Unfortunately, scaup, the remaining host type to be tested, were disinclined to feed normally under laboratory conditions.

Bethel and Holmes (1977) interpreted their results to mean that although these parasitized amphipods are not uniquely vulnerable to the definitive hosts, and many are eaten by unsuitable hosts, the nature of the behavioral changes induced by each parasite does increase exposure to appropriate hosts. Thus, the skimming and clinging behavior of *P. paradoxus*-infected amphipods results in increased predation by mallards and increased accidental ingestion by muskrats. The more variable behavior exhibited by hosts of *C. constrictum* may correspond to its broader definitive host range. In the case of *P. marilis*, which infect scaup, the shift in photophilia alone may be sufficient to bring its host to the attention of diving foragers (fig. 3.3).

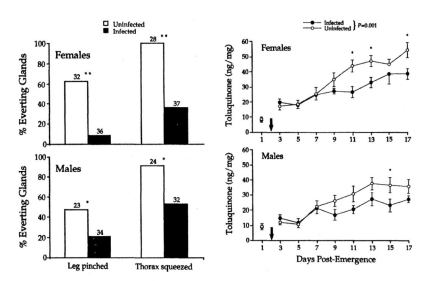

Figure 3.3.   Cestodes also place intermediate hosts in harm's way. When infected with metacestodes of *Hymenolepis diminuta*, the beetle *Tenebrio molitor* was less likely to evert defensive glands when stimulated to do so (left), and these glands contained less noxious substances (right). Glandular toluquinone content expressed as ng/mg body weight of beetles. (Blankespoor et al. 1997; but see Webster et al. 2000). (Reprinted with the permission of the Cambridge University Press.)

Timing

Demonstrating that altered behaviors occur in intermediate hosts that are also subject to increased risk of predation by final hosts is one convincing piece of evidence that these alterations are adaptive for the parasite (but see Edwards 1987; Webber et al. 1987a). In addition, the timing of the behavioral change can be an indirect, but nonetheless compelling, indication (see table 3.4 in the Appendix). In the case of the amphipod G. lacustris parasitized by P. paradoxus, the altered responses only appeared consistently after the acanthocephalan became infective to the final host (Bethel and Holmes 1974). Many of the crustaceans that change color in response to acanthocephalan infection do so only after the parasite becomes infective to the next host. In all these cases, altered behavior has been associated with increased predation (fig. 3.9).

Similar coordination between infectivity and altered behaviors has been shown in systems where enhanced transmission has not been demonstrated but is nonetheless likely. The three-spined stickleback (G. aculeatus) loses its aversion to predators when the tapeworm Schistocephalus solidus reaches an infective size (Tierney et al. 1993). This is in contrast to behaviors that are associated with hypoxia and that change gradually over the course of infection (Giles 1987b), although those may play a role in transmission as well.

Behavioral changes can follow a predictable chronology without necessarily having implications for transmission. They may instead reflect parasite developmental stages and concurrent effects on host physiology (Rau 1983a, 1984a; Berry et al. 1987a, 1988). For instance, anorexia in rats infected with Nippostrongylus brasiliensis is not continuous, but occurs in two phases that reflect parasite development, including a pulmonary phase (Ovington 1985).

What causes these changes? First of all, the behavioral alterations were not a general response to parasitism. Not only did different acanthocephalans elicit different responses, but a larval cestode (*Latiporus*) and uninfective acanthocephalans had no behavioral effect. (This nonconformity of responses across host–parasite associations is not unusual; see chapter 6.) In addition, in at least some cases the cystacanth must be alive in order to influence gammarid behavior (Bethel and Holmes 1973).

Might the parasitized amphipods, with their large parasites, be seeking more oxygen near the surface of the water? Although P. laevis caused G. pulex to be more sensitive to oxygen deprivation (McCahon et al. 1991b, but see Rumpus and Kennedy 1974), oxygen did little to influence the surface-seeking behavior of another amphipod (G. lacustris) containing another infective acanthocephalan (P. polymorphus; Bethel and Holmes 1973).

When injected with serotonin, uninfected amphipods briefly expressed the skimming, clinging, and positive phototactic behaviors found in gammarids infected with *P. paradoxus*. The clinging response, important in vulnerability to surface-feeding predators, was suppressed by octopamine. This response is similar to precopulatory clinging in male amphipods, and when induced under inappropriate conditions it could interfere with escape from predators (Bethel and Holmes 1973; Helluy and Holmes 1990). The precise mechanism is not yet clear, but Maynard and co-workers (1996) found evidence that the ventral nerve cords of infected amphipods contain more putative serotonergic varicosities (or perhaps greater amounts of serotonin in these varicosities) than the nerve cords of uninfected amphipods do. This was not true of amphipods infected with *P. marilis*, which do not cling. As for positive phototaxis, the accessory screening pigment in the eyes of infected gammarids was more light adapted than that in uninfected animals, and this may be related to changes in phototaxis, but serotonin injection did not induce pigment migration (Helluy and Holmes 1990).

The involvement of serotonin in acanthocephalan–amphipod associations raises broad possibilities. Might serotonin mediate changes in the escape response of cockroaches infected with another acanthocephalan (*Moniliformis moniliformis*; Libersat and Moore 2000)? Is the active wandering seen in isopods infected with *Acanthocephalus jacksoni* related to the skimming behavior in *P. paradoxus*-infected amphipods (Muzzall and Rabalais 1975)? A comparison of underlying mechanisms would go far in answering these questions. In general, a comparative approach to mechanisms would certainly contribute to our understanding of the evolution and current function of parasite-induced behavioral alterations. A glance at the primary literature that asks about these mechanisms, however, reveals that it is remarkably easier to suggest such an approach than to gather the necessary data (table 3.1, p. 166).

Acanthocephalans are notorious for changing the color of aquatic crustacean hosts (table 3.2). The parasites themselves may be brightly pigmented; mature cystacanths of *P. paradoxus, P. marilis*, and *C. constrictum* appear as intense red or orange dots under the amphipod's exoskeleton (fig. 3.4). The gaudy cystacanth may not be to blame for all of the increased risk of predation experienced by its amphipod hosts, however. Painted "cystacanths" on uninfected gammarids produced no shift in duck predation, thus indicating that the altered behavior, and not the appearance, was responsible for the increased transmission (Bethel and Holmes 1977). In contrast, artificially "infected" *G. pulex* painted to mimic *Pomphorhynchus laevis* infections (fig. 3.5) were at increased risk of predation by three-spined sticklebacks (Bakker et al. 1997); they also encountered uninfected gammarids at an unusually high rate, a shift in amphipod interaction that is intriguing and difficult to explain (A. F. Brown and Thompson 1986).

The cystacanth color probably comes from host carotenoids, and may be peculiar to acanthocephalans, for colorless cestode cysticercoids co-occur with brightly colored cystacanths. Although not all acanthocephalans in crustacean intermediate hosts are pigmented, the acanthocephalans that do exhibit colors

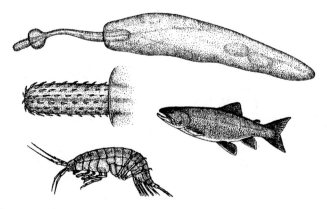

Figure 3.5. *Pomphorhynchus laevis* uses gammarid intermediate hosts, but lives in fish when an adult. Moore 1984a

use crustacean intermediate hosts (Barrett and Butterworth 1968). Bakker and co-workers (1997) noted that colored cystacanths commonly induced increased photophilia in crustacean intermediate hosts, and they suggested that perhaps the pigment shields the parasite from UV-B radiation. Moreover, the carotenoids in the parasite may be attractive to male sticklebacks, which have carotenoid-based breeding coloration. Ironically, because carotenoids can function in health maintenance, they may be an indicator of good health in systems where mate choice is carotenoid-influenced (Houde and Torio 1992; Lozano 1994; see also Shykoff and Widmer 1996).

Pigment dystrophy (loss of color) also occurs in some infected crustaceans. In contrast to bright colors, one suggested mechanism for pigment dystrophy (which does not include eye pigments, by the way) is interference with protein metabolism during critical phases of isopod growth (Oetinger and Nickol 1982a,b). However effective the color change may be in altering predator–prey relationships, it is not a universal characteristic of acanthocephalan–crustacean interactions, nor is it an inevitable outcome even within specific associations where it has been reported (Pilecka-Rapacz 1986; Lyndon 1996).

Many intermediate hosts with altered colors are conspicuous against their normal backgrounds; in other cases, what is altered is not the color of the host, but its background preference (table 3.2, p. 171). Although one might assume that crypticity (or lack thereof) is the result of such a shift, substrate color, as well as integument color, can also affect body temperature, which in turn can have positive or negative effects on the parasite. Color change is one of many alterations that have ambiguous potential effects for parasite and host (chapter 5).

Although terrestrial isopods infected with acanthocephalans are not discolored, they do not escape behavioral alteration, nor its consequences in the food chain. In the isopods that I studied (chapter 1), infected animals spent more time on light-colored substrates, in less humid areas, and away from shelter. In an aviary, European starlings (*Sturnus vulgaris*), the final host, ate more infected than uninfected isopods when given a choice of equal numbers of each in a container that offered dark and humid versus light and dry substrates. In these tests, substrate color seemed especially important in the birds' choices; the starlings took almost all their isopods from the light area, where they were much more conspicuous than they were on dark substrate. Because infected isopods also prefered light substrate, which in nature may very well be drier than dark substrate, such alterations may act synergystically to create even more divergent microhabitat preferences on the parts of parasitized and unparasitized animals (Moore 1983a).

I was surprised to find the *Plagiorhynchus*–isopod association cited in the *Creation Research Society Quarterly* as one "evidence of design," with an attribution to Revelations 4:11, "for thou hast created all things and for thy pleasure they are and were created" (E. N. Smith 1984, p. 124). I can understand how parasite-induced behavioral alterations would interest a deity, and I am pleased to be in such good company, but if *P. cylindraceus* was indeed especially created, that must surely seem a perversion to a starling or an isopod.

Returning to amphipods, acanthocephalans are not the only parasites to alter amphipod behavior in ways that result in increased predation and transmission to the next host. In contrast to the acanthocephalans, which float in the arthropod hemocoel, a digenean trematode, *Microphallus papillorobustus*, encysts in the nervous systems of its second intermediate hosts, *Gammarus insensibilis* and *G. aequicauda* (Helluy 1982). Apparently, only one metacercaria is needed in the protocerebrum to induce positive phototaxis and hyperactivity (when disturbed) in amphipods. Their response to ventral illumination was weaker than that to overhead light, however, indicating that unlike acanthocephalan-infected amphipods, some geotactic influences may also be at work (Helluy 1983a,b, 1984).

The cerebral location on the part of the trematode is no accident; the digeneans follow a chemical attractant to the brain of young amphipods. Predation tests with two captive gulls (*Larus cachinnans*) revealed that infected amphipods were more than twice as likely to be eaten as uninfected amphipods. In addition, infected amphipods in nature occur in clumps on the edge of steep banks and probably encounter birds that use tactile foraging methods.

It is possible that *Maritrema subdolum*, a digenean that cooccurs with *M. papillorobustus* in both amphipods and snails, but is not found in the amphipod nervous system, may hitchhike to the definitive host in gammarids infected by *M. papillorobustus*, sharing the benefits of altered behavior. Amphipods infected with *M. subdolum* alone did not exhibit behavioral alterations (Helluy 1983a,b; see also Fukase et al. 1986, 1987 for a possible cestode hitchhiker). Recall that infected gammarids swim higher in the water column than uninfected ones. Cercariae of *M. subdolum* are more likely to swim than those of *M. papillorobustus* are, and thus are more likely to encounter gammarids infected with the latter species; in other words, the swimming behavior of the cercariae may favor hitchhiking by increasing encounters with infected gammarids (F. Thomas et al. 1997; see F. Thomas et al. 1998a for general discussion.). In contrast, two nonmanipulative trematode metacercariae, *Microphallus hoffmanni* and *Levinseniella tridigitata*, were not positively associated with manipulative *M. papillorobustus* in field-collected *G. aequicauda*; some apparent hitchhikers may simply be "lucky passengers" (F. Thomas et al. 1998b).

This shift in vertical zonation caused by *M. papillorobustus* placed infected gammarids in close proximity and thus had another effect on behavior: infection was responsible for assortative mating, with parasitized animals being more likely to mate with one another. This effect disappeared in shallow water (F. Thomas et al. 1996a,b). Zohar (1993) found a similar phenomenon in gammarids infected with acanthocephalans, which she attributed to changes in proximity due to altered microhabitat preferences. The gammarids' reproductive success may be diminished under such circumstances due to a failure to locate suitable mates, because shallow water is also frequented by small females and juveniles.

Assessing predation is never an easy task. The experiments reported above (see also table 3.3, p. 177) were done in laboratories and were fairly straightforward, involving systems that were amenable to laboratory study. One might argue that these demonstrations are somehow unnatural, and that for all our wishful thinking, altered behavior in parasitized animals has little real consequence, and limited relevance for predator–prey interactions in the field. However, although Bethel and Holmes (1973) did not quantify distributions of acanthocephalan-infected and uninfected amphipods under field conditions, their qualitative observations in the field supported their laboratory behavioral tests. For instance, they noticed gammarids skimming on the surface where people had walked through the water, and were themselves the objects of clinging amphipods when doing field work. Zohar (1993) confirmed that the distributions of infected and uninfected amphipods in the field reflected the altered behaviors observed in the laboratory. Moreover, the results from at least two laboratory systems have been shown to be consistent with field data.

In (and Inspired by) the Field

In the case of the acanthocephalan-infected isopod, I calculated rates of isopod delivery to nestling starlings and measured acanthocephalan intensities in isopods (thousands of isopods) taken from the areas where the parent starlings foraged. (I tried to forage like a starling as well as I could; that is, I did not turn over any rocks or move heavy objects to acquire my sample.) There were far more infected nestlings than I would have expected had they been fed infected isopods in the proportions that occurred in the isopods population. This was true for all my field studies—three consecutive field seasons on an urban campus and one season in a rural area. Although this is an indirect assessment, the analysis is highly conservative. For instance, I counted an infected nestling as one predation event on an infected isopod; given the overwhelming predominance of single-cystacanth infections in isopods, however, multiply-infected nestlings probably had been fed more than one infected isopod.

Moreover, not every infected isopod fed to a nestling would necessarily produce an acanthocephalan in the nestling; for a variety of reasons, including the possibility that some cystacanths may have been swallowed by parent birds, it is probable that some acanthocephalans from infected isopods did not establish in the nestlings. Despite these and additional conservative approaches, I still found more infected birds than one would expect, given the prevalence of infected isopods, probably because of increased encounters between parasitized isopods and parent starlings as a result of behavioral alterations in infected isopods (Moore 1983a).

There are also indications from the field that digenetic trematodes living in Hawaiian corals increase predation by fish on polyps (Aeby 1992, 1998). Predator exclosure experiments revealed that corals exposed to butterfly fish predators had a greater decrease in infected polyps than protected corals did, probably because the butterfly fish ate infected polyps. This was confirmed by laboratory observation. Although the behavior of coral polyps may not seem terribly provocative, polyps infected with trematodes exhibit altered behavior, as well as altered appearance (fig. 3.6). They cannot perform an important polyp behavior—retraction—and are left exposed to predators.

Enhanced predation on infected *Porites* may have some unusual fitness consequences, however. The parasite impaired the growth of the coral colony; in contrast, when infected colonies were exposed to grazers and presumably relieved of their parasites, their growth rates matched those of uninfected corals. The predator, in other words, did not harm the entire colony, and the consumption of infected polyps may have had a beneficial effect, a result contrary to the customary view of parasite manipulation of host behavior as a win/lose situation. Because of shared genes and an ability of individual polyps to regenerate, these colonial animals offer the potential for a different outcome for the intermediate host (Aeby 1992; see also Aeby 1998).

The field tests cited above are unusual in that they are supported by laboratory tests of predation, and, in the case of the isopod–starling system by experimental infections that have also been linked to altered host behavior. However, we have known about enhanced predation upon intermediate hosts in the field for quite a while. W. H. van Dobben (1952) in the Netherlands was among the first to present field data supporting the hypothesis of increased predation on parasitized intermediate hosts. He took advantage of the fact that cormorants (*Phalacrocorax carbo*) eject their meals in rookeries, and he collected the undamaged roach (*Leuciscus rutilus*) delivered by these birds. He then compared the prevalence of the cestode *Ligula intestinalis* in cormorant-captured roach with that in roach that had been taken by fishermen at the same time and in the same location. Fewer than 7% of the roach in the fishermen's catches were parasitized, but 15–30% of the roach taken by the cormorants contained the tapeworm.

This avian parasite *L. intestinalis* has two intermediate hosts: when a fish such as a roach eats an infected copepod (the first intermediate host), it acquires the parasite, which grows to be 20 cm long and 1 cm wide in the roach body cavity. This has fairly noticeable consequences for the fish, which not only behaves differently from uninfected fish, but also develops a rotund shape that is highly visible (see Arme and Owen 1968; also Sweeting 1977).

We know much more about *Ligula*'s close relative, *Schistocephalus solidus*, another diphyllobothriid cestode (Schmidt 1986), and I will tell you this story, which contains many more nuances of mechanism and variation, but lacks such clear predation tests (see Jakobsen et al. 1988). Much of this work is laboratory based, but it is relevant to van Dobben's initial observation. Three-spined sticklebacks (*Gasterosteus aculeatus*) infected with *Schistocephalus* become distended, and they engage in apparently foolhardy behavior: They swim closer to the surface and recover more quickly from an overhead fright stimulus than uninfected fish do, a recovery time that is negatively correlated with the weight (maturity) of the parasites. In addition, infected fish are more likely to feed after such a scare (Giles 1983, 1987a; Godin and Sproul 1988; see also Ness and Foster 1999). Infected threespine sticklebacks in the presence of a cichlid predator (not a definitive host) did not adjust their foraging areas; that is, they foraged at all distances from the predators, whereas uninfected fish stayed far away from predators (fig. 3.7). Fish infected with the protistan *Glugea anomala*, a parasite that does not have to be transmitted by predation, fed at intermediate distances (Milinski 1985). Infected nine-spined sticklebacks (*Pungitius pungitius*) also returned to the surface more quickly than did uninfected fish after a threat from a model of an avian predator (R. S. Smith and Kramer 1987).

Although we lack field evidence such as that of van Dobben's (see above) for *Schistocephalus* and sticklebacks, their risky behavior is likely to transmit the tapeworm. Evidence from a Norwegian lake supports this idea: In the presence of another predatory fish, Atlantic salmon (*Salmo salar*), the number of parasitized sticklebacks decreased to nearly zero, a change that was attributed to selective predation on the hungry, fearless stickleback intermediate hosts (Jakobsen et al. 1988; but see Gilbertson 1980, cited in McPhail and Peacock 1983).

In the laboratory, fish containing infective plerocercoids were most intrepid. They took less time to recover from fright and spent more time engaged in pectoral sculling, a behavior that characterizes undisturbed sticklebacks. Meanwhile, fish with uninfective worms showed a tendency to be even more cautious than uninfected fish (Tierney et al. 1993).

Why are infected fish so audacious? The infected fish's apparent attachment to food (above) offer some clues. Like its relative *Ligula*, *Schistocephalus* is large and produces a rotund fish; the worms can be 50% of the apparent fish biomass (fig. 3.8). The size of the parasite suggests the mechanism that is responsible for "suicidal" sticklebacks, and why it is especially noticeable in fish with large, infective worms: The worm takes up so much space in the fish that it physically cannot feed to satiation and therefore spends more time foraging. Three-spined sticklebacks containing *S. solidus* are poorer competitors than uninfected sticklebacks and are less cautious around predators (Milin-

Fearlessness: One Mechanism

In several host–parasite systems, altered responses to predators have been observed. In a surprising number of cases, parasitized animals are less likely to avoid predators (table 3.5). What causes such carelessness?

Kavaliers and Colwell (1993a,b, 1994, 1995a) have done the most detailed study of the neurochemical correlates of parasitism in mice and have found a possible biochemical basis for some behavioral shifts (Kavaliers et al. 1998, 1999). They found that infection with a coccidian, *Eimeria vermiformis*, resulted in increases in analgesia during the prepatent (pre-oocyst shedding) period; this declined upon patency (see also Colwell and Kavaliers 1993). The analgesia was caused by increases in endogenous opioids; it could be blocked with naloxone, an opiate antagonist. Opioids can mediate many behavioral and physiological attributes that are known to be altered by parasites (e.g., feeding, locomotion, mating, aggression, immune responses, and fearfulness). In addition, when briefly exposed to a cat, asymptomatic (low-level) infected mice failed to exhibit a non–opioid-mediated (5-HT$_{1A}$ serotonin sensitive) analgesia typical of uninfected mice when confronted by a cat. This non–opioid-mediated analgesia is thought to be induced by brief exposures to danger, whereas the opioid-mediated analgesia occurs in response to threats and stresses of longer duration. Kavaliers and Colwell (1994) suggested that the absence of serotonin-sensitive analgesia in infected mice may be relevant to the absence of fear observed in some parasitized animals when confronted by a predator (see also Stibbs 1984; Stibbs and Curtis 1987). Observing *E. vermiformis*-infected mice in a Y-maze, Kavaliers and Colwell (1995a) found that uninfected mice avoided an area with cat odor, but infected mice did not. Infected mice were capable of other olfactory discrimination (e.g., estrous females), so olfactory impairment cannot be invoked easily. This failure to respond to cat odor was not associated with endogenous opioids, but may be related to GABA neurochemistry. Berdoy and co-workers (2000) have found similar fearlessness in rats infected with *T. gondii*, a parasite known to alter personality in humans (Flegr and Hrdý 1994).

Eimeria vermiformis is directly transmitted via fecal contamination from mouse to mouse, and enhanced predation would not benefit either mouse or parasite. However, the reduced anxiety that accompanies this infection can also facilitate intermouse social interactions, and that can enhance *E. vermiformis* transmission (Kavaliers and Colwell 1995a; see also Kavaliers et al. 1997a).

Should some hosts be more prone to loss of predator aversion than others? The failure of some parasitized animals to avoid predation may be subject to the kinds of cost–benefit considerations we see in other behaviors where

continued

Fearlessness: One Mechanism (*continued*)

antipredator behavior is compromised in favor of potential fitness gains (e.g., reproduction, feeding; see Poulin et al. 1994). If this is true, then older hosts may not oppose the relaxation of antipredator behaviors as vigorously as younger ones. In a population of upland bullies (*Gobiomorphus breviceps*) that are usually intermediate hosts for the trematode *Telogaster opisthorchis* (often more than 20 cysts per fish), not only was the relaxation of antipredator behavior intensity dependent, but it was much more apparent in fish at least 2 years of age. Another population of bullies and a population of common river galaxias (*Galaxias vulgaris*) did not exhibit such an effect; they also had fewer parasites. The mechanism by which the trematode exerts this influence is not understood, but the results are consistent with an age-dependent response to behavioral alteration, especially in heavily parasitized populations (Poulin 1993a). The suicidal behavior of pea aphids when parasitized by a braconid wasp was also cost sensitive. Those that were parasitized late in life (as fourth instars) could expect to reproduce and were not suicidal; this was in contrast to those parasitized as second instars that were therefore nonreproductive (McAllister et al. 1990).

ski 1984, 1985; Barber and Ruxton 1998). The size of *S. solidus* and the resources it commands mean that the parasitized fish has energy requirements that probably exceed those of uninfected counterparts and may cause the fish to engage in risky behaviors (Milinski 1990; but see Ness and Foster 1999).

Schistocephalus-infected fish also spend more time at the surface of the water than uninfected fish do. Of course, the distended body might simply appear to be closer to an observer, an illusion manifested by fish infected with *Diplostomum spathaceum*, which become darker than uninfected fish (McPhail and Peacock 1983; Milinski 1990). However, *Schistocephalus*-infected sticklebacks consumed more oxygen than uninfected ones did (see also Walkey and Meakins 1970; Meakins and Walkey 1975; R. S. Smith and Kramer 1987; review in Milinski 1990) and were overrepresented near the shore and underrepresented in pelagic catches. One possible result of such a shift is exposure to a different set of predators, especially piscivorous birds, the final host (Lester 1971). Lester (1971) thought that perhaps the additional oxygen demand was created by the swimming inefficiency resulting from distension. The swim bladder may also be compressed by the large parasite (Meakins and Walkey 1975). In both three-spined and nine-spined sticklebacks, tests under hypoxic conditions revealed that infected fish spent more time at the surface and, again, did so even when frightened by an aerial predator stimulus (Giles 1987a; R. S. Smith and Kramer 1987). This is consistent with the fact that the plerocercoid may be more energetically demanding than the same amount of stickleback tissue would be (Walkey and Meakins 1970).

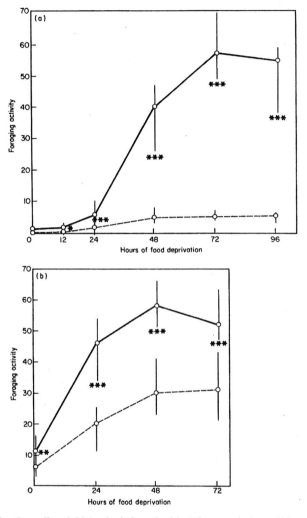

Figure 3.7. Juvenile sticklebacks infected with *Schistocephalus solidus* are highly food motivated. Solid lines indicate infected fish. Fish in the upper graph have also been exposed to a frightening stimulus. * = P < 0.05, ** = P < 0.01, *** = P < 0.001. (Reprinted from Giles 1987a, with permission of Academic Press.)

In a Canadian population of three-spined sticklebacks, hosts of *S. solidus* were more buoyant than uninfected fish. This was associated with a deme-lanization that, in turn, was probably linked to infective stages of the parasite. The resulting fish were white with dark eyes and swollen abdomens—possibly an easy catch, given the small proportions of such fish compared to normally pigmented fish with younger infections. Thus far, these characteristics have been observed in only a few populations (LoBue and Bell 1993; Ness

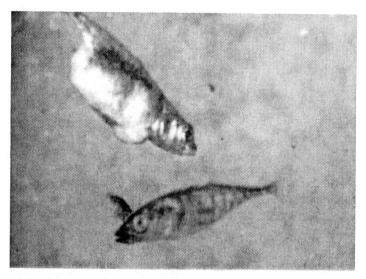

Figure 3.8. *Schistocephalus solidus* is huge in its stickleback intermediate host, creating problems in every area of behavior from feeding to reproduction. (Photo courtesy of Claus Wedekind.)

and Foster 1999) and raise questions about the existence and importance of geographic variation in host–parasite associations (see also Heins et al. 1999).

All of these alterations are not inevitable outcomes of general debilitation. For instance, although debilitation can influence competitive success, much depends on the manner in which the host is debilitated. Thus, the foraging of a blind host (e.g., dace with the trematode *Diplostomum spathaceum* in its eyes) would be compromised, as would the foraging of a stickleback with *S. solidus,* but it would not be compromised in the same way. The infected stickleback is an inefficient forager and may have energetic constraints, but it can see the prey, whereas the dace has trouble seeing food (Crowden and Broom 1980; Milinski 1984, 1985, 1990).

In analogous fashion, the copepod (*Macrocyclops albidus*) first intermediate host of *S. solidus* avoided stickleback-conditioned water that contained competitors in favor of similar water without competitors. Unlike uninfected copepods, they seemed to be more averse to competition than to predation, perhaps because of energy constraints (Jakobsen and Wedekind, 1998). This raises more widespread questions. In general, how might parasites affect zooplankton? Given the involvement of these small, abundant organisms in the life cycles of some parasites, how do parasites affect aquatic ecology? And what of the elaborate defensive morphological responses to predators that are exhibited by many zooplankton? Could a parasite interfere with the development of such defenses?

The examples described here have one thing in common: infected intermediate hosts transmit the parasites they contain by falling prey to final hosts, an event that seems to be hastened by the fact that they are infected. Thus, intermediate hosts are taken in greater numbers than one would predict from their relative abundance in the prey population; in many cases, a predator is probably more likely to encounter a parasitized prey than its uninfected conspecific, and these infected prey may be easier to catch.

Parasites that can accomplish such manipulation will experience an increase in reproductive rate, if all other things are equal, and will have a selective advantage over parasites with no such ability. This is clearly a case where the parasite that kills, weakens, or otherwise deranges its (intermediate) host can be favored by natural selection. In addition, the dynamics of predator–prey interactions will be altered substantially. Such nonrandom interactions may mean that parasitic infections can establish in host populations at densities lower than we would predict in the absence of behavioral alterations.

Examples with More Uncertainty

Increased predation on intermediate hosts has been quantified for many systems discussed thus far (see also table 3.3, p. 177). In other host–parasite systems, such increases are speculative and raise tantalizing questions. One such example is the case of *Echinococcus*-infected moose.

Echinococcus granulosus is an amazing tapeworm. The adult lives in canids, where it is less than a centimeter long. One of its eggs, when eaten by an appropriate intermediate host (many large herbivores, including humans), produces a metacestode that can asexually reproduce an indefinite number of times. This results in increasingly large and abundant cysts that can be debilitating, if not lethal, depending on their location.

All this metacestode reproduction is unusual for a tapeworm, and has unusual life-history implications (Moore 1981). One outcome may well be rendering the intermediate host more vulnerable to predators. While filming in the Brooks Range of Alaska for Walt Disney Studios, Lois and Cris Crisler came upon one possible example (see also Mech 1970). Lois Crisler reports:

> On September 29 1954, two two-year-old caribou cows came through together. When surprised by a wolf, one ran lightly away. The other ran, but remarkably slowly. Five wolf puppies that we were raising lumbered after it, single file. It turned and faced them and sank on one knee, then lay down. Hesitatingly the small wolves surrounded it. It got up and ran. Then, untouched, it faced and voluntarily lay down again. . . . Its lungs were partly deflated and contained eight abscesses, some as big as pingpong balls, half buried in the tissue and full of watery fluid. They seemed to be tapeworm cysts. (Crisler 1956, p. 345).

Infected moose are probably more susceptible to predation by wolves (Mech 1966); we know that in infected humans, for instance, exercise can result in significant chest pain. Predation of moose by wolves is nonetheless difficult to quantify (Messier et al. 1989). Infection does have a surprising consequence for a novel form of predation: Moose that are heavily infected with

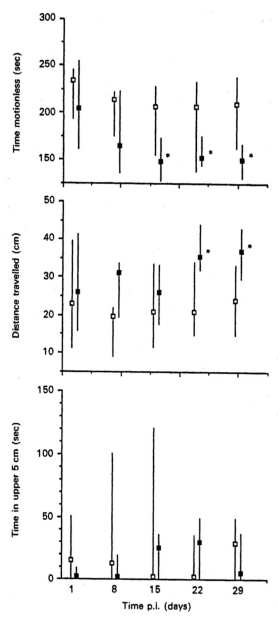

Figure 3.9. Several measures show that the copepod *Cyclops vernalis* infected with a cestode becomes increasingly active over the course of the infection, an increase that is paralleled by rising predation levels. p.i. = post-infection * = P < 0.05 (Poulin et al. 1992). (Reprinted with the permission of Cambridge University Press.)

Echinococcus are more likely to be shot by hunters than uninfected moose are, and infected moose are overrepresented early in the hunting season (Rau and Caron 1979). Again, this is an example of a host–parasite interaction that probably increases predation and transmission, but for which estimates of predation on infected and uninfected hosts are difficult to acquire (Messier et al. 1989).

At least two other well-known examples fall into this category of uncertain outcomes. One of the classic, frequently cited examples of parasite manipulation of intermediate host behavior that results in transmission is as much about appearance as behavior, and concerns a terrestrial snail, *Succinea*, that is host to larval stages of the trematode *Leucochloridium* (e.g., Rennie 1992). *Leucochloridium* sp. has always been a parasitological favorite; the sporocyst stage invades the snail tentacles and is quite conspicuous (fig. 3.10): "The first sight of the 'pulsating broodsacs' in the tentacles of the snails is a thrilling

Figure 3.10. The antennae of snails infected by *Leucochloridium* are abnormally large and conspicuous, full of trematode. The pigments, while contributing to the insectlike appearance and overall conspicuousness of the trematode, are closely associated with the muscles of the worm and may act as light receptors. (Photo courtesy of Paul Lewis.)

sight to a parasitologist" (Woodhead 1935, p. 337). Despite this remarkable advertising campaign on the part of the trematode, I am not aware of any quantitative documentation in the scientific literature of enhanced predation on these infected snails.

Leucochloridium has an unusual life cycle in that there is only one intermediate host, the snail, which becomes infected when it eats parasite eggs from contaminated vegetation. Recall the basic digenean life cycle: egg, miracidium, sporocyst/redia, cercaria, metacercaria, adult. In contrast, *Leucochloridium* sporocysts form cercariae, but these remain in the sporocysts, which develop branched extensions called broodsacs, the stage in the tentacles.

Wesenberg-Lund (1931), one of the few people to investigate predation on these infected snails, penned a delightful account of the snail and its garish parasite:

> It seems as if infested snails seek the light; they are very often seen balancing on the borders of the leaves, or sitting on the underside of the leaves with only the antennae infested with the *Leucochloridium* protruding from the borders of the leaves; in this position, when the sacks are pumping with great regularity, the aspect is very peculiar, and the parasite remarkably conspicuous. (p. 97)

Should *Leucochloridium* join the ranks of *Plagiorhynchus, Polymorphus,* and *Ligula,* parasites that induce behavioral changes in the intermediate host and that increase the probability of transmission to the avian final host? To some extent, Wesenberg-Lund (1931) agrees:

> With regard to the *Leucochloridium* it is beyond doubt that it is to the sporocyst that the problem is committed of attracting the attention of the bird to such animals as do not always belong to their normal diet. For it must be remembered that many of the small songbirds which are to take the sacks of the *Leucochloridium* are never snail-eaters and are quite unable to swallow a *Succinea*.

He is tentative, however (p. 98):

> . . . For my own part I could have snails with *Leucochloridia* before my eyes for hours. The birds were singing only a few metres from them, but nevertheless I never had the good fortune to see the birds take the parasites. I confess that this fact has always troubled me a good deal. For it must be kept in mind that hitherto no one has ever seen a bird in natural conditions attack a snail infested with *Leucochloridia*, nor pick out a sack from the antennae.
> . . . these observations . . . have puzzled me a little, and I confess I have been sorry for all the wasted efforts of the sacks pumping and pumping at a rate of about 70 strokes a minute (pp. 129, 98)

One suspects that a good deal of trematode watching preceded Wesenberg-Lund's sympathy for their predicament.

Despite Wesenberg-Lund's conviction that *Leucochloridium* has become a caterpillar in the eyes of the avian final host, and Ahrens' (1810, cited in

P. D. Lewis 1977) mistaking a broodsac for an immature insect, the empirical verification of enhanced predation in this system has been elusive. We know that the branched broodsacs (sporocysts) become active and pulsate in the presence of light, elevated temperatures, or digestive enzymes and that their appearance (color bands) is species specific. We also know that the sporocysts can spontaneously rupture from snail hosts. It would be a challenge to experimentally establish the cues that the avian predator might respond to—size, color, or caterpillar search image? This difficulty is exacerbated by the fact that snails with broodsacs can be relatively rare and difficult to collect, and some do not survive well in the laboratory. Thus, *Leucochloridium* may well be another example—and a spectacular one—of parasite manipulation of intermediate hosts, but understandably, it has not been particularly amenable to quantitative investigation.

This problem also plagues another well-known host–parasite system, the digenean *Dicrocoelium dendriticum* and congeners in ant second intermediate hosts. These parasites reach adulthood in large herbivorous mammals, which are not usually considered ant predators. Terrestrial pulmonate snails feed on eggs of the parasite and become infected, liberating cercariae in slime balls exuded from the mantle cavity. These slime balls are carried to the ant nest and devoured. When the *D. dendriticum* cercariae enter the body cavity of the ant, they begin an anterior migration that results in one metacercaria forming a thin-walled cyst in a depression in the anterior part of the subesophageal ganglion, between the nerves that serve the mouth parts; the rest of the parasites remain in the hemocoel. The infected ant behaves normally during the day, but when temperatures drop, it climbs to the tops of grass blades, mandibles opening and closing, and enters a kind of torpor, anchored by its mandibles (fig. 3.11, see also the cover). The ant is difficult to dislodge from its perch, and it may be ingested accidentally along with the grass by the herbivores that graze in the cool of the evening or the following morning. As temperatures warm, the infected ants resume normal activity (Anokhin 1966; Carney 1969, and references therein; Holmes and Bethel 1972). Note that accidental ingestion serves the parasite as well as purposeful predation; the motives of the next host are not a concern. Thus, some parasite-induced behavioral alterations may change foraging patterns by introducing items to the diet of the next host that it would otherwise rarely include.

The cercaria that succeeds in reaching the ant's brain—the "brain worm"—loses its infectivity for the mammalian host. It sacrifices its own adult reproductive capability. If one recalls how that cercaria originated, however (from an asexually reproducing mass within the previous host), such behavior can be seen to increase the trematode's inclusive fitness. It is likely that when devouring a slime ball, the ant ingests related, if not identical, cercariae, and the brain worm's infective kin stand to benefit from the altered behavior (Wickler 1976; Wilson 1977). *Dicrocoelium hospes* causes ants (*Camponotus compressicapus*) to behave in a similar, temperature-dependent fashion, with minor differences (Lucius et al. 1980; Romig et al. 1980).

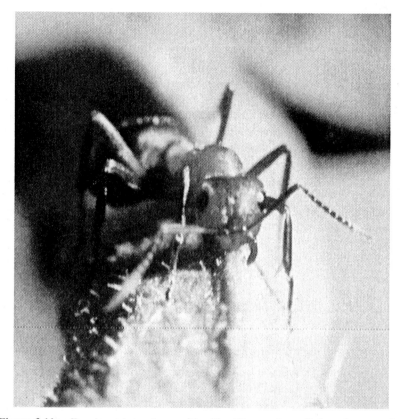

Figure 3.11. *Formica praetensis* parasitized by *Dicrocoelium* climb to the tops of plants, where they are probably inadvertently preyed upon by grazing animals. (Photo courtesy of Richard Lucius.)

Sadly, ant ingestion by herbivorous mammals is not easily quantified. A related digenean in yet other species of *Camponotus* offers more hope, however. The metacercaria of *Brachylecithum mosquensis*, another dicrocoeliid trematode, encysts near the supraoesophageal ganglion and causes its ant second intermediate host to increase time spent in open areas and on rock surfaces. Infected ants exhibit circling behavior when they move at all, and they do not respond to changes in light intensity, nor do they retreat to nests as autumn approaches. In addition, the parasite apparently causes an increase in parenchymatous tissue in the ant abdomen. The result is a larger-than-average ant with a striped appearance caused by the exposure of the intersegmental membranes in the distended abdomen (Carney 1969). These infected ants presumably are more easily noticed and preyed upon by the final host, the western robin, and ants and robins in an aviary might prove more amenable to quantitative studies than ants and herbivores in a pasture.

Romig and co-workers (1980) acknowledged that it was tempting to use the location of these trematodes as an explanation for the behavioral disruptions, but also pointed out that some trematode-infected ants exhibiting infected behavior did not have metacercaria in or near ganglia. In addition, cestodes in ants (*Leptothorax nylanderi*) are able to induce behavioral changes such as lethargy without proximity to major ganglia. [If infected *L. nylanderi* stay in the nest, one would think that predation risk would be minimized. However, when the woodpecker final host destroys the nest in search of food, these ants do not flee rapidly, and it is possible that they are more readily eaten (Plateaux 1972; Plateaux and Péru 1987; fig. 3.12).]

In an intriguing convergence, ants infected with fungi often crawl up on vegetation, also using their mandibles (Bünzli and Büttiker 1959; Carner 1980; Evans 1989; but see Sanchez-Peña et al. 1993). The behavior of fungus-infected ants is remarkably similar to that of ants infected with *D. dendriticum*, and we are left wondering if the fungus is somehow producing similar physiological effects, or if we are observing some other, more generalized response. [Some fungi do send hyphae to the area of the subesophageal ganglion and other neural structures (Evans 1989).] If this elevated location really does enhance accidental predation, as it may with helminth-infected ants, how do the fungal spores fare in their passage through the predator? Does elevation benefit them by enhancing windborne spore dispersal? What can we learn from the [unusual] fungi that do not alter behavior (Sanchez-Peña et al. 1993)? Finally, some insects infected with fungus can defeat the parasite with high enough body temperatures, something that can be achieved at the tops of vegetation and exposed places (e.g., Watson et al. 1993; see chapter 5). This probably does not apply to *F. rufa* and other hosts that seek higher elevations at dusk, but we may still ask if elevation seeking was initially a febrile behavior that has been captured by some parasites (table 3.6, p. 183). In that case, timing may be everything (chapter 5).

Pathology and Adaptation

Multiple approaches—including comparative analysis—may be necessary to sort out the evolutionary influences on parasite-induced behavioral alterations (Moore and Gotelli 1996). The role of adaptation, for instance, can be unclear. One seemingly unambiguous way to categorize parasite-induced behavioral alterations involves partitioning them into behaviors that benefit the host, those that benefit the parasite, and those that are side effects of the host–parasite interaction (e.g., pathology) and of no benefit to parasite or host (Jones 1985, Minchella 1985, Milinski 1990, Moore and Gotelli 1990). The relationship of pathology to altered behavior is substantial and has been reviewed by Holmes and Zohar (1990).

The moose–wolf–*Echinococcus* story seems to be a fairly unambiguous example of debilitation resulting in weakened prey that can then transmit the parasite more efficiently. One could argue that such pathology differs fundamentally from some of the more subtle measures described above, such as parasites floating in host hemolymph and (biochemically?) altering phototaxis or responses to substrates, that pathology is somehow less refined, possibly even accidental. In the case of *Echinococcus*, the asexual reproduction and increase in parasite mass that proves to be so pathological to the moose intermediate host may be a consequence of parasite life-history adaptations, with that asexual reproduction occurring in the metacestode (Moore 1981). In that case, the result of increased predation could be coincidental. Then again, it may be that *Echinococcus* has responded to natural selection that favors the damage the worm causes the moose, speeding it on its way to the wolf.

Another example of behavioral change that has a pathological basis of uncertain evolutionary origin is that of the trematode metacercariae that encyst in fish eyes, causing blindness and a variety of subsequent ecological and behavioral effects (Crowden and Broom 1980; Brassard et al. 1982; Lafferty and Morris 1996). Again, we find ambiguity. Of course, these effects could be adaptive for a parasite that is transmitted in the food chain. Then again, because the eye is a protected place, immunologically speaking, the initial selective advantage associated with encysting in this location may have been escape from host immune defenses, with the concomitant behavioral alterations a fortuitous coincidence (see Szidat 1969). But perhaps this is a side effect of unavoidable pathology—an explanation that is reasonable to invoke when the nervous system is invaded.

In this regard, parasites from several clades have found a home in the host nervous system, with varying outcomes. The protistan *Toxoplasma gondii* invades the brain of rats and is associated with activity increases. These activity changes may not be byproducts of encephalitis; other brain parasites in this rat population (e.g., *Coxiella burnetti*) were linked to various pathologies but did not influence activity. Of six such parasite species, *T. gondii* was the only one associated with increased activity and the only one transmitted by predation (Webster 1994; but see Hutchison et al. 1980a). In fact, *T. gondii* can be transmitted both by predation and congenitally, and mode of infection also influences the way that activity is altered. This fact also undermines any general debilitation explanation for *T. gondii*'s effect (Hutchison et al. 1980b; see also Burright et al. 1982; Dolinsky et al. 1985). [Oddly, *T. gondii* affects social interaction in mice (Arnott et al. 1990), but not in rats (Berdoy et al. 1995), providing yet another example of host-specific behavioral alterations.]

Thus, it can be difficult to disentangle pathology that occurs because of, say, evolutionary or physiological constraints on the parasite–host association and pathology that is actually part of an adaptive complex. Dawkins acknowledged this problem and used it as a platform from which to illustrate extended phenotypes (1982, 1990). Read (1994) grappled with the ideas of pathology and adaptation and suggested, among other things, that pathology be considered a null hypothesis against which adaptive hypotheses can be rigorously tested; he warned that evidence refuting such a null hypothesis "may be difficult or impossible to obtain" (p. 120). Poulin (1995) proposed four characteristics of an adaptive behavioral alteration: complexity (admittedly subjective and difficult to gauge); purposiveness of design—conformity between a priori design specifications and the modification; convergence—similar adaptations in several independent lineages; and fitness effects. He went on to suggest some useful general guidelines that could assist in determining whether a behavioral change is adaptive. Endler (1986) has also provided guidelines for discerning natural selection in natural populations. We do not know enough about most host–parasite associations to apply many of these criteria, a reminder that our assumptions about adaptation may be just that—assumptions.

What about the effects of parasites on novel hosts? Behavioral alterations in a novel host are unlikely to be adaptations, although in some cases they may have been adaptive in an ancestor of that host—say, the progenitor of a clade in which the parasite has been retained by some, but not all, descendant species. In the absence of such history, and in a novel host, behavioral alterations could reasonably be viewed as nonadaptive.

In this vein, while gathering data for a comparative study of behavioral alterations in seven species of cockroaches parasitized by the acanthocephalan *M. moniliformis* (see Moore and Gotelli 1996), we were surprised to find one cockroach that was unaffected by the parasite in any of our behavioral tests (Allely et al. 1992). This cockroach, *Diploptera punctata*, a blaberid, was ecologically and phylogenetically far removed from the others in the study, and we wondered if such divergence might help explain the absence of behavioral modification. Unfortunately, the blaberids as a group are often stubbornly resistant to *M. moniliformis* (Moore and Crompton 1993), eliminating opportunities to behaviorally test most other infected blaberids.

It is a mistake to view every altered behavior as adaptive when we are ignorant of its fitness consequences or, for that matter, if we have not experimentally identified the linkage between infection and alteration. Some animals may differ from others behaviorally to begin with, and that may be the origin of both "altered" behavior and infection. In addition, although characteristics such as sudden shifts in behavior that correspond to parasite infectivity or a specialized feature uniquely linked to the behavioral modification do support an adaptive hypothesis, the absence of these traits need not preclude an adaptive function. Finally, the absence of demonstrable differences in, say, reproductive success does not necessarily mean the absence of any difference of evolutionary importance. Differences that are too small to be revealed in the sample sizes of most experiments or observations can nonetheless have evolutionary consequences. Thus, it may also be a mistake to be too willing to view pathology and its behavioral consequences as an accident.

Some of this willingness to evoke an "accidental" association may be a holdover from the traditional idea that pathology is an almost unnatural outcome of host–parasite interaction. If given enough time, says the Panglossian ideal, these associations would naturally evolve toward some benign state. All heritable aspects of host–parasite interactions are potentially subject to natural selection, however, and pathology is no exception. Returning to May and Anderson's basic model, if pathology increases reproductive ratio of host or parasite, then it will be favored by natural selection.

Viewed in this light, the distinctions between pathology and adaptation become blurred indeed. The evolutionary origins of parasite-induced behavioral modifications may be lost to us in many, even in most, cases, just as the evolutionary origins of many other traits are unclear. If a trematode accidentally wandered into a nervous system and then found itself dispatched to a final host more quickly, and if that tendency to wander into a brain could be inherited by that trematode's offspring, then there would be no reason to disclaim its adaptive nature. After all, mutations arise by accident, and if they prove beneficial and are favored by natural selection, they are said to be adaptive. To make things worse, we are ignorant of the genetic basis of any behavioral alteration in an intermediate host (see Hinnebusch et al. 1996 for genetics, fleas, and plague). Perhaps we can assume that if a kind of pathology is dependably part of a host–parasite complex, then some genetic basis for the pathology is likely in at least one of the participants. Such assumptions risk lack of rigor; on the other hand, in that (heritable) case, pathology may be seen as another host or parasite trait, potentially a product of selection. Finally, despite our ignorance of history, we can at least ask about the current beneficiaries of the modifications. Does the pathology affect parasite or host fitness? As will become apparent, even that question can yield multiple, sometimes ambiguous, answers.

Behavioral alterations are abundantly associated with parasitism, and given the examples reported in this chapter, one might nonetheless conclude that, in general, parasites can be depended upon to produce almost ubiquitous increases in transmission rates wherever they are found. Despite the rather disconcerting complications that this scenario holds for ecologists, who can no longer assume that relative abundance of prey is necessarily related to encounter rates with predators, a world infested with manipulative parasites would still not be overly complex. One can envision (if wildly optimistic) that enough research could yield a kind of corrective behavioral index for a given type of parasite–host system. At the least, we would know that transmission rate is likely to be higher than it seems based on population estimates alone. Indeed, in at least two studies, altered behavior seems to be almost fine tuned so that not just any predator, but final hosts in particular, are more likely to be involved in the predation event (Bethel and Holmes 1973; Webber et al. 1987a).

The nature of the predator–prey interaction that transmits the parasite may advise us in some cases about the types of alterations that could be expected. If the final host is a predator that chases prey, then debilitating alterations are likely to increase transmission (table 3.7, p. 191). Prey that depend on crypticity to avoid predators may be modified in other ways (e.g., color changes; Holmes and Bethel 1972; Moore and Gotelli 1990; table 3.2, p. 171). A parasite that debilitates its intermediate host may increase its likelihood of being taken by some predators (e.g., inefficient ones) more than others (Tiner 1954; Milinski and Löwenstein 1980). For instance, the genus *Diplostomum* (in the nervous systems of fish) exhibits great variation in the intermediate host debilitation it causes, and the variation correlates with the efficiency of the definitive host predator (Szidat 1969).

Poulin (1994) wondered how parasite manipulation of hosts might evolve. He noted that in some cases, the ability to alter behavior may not be costly to the parasite. This will be true when the altered behavior occurs as a corollary of something the parasite might do under any circumstance. For instance, encysting in muscle or nerves will affect behavior, but these encystment sites may be favored for reasons other than or in addition to the fortuitous behavioral change. On the other hand, he noted that if manipulation requires additional "manipulative effort" on the part of the parasite, then the optimization of this effort may be subject to other influences, such as parasite infrapopulation size, transmission rates, fecundity, or longevity. Thus, not only do we expect different types of manipulation from different parasites, we also expect differences in manipulation effort among parasites. Unfortunately, we know little about manipulation effort in most host–parasite systems.

Even if we did understand far more than we do about transmission dynamics and costs of manipulation, the assumption that altered behavior is always adaptive is flawed. For instance, although a trait may be favored in some interactions because it enhances transmission, it may also be present because it is a corollary of another highly favored aspect of the interaction. It may therefore be a constraint of the host–parasite interaction (see Read 1994). One possible outcome of such constraints would be increased risk of predation without concomitant transmission. In some cases, parasite survival may even be cut short if, say, unavoidable pathology increases transmission to an inappropriate host (table 3.8). One would be hard pressed to argue that such modification was adaptive.

This is probably the case with red grouse (*Lagopus lagopus*) infected with the cecal nematode *Trichostrongylus tenuis* (chapter 2), a parasite that can be instrumental in host population cycles (Hudson and Dobson 1991; Dobson and Hudson 1992; Hudson et al. 1992). Peter Hudson and co-workers (1992) realized that as predator control measures intensified, the number of *T. tenuis* in grouse increased. In addition, birds that were killed by predators had higher nematode intensities that those that were shot by hunters. By treating infections in some birds, Hudson and co-workers showed that dogs could locate infected birds more readily than they could locate uninfected ones. *Trichostrongylus* causes cecal pathology and may well hinder incubating hens from reducing the scent associated with cecal feces (Hudson et al. 1992). Predation on odoriferous, infected grouse would appear to be the adaptive result of parasite manipulation and enhanced transmission if it were not lethal for the worms. Instead, this hints at a linkage between pathology and increased predation that benefits neither host nor parasite. Nonetheless, it has an ecological impact.

Do we then assume that without selection to the contrary, host–parasite associations will lead to increased predation simply because of a weakened or somewhat deranged host? Perhaps all those embattled moose with golf ball-sized *Echinococcus* cysts in their lungs support such a notion. Then again, such pathology–transmission linkages may not be as unquestionable as they initially appear. Monogenean infections caused guppies to swim in a lethargic and unusual manner; this attracted other guppies, favoring guppy-to-guppy transmission of monogeneans (Scott 1985). When largemouth bass (*Micropterus salmoides*) were hosts to monogeneans (*Dactylogyrus*), they also became sluggish, but in this case they were more vulnerable to bowfin predators (*Amia calva*; Herting and Witt 1967), a lethal result for both parasite and host. In contrast, fish with other ectoparasites were not sluggish, neither were they at risk of increased predation (Vaughan and Coble 1975; but beware of possibly confounding population effects).

Thus, parasites, even ones that are superficially similar, do not inevitably enhance transmission, nor do they inevitably cause the kind of pathology that results in increased predation, with or without transmission (table 3.8, p. 201; Webster et al. 2000). Recall that *Gammarus lacustris* containing *Lateriporus* metacestodes behaved normally (Bethel and Holmes 1973), despite the advantages that would accrue to the parasite if predation were favored, despite the fact that other parasites behaviorally alter *G. lacustris*, and despite the fact that other tapeworms can alter the behavior of crustacean intermediate hosts in ways that enhance transmission (e.g., Poulin et al. 1992).

And what about *Echinococcus*? Many other taeniids (using similar, if not identical, hosts) do not have such a debilitating effect on their intermediate hosts (Moore 1981). Asexual reproduction of larval *Echinococcus* results in large lesions. Are these favored because they enhance transmission? Because, by becoming large, they can escape some immune responses? Because such cysts, when consumed, give rise to thousands of tiny adult cestodes with the combined reproductive ability of a large *Taenia*, but capable of exploiting different intestinal environments? We do not know.

Meanwhile, the case of the grouse and the dogs that were able to locate the parasitized birds brings up another caution: many of the behavioral alterations that are best investigated are those that seem most obvious to human observers. Probably because of our visual bias, alterations that would increase nonvisual conspicuousness have been virtually ignored. Parasites can alter host odor (Hudson et al. 1992) and responses to chemical cues, be they pheromones (Hurd and Fogo 1991; Carmichael et al. 1993) or food (Etges 1963). They can also change the sound a flying insect makes (bees: Lundberg and Svensson 1975; midges: Wülker 1985) and response to sound (Morris et al. 1975; see also Zuk et al. 1998a). The natural world may abound with parasite-induced changes that are not conspicuous to human senses, changes that scientists do not even imagine, much less seek.

TRANSMISSION BY ARTHROPOD VECTORS

Although predation upon intermediate hosts is one of the most dramatic methods of getting from one host to another, it is by no means the only route. In fact, most of the intestinal parasites that humans acquire from consuming intermediate hosts cause relatively little damage. Our role in the lives of these parasites is nothing more than to provide food and a place to live while they disseminate eggs for the next intermediate host, and in that case, minimal virulence will usually enhance transmission and maximize parasite reproductive ratio.

Different evolutionary pressures confront the parasites that live in the bloodstream or tissues, especially if they enter and exit that environment through the good graces of a vector. Partially because of these evolutionary pressures, at least 2 million people die of malaria annually, a tiny fraction of those debilitated by the disease (>400 million cases; Marquardt et al. 2000). Despite these numbers, quantitative estimates of transmission success in vector-borne parasites are in many cases as difficult to come by as are estimates of ant predation by sheep.

Transmission from Arthropod to Vertebrate

What alterations in host behavior might we expect from vector-borne parasites that maximize reproductive rate? Again, we see enhanced transmission in expected, and unexpected, ways. Although almost every type of blood-feeding organism is capable of transmitting parasites [even leeches are elements in the life cycles of some blood parasites (trypanosomes) that infect aquatic hosts], we know the most about hematophagous insects.

Foraging is central to the transmission of parasites by these insects and is often modified by parasites (James and Rossignol 1991; table 3.9, p. 203). Ribeiro and co-workers (1984) discussed possible parasite strategies for insect-to-vertebrate transmission. These include the promotion of salivary pathology, an increase in probing, and a decrease in persistence and blood location ability. The impaired feeding of hematophagous insects may reduce their fitness: Infected insects may be malnourished, produce smaller (or no) egg clutches, and incur more risk from defensive hosts as they attempt to feed. In addition, insects that have trouble feeding may probe numerous times, spend a long time feeding, and may be interrupted and continue foraging on a different host. Several parasite–insect associations have converged upon altered foraging, but the mechanisms that induce it vary and may range from physical blockage and salivary gland pathology to sensory interference and food depletion (Schaub 1992). In general, the complexity of the blood-feeding process provides numerous opportunities for interference by parasites (Ribeiro et al. 1985; Rossignol 1988; James and Rossignol 1991).

In this section, I must abandon my earlier claim for emphasis on examples in which increased transmission has been documented. Although demonstration of preferential predation is difficult in many predator–prey interactions, proof of behaviorally increased vector capacity is even harder to generate and laboratory model organisms may not always be critical in field transmission. What we do know is that vector behavior is frequently altered in ways that would seem to enhance transmission, and that it is physically possible for the behaviors (e.g., increased probing) to result in multiple infections.

Popular lore to the contrary, medical entomologists have known for decades that the parasites transmitted by insects are not benign in their insect hosts. The ways in which they are not benign invite notice. As early as 1914, A. W. Bacot noted that the plague bacillus (*Yersinia pestis*) entered vertebrate hosts when regurgitated by a flea as it attempted to feed. A valve in the flea's proventriculus is covered with spinelike cells, and these catch plague organisms. Bacot reported that the bacteria reproduce there and can clog the foregut, thus causing regurgitation of parasites into the wound when the flea attempts to feed. Even if the valves rupture, infectious organisms can be transmitted in the backflow of blood from the stomach (see also Bacot and Martin 1914).

We now know that impaired feeding is a common result of parasitism in hematophagous insects (Molyneux and Jefferies 1986; Moore 1993; see table 3.9). In the case of the plague bacteria (*Yersinia pestis*) blocking the flea proventriculus, the underlying mechanisms are still being elucidated (fig. 3.13; Hinnebusch et al. 1998), but the blockage is visible under low power (10×) magnification (Holdenried 1952). Bacteria multiply in this block, and some are washed back into the host as the flea tries to feed with limited success. The blockage is inversely related to temperature; a 110-kilobase plasmid must be present for the bacterium to colonize the flea (Hinnebusch et al. 1998). Inside the vertebrate host, other temperature-dependent gene expression occurs,

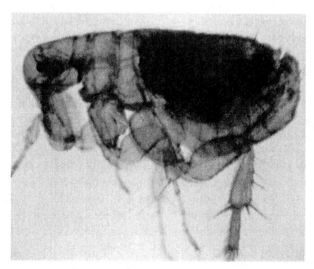

FIgure 3.13. A flea with its proventriculus blocked by a plague infection. (Photo courtesy of B. Joseph Hinnebusch, NIH.)

which results in a variety of events that are both pathogenic and that impede host defensive responses. The death of that first host causes the flea to seek another host—a kind of flea manipulation by *Y. pestis* with the dead vertebrate as intermediary—and to repeat the transmission process (R. E. Thomas 1996). (NB: Flea species are not equally susceptible to blockage.)

In addition, another chromosomal segment contains the hemin storage locus (*hms*). *hms+ Y. pestis* block the flea proventriculus and eventually starve the flea; *hms− Y. pestis* do not colonize the proventriculus, but remain in the flea stomach, where they do little harm. Hinnebusch and co-workers (1996) noted that from a human perspective, this locus influenced plague bascilli transmission, but from the perspective of the flea, it influenced virulence.

The first report of a protistan parasite affecting the feeding of its insect host came from *Leishmania mexicana amazonensis* and the sandflies they inhabit (*Lutzomyia longipalpis*; Killick-Kendrick et al. 1977; Killick-Kendrick and Molyneux 1990). The infected flies probed multiple times, a condition that correlated with the parasites' presence in the cibaria, not the head or abdomen. The parasites may interfere with receptors; they could also physically block flow, similar to the situation in fleas (Shortt et al. 1926; Killick-Kendrick et al. 1988; Schaub 1992). Infected sandflies fed on more mice than their uninfected counterparts did (Beach et al. 1985, observing *Leishmania major* and *Phlebotomus doboscqi*), a phenomenon linked to the subsequent appearance of multiple leishmania lesions (Killick-Kendrick and Molyneux 1990). Salivary gland products of sandflies may also assist parasite development, possibly by interfering with immune function (Titus and Ribeiro 1988, 1990).

The causative agent of Lyme disease (*Borrelia burgdorferi*) was transmitted more effectively by ticks than by syringe, probably because tick saliva inhibits neutrophils and interferes with complement. Microfilariae may actually be attracted to mosquito saliva and sites where they are likely to be ingested by mosquitoes (Stark and James 1996; see also Jones et al. 1989; Wikel 1999).

Indeed, the saliva of hematophagous arthropods is a pharmacopeia of biologically active compounds, well stocked with anticoagulants, vasodialators, anti-inflammatory agents, and perhaps even immunosuppressants. Titus and Ribeiro (1990) argued that because of this, experimental infection using syringes may be less natural than one might expect: "blood-sucking arthropod vectors are not simply flying syringes" (p. 157; see also Ribeiro 1995; Nuttall 1998).

Saliva is essential for efficient foraging by mosquitoes; the apyrase it contains serves an important function in blood vessel location and inhibits platelet aggregation, thereby facilitating blood uptake (Ribeiro 1984; Ribeiro et al. 1984). Salivary gland pathology is instrumental in the unsuccessful probing that characterizes *Plasmodium gallinaceum*-infected *Aedes aegypti*. In uninfected mosquitoes, salivary apyrase is injected into the host during feeding and inhibits the aggregation of platelets, which in turn inhibits clotting. As anyone knows who has gotten to the end of an old-fashioned malt or milkshake made with real ice cream, the efficiency of sucking liquid through a straw plummets when such delicious "clots" clog the soda straw. Without salivary apyrase, the mosquito encounters an analogous problem: the hematoma that usually forms in the area of a pierced blood vessel, increasing mosquito feeding efficiency, does not form when the blood is allowed to clot. The *P. gallinaceum* sporozoites do not damage the mosquito salivary glands indiscriminately. Instead, they concentrate in the areas of the salivary gland devoted to apyrase production; this leaves the rest of the gland able to continue to make normal amounts of saliva, thus allowing the entrance of infectious parasites into the vertebrate host. The saliva is apyrase deficient, however, and feeding is much more difficult, requiring prolonged and repeated probing, perhaps on several hosts (Rossignol et al. 1984, 1986; fig. 3.14).

There are sure to be limits to the ways and extent that host behaviors can be altered, based on the available range of possible behaviors and physiologies. In anopheline mosquitoes naturally infected with *P. falciparum*, probing was increased, but there was no evidence of impaired feeding among infected mosquitoes, which were able to acquire full blood meals with the same frequency of uninfected mosquitoes (Wekesa et al. 1992). The failure of one malarial organism (*P. berghei*) to affect feeding in its mosquito host (*An. stephensi*) illustrates another constraint on how parasites may manipulate hosts. It appears that infected *An. stephensi* experience no difficulty in feeding. Salivary apyrase levels in *An. stephensi* are naturally low, certainly lower than those of *Ae. aegypti* or *Ae. freebornii*, so that *An. stephensi* may not be subject to this type of manipulation of feeding behavior (Li et al. 1992).

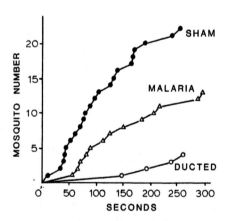

Figure 3.14. Rossignol and co-workers (1984) compared probing time of saliva-deprived (ducted) mosquitoes, sham-operated mosquitoes, and mosquitoes infected with malaria. The infected mosquitoes took longer to probe than sham-operated ones, although not as long as mosquitoes that had been completely deprived of saliva. (Reprinted with permission of *American Journal of Tropical Medicine and Hygiene*.)

Although saliva is important, there are a variety of other ways to impede feeding. Protists such as *Trypanosoma congolense* may interfere with the feeding of their tsetse fly host (again, causing repeated probing and difficulty in engorging) by gathering around specific mechanoreceptors in the labrum of *Glossina morsitans*. These sensilla monitor the flow of incoming blood, and rosettes of trypanosomes have been observed to be almost tangled around the receptors. They may not only compromise the function of the sensilla, but could also physically slow the ingestion of blood (Molyneux and Elce 1979; Molyneux 1980; Molyneux and Jenni 1981). Jenni and co-workers (1980) observed a similar phenomenon in *G. morsitans* and *G. austeni* infected with *T. brucei* (see also Roberts 1981). The composition of saliva is also affected in infected flies, and in addition, their salivary glands leak cholinesterase, a condition that progresses with the development of the infection. This, too, could affect feeding efficiency (Patel et al. 1982; Golderm et al. 1987). In contrast, some workers have not observed significant feeding problems in infected tsetse, so these phenomena may not be universal or may not have equally dire effects on all flies (see Moloo 1983; Moloo and Dar 1985; Chigusa and Otieno 1988; Makumi and Moloo 1991). As is true with most hematophagous insects, condition of the vector and intensity of infection may be partially responsible for varying results. These parameters deserve careful monitoring (Molyneux and Jefferies 1986). Such variation is probably not uncommon in many host–parasite associations that show behavioral alterations, and may even be expected, along with the variation we anticipate in much of the natural world.

In general, flagellate–insect associations may offer rich opportunities for comparative study. Although many insects have flagellates, the effects of the flagellates on their hosts vary widely, as do their locations within hosts (Grewal 1957). In some cases, the flagellates are involved in life cycles that alternate between animal hosts; in some cases they alternate between animal and plant hosts, and in others, the insect may be the only host. Is insect foraging affected in all cases? How are epidemiologic considerations affected across such an array of host taxa?

Returning to arthropod vectors, there is less information about chelicerates than about biting flies, but ticks (*Ixodes trianguliceps*) that feed on rodents infected with *Babesia microti* experience greater feeding success and molting success than other ticks (Randolph 1991). This could occur if *B. microti* had antihemostatic effects on the host circulatory system and/or if immunity to ticks was suppressed, and could well have salubrious effects on tick fitness. Because of the difference in feeding habits and schedules between ticks and many hematophagous insects, pathogens that are transmitted by ticks could be especially sensitive to the repercussions from impaired feeding (and reproduction) on the part of their vectors (Randolph 1991).

Altered probing is therefore thought to characterize a wide range of arthropods that can transmit bloodborne parasites. What are the epidemiological consequences of such probing? Increased transmission is certainly a possibility. Difficulty in feeding and the increased probing that accompanies it can result in parasite transmission, and perhaps transmission to more than one host, if difficult feeding means that vectors are more easily interrupted in their meals, which are then resumed later elsewhere (e.g., Gargan et al. 1983; Rossignol and Rossignol 1988). In houses in Namawala, Tanzania, *Anopheles gambiae* infected with *Plasmodium falciparum* were more likely to have fed on multiple hosts than were uninfected mosquitoes, based on DNA from host blood (Koella et al. 1998a). In addition, probing alone can transmit *Leishmania mexicana* and *Trypanosoma vivax* (Strangways-Dixon and Lainson 1966; Makumi and Moloo 1991), and one vector can infect multiple hosts with some parasites (e.g., *L. major, P. gallinaceum, T. brucei,* and dengue-2 virus; Corson 1932; Beach et al. 1985; Killick-Kendrick and Molyneux 1990; Kelly and Edman 1992; Putnam and Scott 1995). Given these observations, if multiple feeding occurs, then models that assume one engorgement per gonotrophic (feeding and subsequent egg production) cycle underestimate transmission potential (Klowden 1988; Kelly and Edman 1992). In fact, mosquitoes (*Culex pipiens pipiens*) that feed multiple times but that do not acquire enough blood to reproduce may nonetheless transmit St. Louis encephalitis virus or *B. pahangi* (a nematode) (Mitchell et al. 1979; Klowden and Lea 1981). Innoculation of more than one clone of parasite can increase outcrossing and spread of genes such as those for drug resistance (Dye and Williams 1997). In contrast, it is possible that feeding on multiple hosts may reduce the size of inocula and the probability of transmission (Burkot et al. 1988). Infection may increase the tenacity of some mosquitoes, which can also affect biting rate (Koella and Packer 1996; Anderson et al. 1999). While we tend to categorize hosts as "infected" and "uninfected," this masks ignorance about the influence of inoculum or, for that matter, manner of exposure, on subsequent infection (Scott et al. 1988; Randolph and Nuttall 1994).

This apparent enhancement of transmission is consistent with the observation that infected vectors may be rare in areas where infected vertebrate hosts are more common, just as infected intermediate hosts can be rare in areas where infected final hosts are not (e.g., May and Anderson 1979; Jenni et al. 1980; Roberts 1981; Beach et al. 1985). This seeming imbalance can be explained by the fact that only a few vectors are necessary to transmit parasites to several hosts (Moloo 1983) or that vertebrate hosts are long-lived reservoirs, requiring few vectors at any one time. Meanwhile, transmission enhancement may carry a price. The engorgement problems that lead to increased probing may reduce fecundity (Freier and Friedman 1976; Grimstad et al. 1980), and one might expect the evolution of resistance to the parasite on the part of the vector (Moloo 1983). Rossignol and co-workers (1985) addressed this problem by noting that for some parasites (e.g., *Plasmodium*), the developmental time needed before becoming infective to the vertebrate host exceeds that required for one gonotrophic cycle. Thus, if feeding on an infected vertebrate is somehow easier than feeding on an uninfected host, or if infected hosts are more readily located (see below), the mosquito will gain that initial advantage (ease and safety of foraging) during at least one oviposition cycle before the parasite is ready for transmission and associated foraging problems

(Daniel and Kingsolver 1983; Day and Edman 1983; Ribeiro et al. 1985; Rossignol et al. 1985; see also Fialho and Schall 1995; Koella 1999).

Meanwhile, fundamental differences between chelicerate and insect life histories may have profound implications for transmission potential. As Randolph (1998) has cautioned, compared to insects, ticks are long lived, with relatively low mobility, few hosts, few oviposition bouts, and infrequent blood meals. We are accustomed to thinking of biting flies as ephemeral links from one vertebrate host to another, but ticks may affect transmission dynamics in very different ways. Not all vectors are alike.

In summary, it is clear that many bloodborne parasites influence feeding behavior of vectors. We know little about how this affects transmission, although it is reasonable to expect that it would.

Transmission from Vertebrate to Arthropod

The return trip (vertebrate-to-insect transmission) also can be enhanced by parasites if they induce the following alterations (Scott et al. 1988, 1990): decreased defensiveness (Day and Edman 1983; Scott et al. 1988); altered blood characteristics (Rossignol 1988); and increased host attractiveness (Mahon and Gibbs 1982). Although lethargy may conserve the energy of an infected host (Hart 1988), it may also produce a docile host and an easy meal (and parasite source) for an insect. *Defensive behavior* is an important component of host choice by vectors and contributes to interrupted feeding; vectors are understandably discouraged by defensive hosts, which can be life threatening (chapter 4).

Some bloodborne parasites reduce defensive behavior on the part of infected vertebrates, thus enhancing transmission to vectors. This means that yet more uninfected vectors can acquire the parasites as the vectors feed on blood sources whose acquiescence favors the transmission of bloodborne parasites to the next vector. Because mice infected with *Plasmodium berghei* and *P. chabaudi* cease defensive behavior when gametocytes are in the peripheral blood, they are most infective to mosquitoes when mosquitoes are most able to feed (Day et al. 1983; Day and Edman 1983). In this way, not only do pathogenic forms of *Plasmodium* find an entry to the next vector, but less virulent species of *Plasmodium* may hitchhike with those that reduce defensive behavior (Day and Edman 1983).

Not all infections interrupt host defensive behavior; asymptomatic infections by either filarial worms or malaria in humans did not result in foraging bias (Burkot et al. 1989). In addition, host defensiveness does not inevitably impede transmission. Squashing a feeding, infected sandfly may actually allow entry of the parasites it contains (*L. mexicana*; Strangways-Dixon and Lainson 1966).

Again, much less is known about tick-borne parasites, but in the case of such parasites, increased host activity would increase exposure to questing vectors. Randolph (1998) alludes to some evidence for this with *Babesia* in rodents.

Infected vertebrates also may provide easier and safer meals for vectors because of *changes in blood characteristics* (Rossignol 1988). *Aedes aegypti* did not feed as long on mice infected with *P. chabaudi* and on hamsters infected with Rift Valley virus as they fed on uninfected animals. They seemed to be able to locate blood and feed more efficiently. Rossignol and co-workers (1985) suggested that such parasites might disrupt host hemostasis. Such disruptions may not be side effects of disease. They may play an active role in transmission. In fact, one explanation for the anemia that accompanies *P. berghei* infection in mice is that it could enhance vector feeding (Day and Edman 1983; but see Freier and Friedman 1976). In the case of the sandfly *Lutzomyia longipalpis,* feeding was random on uninfected mice, but probing increased when the sandflies were on the parasite-rich lesions that characterize mice infected with *Leishmania mexicana* (Coleman and Edman 1988). Blood is close to the surface in these lesions and that can be important to sandflies, some of which are pool feeders and lack long mouthparts.

In some cases, hosts infected with other parasites are actually *more attractive* to mosquitoes (e.g., Mahon and Gibbs 1982). Qualitative changes in blood as a result of parasitism could favor foraging on infected hosts (Daniel and Kingsolver 1983).

Many lines of evidence point to parasite influence on vector feeding. Models of vector-borne disease transmission show that such influences may have profound epidemiological consequences (Kingsolver 1987; Rossignol and Rossignol 1988; but see Dye and Hasibeder 1986; Burkot 1988; Randolph and Nuttall 1994). In general, the likelihood of parasite transmission and establishment increases when host–parasite contact is not random (Dye and Hasibeder 1986).

Blood-feeding arthropods do not forage randomly. Their choice of hosts is influenced by the presence of parasites and by other cultural and demographic factors, and all of these can bias transmission rates for bloodborne parasites (Burkot 1988).

Behaviors Other Than Feeding

Other behaviors can be affected by bloodborne pathogens, and a shift in arthropod elevation preferences is among them. Although little of this information is available for biting flies, the bacteria that causes Lyme disease (*Borrelia burgdorferi*) alters a variety of behaviors in black-legged ticks (*Ixodes scapularis*) in complex, age-related ways. Some of these include changes in the elevations that the ticks seek in the laboratory (Lefcort and Durden 1996; fig. 3.15). Ginsberg and Ewing (1989) found that in the field, infection did not seem to alter tick habitat preference. Within a habitat, however, Lefcort and Durden's findings may well have relevance.

Mather and co-workers (1993) tested the idea that burning an area should reduce Lyme disease risk. They burned a woodland understory, and found that although tick abundance was generally reduced in the burned area, the proportion of infected ticks in that area was greater than in an unburned area. Thus, the risk of encountering an infected tick in the burned area may not dif-

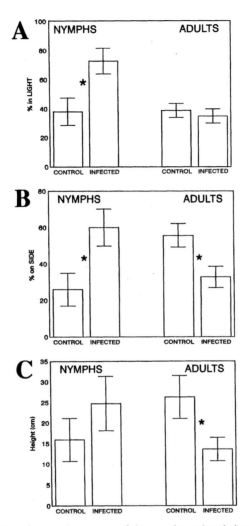

Figure 3.15. These figures show some of the age-dependent behavioral alterations that Lefcort and Durden (1996) found in ticks infected with the causative agent of Lyme disease. *$p < 0.05$. (Reprinted with permission of Cambridge University Press.)

fer from that in the control site, despite fewer ticks in the former location. Based on lower prevalence of a nematode (possibly from deer) in ticks from the burned area, Mather and co-workers speculated that the survival of ticks infected with *B. burgdorferi* in the burned area may reflect a greater impact of the fire on deer-derived (adult) than on rodent-derived (nymphal or larval) ticks; the latter, perhaps living in burrows, may have been protected from the fire. Coincidentally, they were also more likely to harbor *B. burgdorferi*, which is transmitted to ticks from rodents, not deer. If *B. burgdorferi*-infected ticks also differ from uninfected ticks in other microhabitat preferences, this could influence burn survival as well. Such behavioral differences would not necessarily be apparent from comparisons of different habitats.

Parasites also affect activity levels in hematophagous insects (table 3.7, p. 191). Although one might argue that reduced activity could keep vectors near host kin (i.e., other susceptibles?), it is difficult to see how vector lethargy could routinely benefit parasite or insect host, unless, of course, the parasite was also transmitted by predation (e.g., Edman and Scott 1987; Dirie et al. 1990).

No matter what the cause, activity differences might be expected to affect dispersal probabilities. There is surprisingly little information about how parasites might influence vector dispersal in the field. Moreover, dispersal may be confounded with shifting microhabitat preferences that could also result from parasitism.

In summary, behavioral alterations of the vector may increase parasite transmission, enhancing basic reproductive rate. In addition, virulence in the vertebrate host can result in decreased defensive behavior. When this decrease coincides with occurrence of infective stages in the peripheral blood, such virulence, and the accompanying change in behavior, can increase the probability of transmission to new vectors—again, enhancing basic reproductive rate. This may be augmented by changes in habitat selection and other behavioral shifts that have yet to be explored.

Reduction in defensive behavior is not without cost. Hosting a blood-feeding arthropod can be an expensive proposition. Hosting an infected one can be lethal, with costs as high as those incurred by intermediate hosts. In response, potential hosts have developed avoidance behaviors, which are described in chapter 4.

TRANSMISSION OF PARASITE PROPAGULES

Not all parasites are transmitted from within hosts; in fact, a majority spend at least one portion of the life cycle as a "free-living" stage, outside the confines of a host. I call this aspect of transmission "propagule dispersal"; it covers a multitude of situations, from egg deposition to the host-seeking activities of digenean cercariae to direct host-to-host transmission of infective agents (table 3.10, p. 210). Like the altered behavior seen in vectors, it is difficult to verify that behavioral changes associated with propagule dispersal increase transmission. Given the nature of some of the changes, however, it does seem likely that many parasites use hosts to disperse propagules in an especially propitious manner.

The Right Place

"Being in the right place at the right time"—this well-worn phrase is as relevant to larval trematodes and bloodborne protistans as it is to aspiring actors and rock stars. For the sake of convenience, I partition the next few examples of propagule dispersal into one of those two categories—place and time—though one is rarely more important than the other.

For instance, the ciliate *Lambornella clarki* affected the feeding of its mosquito host (*Aedes sierrensis*), with results that differed considerably from those induced by many bloodborne parasites but that nonetheless put *L. clarki* propagules in the right place for transmission (see below). Mosquito hosts of *L. clarki* sought and obtained blood meals with notably reduced enthusiasm, despite no apparent effect of the ciliate on the speed of engorgement (Egerter and Anderson 1989). Mosquito feeding is high-risk behavior and does not transmit this ciliate.

Unlike parasites that are transmitted to other hosts by mosquito vectors, *L. clarki* depends on the mosquito for a return to the water, where it can infect other mosquitoes. To this end, infected mosquitoes mimic oviposition behavior and disseminate the ciliate at oviposition sites, where a new generation of larval mosquitoes awaits infection. Likelihood of death of infected mosquitoes in water is also correlated with infection, and host death can disperse the parasite (Egerter et al. 1986). This particular ciliate is highly versatile and also pursues a free-living existence in the absence of mosquito larvae. Treehole mosquito larvae (*Aedes sierrensis*) consume these protists; the insects also produce a chemical factor that signals the ciliates to undergo a metamorphosis and to become parasitic on their predators. They encyst on the mosquito cuticle, invade the hemocoel, and multiply within the larva, ultimately killing it (Washburn et al. 1988; see also Laird 1959).

Lambornella clarki illustrates how, in addition to enhancing transmission by affecting vector or intermediate host behavior, parasites themselves can be distributed or can move under their own volition to places that result in increased encounters with the next host. In the interest of moving to a favorable location, some cestode proglottids, or segments, crawl away from the feces in which they were deposited. The impact of this behavior on transmission is not clear, but it may serve to position the eggs in areas where they can be fed upon by hosts that do not normally ingest feces (Mackiewicz 1988). Juveniles of the bovine lungworm, *Dictyocaulus viviparus*, accomplish the same end when they crawl to the surface of feces. Fungi (*Pilobilus*) often grow there, and when a juvenile nematode encounters a sporangium, it crawls to the upper surface, where it coils and awaits the discharge of the sporangium. The fungus can fling the young nematode as far as 10 feet away from the fecal pat (Robinson 1962), a probable boon for transmission since cattle tend to avoid fecal deposits when grazing (see also Durie 1961; Bizzell and Ciordia 1965).

In some cases, a parasite propagule is not necessarily dispersed into a type of habitat, but may move directly to another host. Guppies (*Peocilia reticulata*) are hosts to the monogenean *Gyrodactylus bullatarudis*, a pathogenic ectoparasite that has omitted the oncomiracidium stage. Instead, a young monogenean leaves the parental uterus and carries three generations of parasites. If it invades a new host, it must move there directly. It is assisted in this task by the lethargy and odd swimming behavior of heavily infected fish, which attract other guppies that can then be colonized (Scott 1985).

Movement of mobile parasites and parasitoids as they respond to hosts is a vast area of study (e.g., Van Alphen and Vet 1986; Pike 1990; Godfray 1994; Combes 1994; Combes et al. 1994; Haas 1994; Lewis et al. 1995; Campbell and Kaya 1999), as is the behavior of parasites within the hosts. The latter is usually not directly related to transmission, although it can be if it causes pathology or conspicuousness that puts an intermediate host at risk of predation or if the parasite moves to places of greater vector accessibility. Nevertheless, the intricacies of both site selection (Sukhdeo 1990, 1997; Sukhdeo and Sukhdeo 1994) and mate selection (MacKinnon 1987; Lawlor et al. 1990) reinforce the view of parasites as organisms responsive to evolutionary forces, for which the notion of random contact, be it with host or other parasites, is surely oversimplified.

In addition, at least three special types of transmission are closely linked to site selection within hosts: parasites can certainly minimize the probability of entering the wrong host and overcome the vagaries of sparse host populations if they are transmitted through milk (transmammary transmission, Stone and Smith 1973; Miller 1981; Shoop 1988), directly into embryos (transovarial transmission; J. E. Smith and Dunn 1991; Dunn et al. 1993, 1995), or via sexual contact. Transmission via sexual contact has been incorporated into a wide-ranging investigation of host behavior and parasitism, that of sexual selection (chapter 4).

Vertical transmission—transmission from parent to offspring—has different implications for the evolution of pathogenicity than the more routine horizontal transmission does (Turner et al. 1998). On the average, vertically transmitted parasites are expected to be less pathogenic because their fitness depends on the fitness of the host. The hypothesis that vertical transmission is associated with reduced virulence has been tested and supported with such diverse parasites as avian lice (Clayton and Tompkins 1994, 1995), bacteria and phages (Bull et al. 1991), mosquitoes and microsporidian protists (Koella and Agnew 1997; Koella et al. 1998b; Agnew and Koella 1999; fig. 3.16), and fig wasp nematodes (Herre 1993). Not surprisingly, feminization is a common occurrence among hosts of maternally inherited parasites (Dunn et al., 1993).

Lipsitch and co-workers (1995, 1996) have wisely cautioned that for many parasites, transmission is not neatly ordered into two possibilities—vertical and horizontal—but may be a mixture of the two, tempered by great heterogeneity in host population structure. Thus, possible outcomes may (to echo Anderson and May) be numerous. The thing to note here once more is the strong relationship between transmission mode and parasite virulence, between transmission mode and parasite propagule dissemination.

False oviposition (i.e., "ovipositing" parasites instead of eggs at oviposition sites where the parasites may then encounter host larval stages) might be seen as a modification of the mother–offspring transmissions summarized above, but one that occurs outside the host, and frequently transmits parasites to conspecifics' offspring. In addition to *L. clarki*, a variety of other parasites use this transmission mode. (NB: For false oviposition to occur, parasites do not have to exit through the ovipositor; false oviposition simply refers to be-

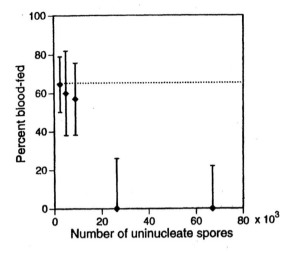

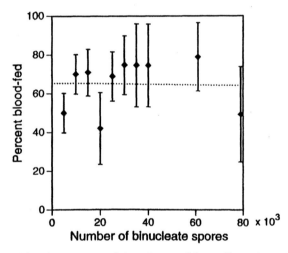

Figure 3.16. Uninucleate spores of the microsporidian *Edhazardia aedis* are strongly negatively correlated with successful blood feeding in the mosquito; these spores are transmitted horizontally. Binucleate spores, transmitted vertically, have no such negative impact. (Koella and Agnew 1997; reprinted with permission from *Oikos*, vol. 28, p. 315.

havior that brings the parasite-shedding host to oviposition sites.) Even parasitic fungi take advantage of oviposition to enhance transmission and dispersal (Undeen and Nolan 1977; Yeboah et al. 1984).

Although few quantitative studies have been performed, mermithid nematodes have been reported to disrupt sexual development in a wide range of insects, which nonetheless display oviposition behavior (table 3.11, p. 211). In doing so, they return the mermithids to an aquatic habitat, where they can emerge and produce young that will infect larval insect hosts. Even parasitized males may fly upstream to "oviposit" (Vance and Peckarsky 1996; see also Wülker 1964, 1985; Poinar and Benton 1986; fig. 3.17). Other nematodes are transmitted by terrestrial "oviposition" (Poinar 1965; Poinar and van der Laan 1972; Nappi 1973; Lundberg and Svensson 1975).

Perhaps one of the most famous examples of a parasite manipulating host behavior in an attempt to place propagules in a felicitous environment is that of nematomorphs, or horsehair worms. Their colloquial name arises from their presence in pools, puddles, and horse troughs. These large, cylindrical invertebrates superficially resemble nematodes, and the life cycles of nematomorphs and mermithid nematodes have independently converged on similar attributes; developmentally and morphologically, nematomorphs are quite different from nematodes, and their phylogenetic affinities are currently unknown. (Nematomorph larvae look intriguingly like little acanthocephalans, however!) Because nematomorphs have not been a hotbed of systematics research, there is little evidence to support or refute proposed relationships between this and other phyla (see Winnepenninckx et al. 1995; Wallace et al. 1996).

Nematomorphs have typical parasitoid life cycles, developing in arthropods until they are ready to exit the host. They are legendary for then forcing the hapless arthropod into water, where adult nematomorphs can emerge and mate, leaving little more than the arthropod exoskeleton (e.g., Poinar 1991). Nematomorphs differ from the "oviposition" examples cited above because in this case, adult parasites are emerging into the water, preparatory to mating. Despite numerous reports of frantic hydrophilia, including those of insects flinging themselves into toilets and dog watering dishes, to my knowledge, choice tests have not been performed. The task is more difficult than one might assume, because nematomorphs have lengthy life cycles and most are not amenable to laboratory cultivation. Again, without the ability to experimentally infect hosts, behavioral tests are severely compromised; naturally infected hosts may have differed from uninfected conspecifics before infection, and, in fact, that difference may have contributed to—not resulted from—their exposure. Nonetheless, if the stories are correct and if hosts in terminal stages of infection indeed express a monomanical attraction to water, the experimental possibilities are fascinating: What cues are hosts responding to? What physiological/neurochemical shift induces such a clear-cut, specific change?

Some parasites need to move from water to terrestrial hosts. For a digenean that must move from an intertidal gastropod first-intermediate host to a semiterrestrial crustacean second-intermediate host, a shift in gastropod location has the potential of enhancing transmission a great deal (e.g., Curtis 1993). In some cases, the cercariae themselves go ashore: the cercariae of *Maritrema misenensis* swim to the surface of the water, where they remain until they are washed onto the beach, into the habitat of the second-intermediate host amphipod (Bartoli and Combes 1986).

Entomopathogenic fungi often depend on hosts to disseminate spores. Two of the most common methods depend on the insect's continued ability to move,

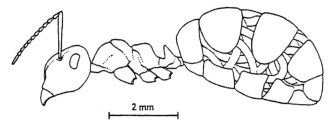

Figure 3.17.   An ant (*Colobopsis* sp.) containing a mermithid. Such ants were observed to repeatedly enter the water, even if removed, and to eventually drown. Mermithids emerged 5–10 minutes later. Normal *Colobopsis* are not so hydrophilic. (Figure reprinted from Maeyama et al. 1994, with permission from *Sociobiology*).

thus broadcasting spores, and on the insect seeking elevated places, from which wind currents can broadcast spores. The result is reminiscent of Hollywood at its most gruesome, for transmission depends upon a living insect, consumed by the parasite, moving around (or up) as the fungus breaks through the body wall (fig. 3.18). Thus the 17-year cicada, *Magicicada septendecim*, loses abdominal segments sequentially as *Massospora cicadina* penetrates the intersegmental membrane. The thorax and head finally are left to transport the fungal hyphae, which by now have reduced the abdominal remnants to a "lump of clay" (Peck, 1878, cited in Goldstein 1929).

> The appearance of insects crawling and flying about with but two or three abdominal segments attached to the thorax, is indeed sufficiently striking to attract the attention of any one . . . the movements of the insect from place to place serve to disseminate the conidia in a way that could scarcely be improved by any natural method. . . . Although the conidia [*that emerge after host death*] are thrown to some distance, such a method seems inefficient when it is compared with the process which takes place in the present instance, in which the live, actively moving infected host mingles promiscuously with its fellows. (Speare 1921, p. 76)

In addition to engaging in "promiscuous mingling" (see also Soper et al. 1976), a wide variety of insects climb to elevated locations, not only as intermediate hosts (see above), but also in response to infection by fungi, viruses, and rickettsial organisms. This tendency was observed more than 100 years ago and given the name *Wipfelkrankheit*—"tree-top disease" (Goulson 1997), also called "summit disease." This behavior can be quite unusual, and subterranean insects may move to the surface, or insects that do not frequent vegetation may be found at the tops of plants (Watanabe 1987; Horton and Moore 1993; but see *Beauveria bassiana* in Krasnoff et al. 1995). Elevation-seeking behavior on the part of parasitized hosts can be both intriguing and highly ambiguous in its function. In the case of ants, while elevation seeking may assist spore dispersal, it also moves the infected ant away from the home nest (Evans 1982). In addition, negative geotaxis and positive phototaxis can be

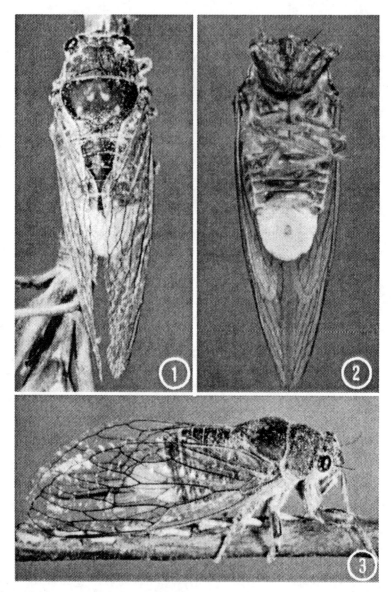

Figure 3.18. A cicada infected with *Massaspora levispora*. The fungal mass has replaced some abdominal segments. (Soper 1963; Reprinted with permission from *Canadian Journal of Botany*, vol. 41, p. 877.)

confounded. In the natural world, the two are rarely independent, and some of the investigators cited in table 3.6 have not examined them separately (see also table 3.12). Although the taxis responsible for surface seeking is of interest, it does not alter the behavioral outcome—the animal is nearer the surface, or in the case of terrestrial animals, in a more elevated position, than uninfected counterparts.

As with altered behavior of vectors, tests of the efficacy of behavioral correlates of parasite dissemination are rare (see Dwyer 1991; Goulson 1997).
One notable exception to this generality is the transmission of the fungus *Entomophthora muscae* from dead to living flies (*Musca domestica*). The dead
fly's behavior is not modified, of course, but the fungus exploits the mate-
seeking behavior of potential hosts as they respond to the dead flies. In choice
experiments, male flies were more attracted to the swollen abdomens of infected female flies than to those of uninfected females, even when the infected
abdomens were attached to the bodies of uninfected flies, and the males were
even more strongly attracted when infected abdomens were attached to infected flies. Much to their misfortune, they attempted to copulate with these
swollen abdomens, and the proportion of resulting infections far exceeded the
4% that became infected in the absence of contact (Møller 1993; fig. 3.19).
In this case, the parasitized host is dead and not behaving; potential hosts, on
the other hand, are being attracted to a place where transmission can occur.
Because transmission does not depend on the behavior of the host itself, the
cost of virulence for the parasite is low (Møller 1993), although we do not
know what, if any, investment the fungus makes in, say, chemical attractants.
Transmission is nonetheless enhanced as future hosts are lured to the contaminated area and is increased far above the rate that would result from random encounters between flies and fungal spores. The positive relationship between abdominal size and fecundity is part of the reason that the males find
these lethal abdomens so attractive; this is not the entire story, however, for

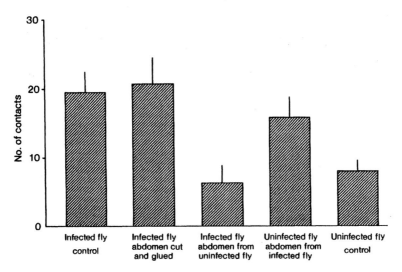

Figure 3.19. Female flies infected with a deadly fungus were much more attractive
to males than uninfected flies, although uninfected flies with infected abdomens
(glued on) also excited interest. (Møller 1993; Reprinted with permission from *Behavioral Ecology and Sociobiology*.)

infected abdomens were attractive to males even when size was experimentally controlled (Møller 1993).

Although little is known about the dispersal and transmission of helminth ova (Mackiewicz 1988), we do know that *E. muscae* is not the only parasite to attract potential hosts to propagules. The rat tapeworm *Hymenolepis diminuta* is transmitted to its beetle intermediate host when the beetle eats tapeworm eggs that are dispersed in rat feces; there is some evidence that beetles find infected feces exceptionally attractive, although this result may depend on the behavioral measure that is used (Evans et al. 1992; Pappas et al. 1995; Shostak and Smyth 1998; fig. 3.20).

Mimicry is not in short supply among host–parasite associations and is used to exploit foraging behavior of future hosts in a bid for transmission. Some cestode eggs resemble food that intermediate hosts find attractive. For instance, false diatoms (the eggs of *Diorchis stefa'nskii*), complete with stripes, appeal to some diatom-eating ostracods. Other *Diorchis* spp. eggs exhibit shapes that increase their likelihood of settling on ostracod food such as *Ceratophyllus*. Even properties such as specific gravity may function in transmission. If the intermediate host is benthic, a cestode egg is likely to be heavy and sink; other species of cestodes that use swimming intermediate hosts may produce floating eggs. The size of the egg may make it more likely to be consumed by some hosts than others (Jarecka 1961; fig. 3.21).

The Right Time

Just as timing of behavioral shifts in intermediate hosts is thought to assist transmission to the next host, so timing of propagule release may be important to transmission (Tinsley 1990; table 3.4, p. 179). In the field, noctuid caterpillars (*Mamestra brassicae*) that are infected with nuclear polyhedrosis virus disperse less than uninfected caterpillars early in the infection, when they are not distributing viral particles, but disperse more than uninfected animals do later in the infection. This result differs from similar comparisons in the laboratory, which revealed increased activity on the part of infected animals throughout the infection (Goulson 1997). This sounds an important cautionary note when comparing results from field and laboratory.

Because cercariae have limited energy reserves, timing of release may be at least as important as location. They tend to invade active hosts, so being in the right environment at the right time may be more profitable than active pursuit of an individual host (reviewed by Combes et al. 1994). The gastropods that liberated *G. adunca* performed most of their migrations in the tides that precede nocturnal low tides. Moreover, their nighttime tracks contained numerous cercariae, whereas trails made during the day had few. This periodicity both protects the cercariae from desiccation and maximizes overlap with the nocturnal crustaceans that are the next host (Curtis 1993; see also Sindermann 1960).

An elegant inference of natural selection involves a bane of human existence, the blood fluke (*Schistosoma* spp.). Between 200 and 400 million peo-

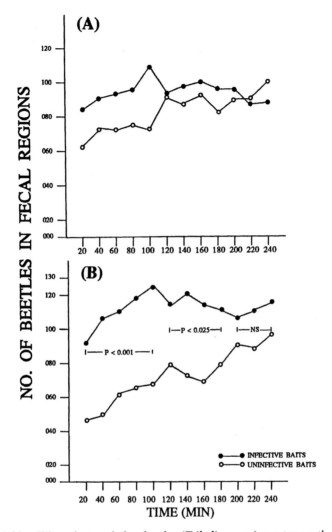

Figure 3.20. When given a choice, beetles (*Tribolium* spp.) spent more time in the vicinity of feces from infected rats than those from uninfected rats (Evans et al. 1992; Pappas et al. 1995; but see Shostak and Smyth 1998). This is probably mediated by a volatile attractant (Pappas and Wardrop 1997, Evans et al. 1998). (A) Well-fed beetles. (B) Beetles from which food has been withheld for 48 hours. (Figure reprinted from Evans et al. 1992, with permission from *Canadian Journal of Zoology*.)

ple are final hosts to this trematode (Combes 1990), which does one of several untrematode-like things when it invades its final host directly, as a cercaria, and circumvents the second intermediate host. In the French West Indies (Guadeloupe), Andre Théron and co-workers studied two populations of *S. mansoni*. One produces cercariae that emerge from snails around midday,

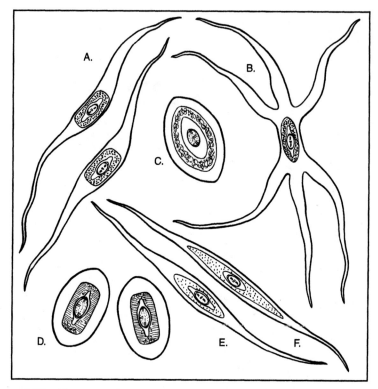

Figure 3.21. These cestode eggs (*Diorchis* spp.) have physical properties that increase their likelihood of being ingested by an ostracod intermediate host. They are likely to settle on aquatic plants such as *Ceratophyllum*, which is eaten by ostracods. (Redrawn from Jarecka 1961 by Conery Calhoon.)

the other in late afternoon. The earlier peak occurs near urban areas, where humans are the primary host; the later peak characterizes worms from *Rattus rattus*, a crepuscular animal that is more likely to frequent water later in the day (Théron 1984). Moreover, an intermediate peak was found in mangrove swamps where both humans and rats occur (Théron 1985). Crossbreeding experiments involving early and late-shedding populations produced an intermediate F_1 generation and an F_2 generation that contained representatives of all three patterns (Théron and Combes 1988; see also Théron 1989; fig. 3.22). Thus, the timing of *S. mansoni* emergence appears to be genetically based, and it favors temporal overlap with the host. Such chronobiological differences may play an important role in schistosome speciation (Théron and Combes 1995).

In contrast to *S. mansoni*, the cercariae of *Fasciola hepatica* do not penetrate hosts directly, but encyst on substrate such as vegetation, where they await the final herbivorous host. Although some ambient conditions can in-

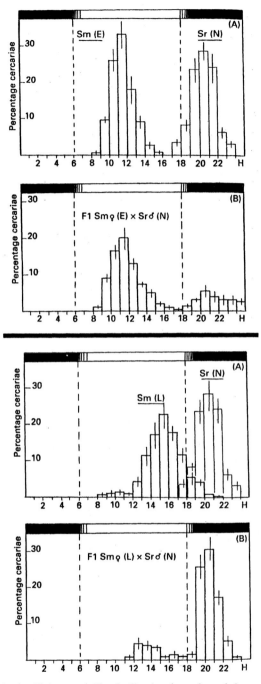

Figure 3.22. Andre Théron and Claude Combes investigated the genetic basis of shedding time in schistosome cercariae. They discovered that cross-breeding early- and late-shedding populations produced an intermediate F_1 generation (Théron and Combes 1988; Théron 1989). Both A panels show the emergence patterns of the parental strains; B panels show the patterns of the offspring. (Reprinted with permission of Cambridge University Press.)

fluence cercarial emergence, there is no evidence of circadian rhythm. After all, the substrates are there around the clock, provided there is water (Kendall and McCullough 1951; Bouix-Busson et al. 1985; see also Shostak and Esch 1990).

Recall the predicament of *Dicrocoelium*, the trematode that causes its ant second-intermediate host to remain at the top of grass blades, awaiting herbivory. This life cycle is entirely terrestrial, so how do the cercariae invade the ant? As noted, they are released in slime balls—and again, timing is important. In this case, the slime balls are secreted just after sunrise, in response to light (Lucius and Frank 1978; Frank et al. 1984). Thus, they are fresh (perhaps an appealing trait in slime balls) and available when ants emerge to forage. Likewise, the parasitized ants do not remain permanently elevated on grass blades, but climb down during the day, when conditions are harsh and the definitive host is not grazing.

For reasons we have yet to understand, timing may be important to the dispersal of the fungus *E. muscae*, which seems to motivate its dying dipteran host to ascend vegetation, all the better for windborne transmission. There is a daily rhythm to both the elevation-seeking behaviors and the deaths associated with this and several other fungal parasites (Skaife 1925; Rockwood 1950; Roffey 1968; Newman and Carner 1974; Mullens 1990; Hajek and St. Leger 1994).

Most fungal deaths of aphids occur during the day, but different parasite species are lethal at different times of day. Given the heterogeneity of the natural environment, this could reflect adaptation to different geographic or seasonal conditions on the part of the fungi. Regardless, it may be an important consideration for field workers who are interested in documenting the action of fungal control agents (Milner et al. 1984). The regularity of fly (*Musca domestica*) deaths from *Entomophthora muscae* could be evidence of a biological clock operating in the fungus which free-runs under constant conditions (Krasnoff et al. 1995).

Bloodborne parasites also exhibit rhythmicity that is often closely linked to the feeding behavior of vectors. The circulatory system is not a uniform environment, and many blood-inhabiting parasites that are transmitted by vectors are only found in the peripheral blood at times the vector is most likely to be feeding (e.g., Hawking et al. 1968). For thousands of years, malaria has been known to cause cyclical host responses; now we understand that its availability to its vectors is also periodic (Hawking 1970). Microfilarial nematodes such as *Wuchereria bancrofti* and *Brugia malayi* show a geographic variation in the timing of their occurrence in peripheral blood that matches the foraging behavior of local vectors (Pichon 1981). It is well known that many parasites shed propagules with a strong enough periodicity that diagnosis can be compromised if samples are taken at importune times. In some cases, such periodic parasite activity can have broad effects on host defensive behavior (e.g., Day and Edman 1983; see above), further increasing opportunities for transmission.

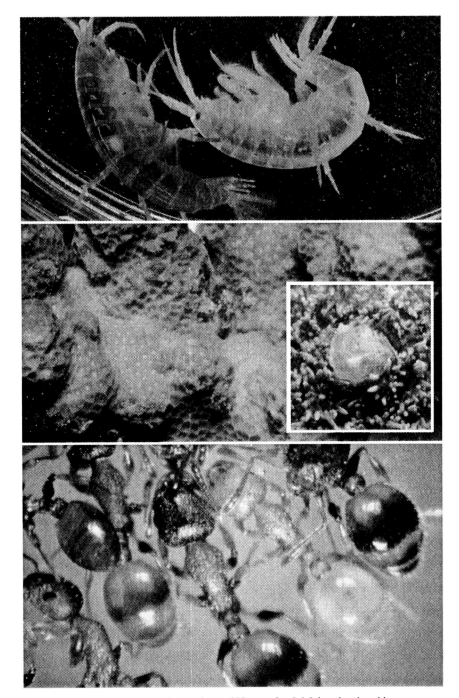

Figure 3.4. Cystacanths, such as these polymorphids, are often brightly colored, and in some cases, may play a role in increased transmission. (Photo courtesy of Barbara Maynard.)

Figure 3.6. Infection with the digenean *Podocotyloides stenometra* causes these coral polyps to become swollen and pink and more likely to be eaten by butterfly fish, the definitive host. (Photos courtesy of Greta Aeby; inset by Greta Aeby and David Gulko.)

Figure 3.12. *Leptothorax nylanderi* parasitized by cestodes become lethargic. They are also yellow compared to their brown conspecifics. Is this a result of endocrinological interference? Does it make the afflicted ants easier for a woodpecker to see? (Reprinted with permission from *BioScience*, Vol. 45, p. 90, 1995. © 1995 American Institute of Biological Sciences. Photo by Laurent Peru.)

Ectoparasites can also be exquisitely adapted to their hosts' reproductive cycles, and the European rabbit flea (*Spilopsyllus cuniculi*) is among the best known in this regard. Once on a rabbit, the fleas attach to the ears and feed. They respond to cues that accompany mating, such as a rise in temperature, and to the hormonal changes that follow as the female rabbit ovulates. The fleas become sedentary at this point, and do not leave their pregnant host, resulting in relatively high flea infestations for pregnant does. When adrenocorticotrophic hormones and corticosteroids are released in the rabbit 10 days before parturition, those hormones also promote maturation of eggs in the rabbit flea. Shortly after the birth of the young rabbits, the fleas move from the ears to the face of the doe. They then transfer to the young, which are being cared for by the doe, and the fleas feed and mate. Thereafter, they leave the young rabbits, lay eggs in the rabbit nest, return to the doe, resorbing any remaining eggs or sperm. [Not all fleas are so specialized. The rat flea (*Xenopsylla cheopis*), for instance, simply requires a blood meal (Rothschild 1965a,b; Rothschild and Ford 1966).] Despite such fine tuning on the part of the rabbit flea, mermithid parasites of fleas can apparently derail the flea–rabbit coordination (Rothschild 1969).

For a monogenean (*Pseudodiplorchis americanus*) of desert toads (*Scaphiopus couchii*), timing is critical. The parasite oviposits when the toads enter the water to mate, taking advantage of an annual 24-hour opportunity for aquatic transmission to another toad (Tinsley and Jackson 1988; Tinsley 1989).

Behavioral alterations that influence propagule dissemination are frequently seen associated with directly transmitted parasites other than eukaryotes, but the identity of the beneficiary of the alterations is frequently unclear. For instance, sneezing and coughing often occur with respiratory infection. These behaviors may serve to clear host passages of some pathogens and associated defensive secretions, but they may also disseminate parasite propagules with little cost to the host (Williams and Nesse 1991). The same ambiguous situation can pertain to gastrointestinal infections and concomitant nausea or diarrhea. In the case of cholera (caused by *Vibrio cholerae*), however, diarrhea occurs in response to an enterotoxin, choleragen (Moss and Vaughn 1979). Because this compound is produced by *V. cholerae*, the diarrhea that accompanies cholera may be seen as a parasite device, although the possibility of a host-adaptive explanation—casting this as a host response to the compound—is not eliminated.

There is a legendary infection that affects the host in less ambiguous ways: rabies. Canine–human transmission of rabies was recognized by Babylonians more than 2000 years ago (Steck and Wandeler 1980). The virus attacks the nervous system, especially the parts of the brain that influence aggressive behavior. As fans of the movie *Old Yeller* vividly recall, the host wanders "aimlessly . . . attacking and biting anything encountered" (Hart 1990, p. 275; see also Steck 1982). The virus uses the nervous system to invade salivary glands and saliva, which becomes conveniently abundant because the nerves associated with swallowing are also impaired. It is thought that the virus is not lethal among organisms in which frequent biting or airborne transmission can occur (e.g., vampire bats; see Hart 1990 for review.)

In a review of viral infections of the central nervous system, Hatalski and Lipkin (1997) cited rabies as the only such virus that elicits behavioral changes favoring transmission. They speculated about the possibility of sexually transmitted viruses affecting both the nervous system and degree of sexual activity, but there are no reliable measures of sex drive, and self-reported sexual activity is prone to many biases. It is notable that many sexually transmitted diseases are not immediately debilitating and afford hosts the opportunity to transmit the parasites to additional sexual partners.

In summary, parasites are capable of altering host behavior in ways that have been shown to enhance transmission, in the case of predator–prey interactions, and in ways that are likely to enhance transmission when vectors are involved. In addition, the dissemination of parasite propagules themselves has behavioral aspects that probably serve to increase the probability of entering the next host. If everything else is equal, then any of these alterations is likely to have a positive effect on R_0.

BEHAVIORS THAT ARE DIFFICULT TO INTERPRET

The potential for parasites to affect mental ability or intelligence has been of great interest (see Hay and Aitken 1983; Nokes and Bundy 1994 for reviews), but as with most studies of this component of behavior, results are often difficult to interpret. Animal studies are limited in scope, and human studies are often confounded by other variables. Nonetheless, every major group of parasite has been implicated in this area of behavioral alteration. Some of these studies are summarized in table 3.13.

In some situations where parasitism is associated with learning difficulties, the difficulties may be attributed to negative effects of the parasite on condition of the animal; in others, no negative physical effects are measurable, but learning may still be reduced. Although it may be argued that parasites requiring predation as part of their life cycles could benefit from an intermediate host that was slow to learn or slow to remember refuges, it is not immediately clear how the fitness of other parasites would be changed by negative effects on learning. As for hosts, it would seem the effect would often be negative, especially if it made food caches or shelter more difficult to locate. Moreover, the results of experiments linking learning to parasitism are not uniform; for instance, acute schistosomiasis has a negative impact on rodent learning, whereas chronic schistosomiasis may not (Stretch et al. 1960a,b,c). Extrapolating to fitness effects is therefore probably premature. What many of these studies do reveal is that subclinical infections may nonetheless have behavioral and probably neurological consequences.

The data from studies of humans are particularly difficult to interpret because of the confounding influences that are necessarily present, ranging from polyparasitism to initial behavioral and socioeconomic differences that resulted in differential parasitism. The latter seems important in the association of *Toxocara canis* infection and mental development (Nelson et al. 1996).

Activity is also thought to influence exposure, in this case, to some of the most pathogenic and prevalent parasites we know. While it is clear that children infected with *S. haematobium* increase their activity levels when treated, it is also likely that exposure itself is activity related, with more active children having a greater tendency to play and work near or in water (Kvalsig 1986; Kvalsig and Becker 1988). Moreover, such interaction may go unnoticed if the children, being resourceful and wanting to avoid censure, are not forthcoming about visiting forbidden swimming locations (Kvalsig and Schulte 1986). Over time, there have been several studies that (more or less convincingly) have linked schistosomiasis to mental ability (table 3.13). One factor that may be especially confusing is the effect of activity on the probability of acquiring schistosomiasis, the subsequent effect of the parasite on activity levels, and the potential overall linkage of mental achievement to activity. It may even be that if more active children are more exposed, then infection-derived lethargy may be difficult to demonstrate. Infected and uninfected children may not differ in activity simply because infected children, who may have initially been more active (and hence more exposed), have reduced their originally higher activity levels (Jordan and Randal 1962; Kvalsvig 1986; Kvalsvig and Becker 1988).

Nonetheless, parasites may, through whatever mechanisms, limit the ability of infected humans to learn (Kvalsig et al. 1991). The list of culprits includes helminths common in children in some parts of the world (hookworm, *T. trichiura, A. lumbricoides*), helminths that are often mistakenly viewed as relatively benign, if only by comparison with co-occurring parasites that are even more pathogenic. Often the cognitive effects appear to be reversible—all the more reason for intervention (Nokes et al. 1992). As Nokes and Bundy (1994) point out, in many cases, children in developing countries may not have educational opportunities that extend beyond primary schooling. The possibility that even that brief exposure to education is compromised for these children by a preventable, curable condition should not be ignored. As it stands, more than 1 billion people have ascariasis, trichuriasis, and hookworms, either singly or in combination. These parasites are not rare (Crompton 1999).

These studies and many more show us that parasites alter behavior in subtle as well as in almost fantastic ways. As Anderson and May warn, we must know much more than we usually do about the biology of a given association before we can predict how transmission, pathology, survival, susceptibility, and other host–parasite traits interact to influence R_0. As it is, there are few predictable outcomes, other than the near ubiquity of behavioral alterations themselves. The same parasite in different host species may have widely varying influences, as may different parasites in the same host, or for that matter, even different intensities of parasites in the same host. Although the behavior of some infected hosts is remarkably different from uninfected conspecifics, in the case of other host–parasite associations, the mean value of the behavioral characteristic may be shifted significantly by parasitism, but the values in the parasitized population as a whole may be more variable and overlap those of uninfected animals. In addition to all this variation, we know virtually nothing about the role of geographic differences in any of these changes. Moreover, the behavioral alteration does not always unambiguously reveal its effect on either parasite or host fitness. The obvious expectation—something like effect on transmission—is generally difficult to assess, and the subtle outcome (e.g., assortative mating because of changes in habitat use) may surprise us, or even go unnoticed. Experimental infections are crucial to determining the reality and timing of behavioral changes, but in most systems, they are easier to suggest than to accomplish. What is clear is that although altered behavior may assist parasite transmission in diverse and wonderful ways, this is nowhere near the only outcome of the changes in host behavior that are associated with parasitism.

4

Behavioral Alterations
and Avoiding Parasites

If you are ever in India or Nepal during the fly season and have occasion to watch captive elephants (of the sort that take tourists into the jungle), you may have the opportunity to see yet another kind of behavior imposed by the presence of parasites. Elephants that swat their backs and sides with branches are not performing stereotypic behavior out of boredom. They are not necessarily cooling themselves. They are using tools to repel blood-feeding flies (fig. 4.1). Switching intensifies with fly activity, and elephants that are given branches can reduce median fly count by almost half (Hart and Hart 1994).

Many potential hosts actively avoid exposure to parasites. These animals also exhibit behavioral alterations as a result of parasites, for in the absence of parasites, the avoidance behaviors may be reduced or even lacking, and some of these behaviors are quite energy intensive. As Hart (1990, 1997a) pointed out in comprehensive reviews, to be rigorous about attributing the function of parasite control to a behavior, we must show that the parasite is harmful and that the behavior mitigates this harm. Be warned that not all the behaviors in this chapter conform to this strict definition; in many cases, at least one of these criteria has been inferred.

Behaviors that are most commonly involved in parasite avoidance run the gamut from movement (e.g., escape) and changes in habitat choice to alterations in the timing or location of foraging and changes in aspects of social behavior (see Murray 1990; Hart 1992, 1994; Møller et al. 1993; Oi and Pereira 1993 for review). These behaviors can have multiple functions (e.g., avoiding parasite exposure and exploiting new food supplies), but they have one thing in common: They have the effect of removing susceptibles from the host population, with the same result as immunity might have. In so doing, the avoidance behaviors can reduce R_0. The host that successfully avoids parasites is unavailable for transmission.

Figure 4.1. Elephants use tools (branches) to repel blood-feeding flies. (Photo courtesy of Ben Hart.)

Avoidance behaviors may influence parasite distribution as well as transmission rates within a population of hosts. If otherwise comparable hosts differ in the efficacy of their avoidance behaviors, or in any behaviors associated with exposure to parasites, then initial behavioral variation within a host population can lead to different infection densities among hosts that appear to be otherwise similar. This ultimately can result in the clumped distribution that characterizes many parasite populations within host populations (e.g., *Transversotrema patialense*; Anderson et al. 1978) For instance, initial behavioral differences among brook trout (*Salvelinus fontinalis*) fry may predispose some of them to parasitism by the ectoparasitic copepod *Salmincola edwardsii*. More active fish are more likely to acquire the parasite, and parasite acquisition itself increases activity. In this case, inactivity would be a good strategy for avoiding parasites because these parasites are attracted to moving shadows and disturbance (Poulin et al. 1991a,b).

In fact, exposure is a mirror image of avoidance. Combes (1991, 1998, Combes et al. 1994) saw exposure as one of two "filters" through which parasites must pass to reach a host, the other filter being susceptibility. Our ignorance of exposure levels in most natural systems is a major flaw in many field studies addressing a variety of ethological and ecological questions about parasites. For example, differences in infection levels between male and female hosts are often attributed to sex-related differences in physiology. But the higher levels of monogeneans in male spadefoot toads also result from the fact that males stay in the water longer than females; physiological differences based on sex do not have to be invoked (Tinsley 1989). Bundy (1988) has summarized literature that shows that humans around the world are subject to the gender-based differences in infection levels, much like those in spadefoot toads, that in many cases can be traced to gender-based differences in behaviors leading to exposure.

Foraging is one avenue of exposure, and well-fed animals may have higher levels of parasites that are transmitted through food simply because well-fed animals eat more food. This seems to be the case among starling nestlings that acquire the acanthocephalan *Plagiorhynchus cylindraceus* when fed infected terrestrial isopods. The heaviest prefledging nestlings tended to have the most parasites (Moore and Bell 1983). From *Daphnia magna* to reindeer, animals that eat more increase their exposure to trophically transmitted parasites (Halvorsen 1986a,b; Ebert 1995). In addition, there are several reports that correlate parasitism and social status in several species (e.g., Jenkins et al. 1963; Hausfater and Watson 1976; Schmid-Hempel and Tanner 1990; Hudson and Dobson 1991; but see Müller-Graf et al. 1996). As early as 1955, Barrow had indicated a possible immuno-hormonal relationship that could affect both parasitism and host social status.

Human infection patterns also reveal the importance of exposure. Many people who have heavy hookworm infections regain them after treatment; it may be that this reflects greater exposure as well as a genetic predisposition to infection (Schad et al. 1983; Schad and Anderson 1985). People who engage in water-related activities (e.g., washing, fishing, water provisioning) are more likely to be exposed to (and acquire) *Schistosoma* in areas where it is prevalent (Bundy and Blumenthal 1990; Bundy and Medley 1992; see Kightlinger et al. 1998 for a similar study of *Ascaris lumbricoides* epidemiology). We can even use examination of parasite eggs from ancient latrines to infer exposure levels (behaviors) across social classes in, say, the eighteenth-century United States (Reinhard 1990).

In addition to epidemiological effects, the behaviors that contribute to avoidance of parasites can affect other aspects of animal ecology, much as predator avoidance behaviors do. Both predator avoidance and parasite avoidance can influence when and where animals feed, when they resume feeding, what they eat, how they handle food, vigilance, group size and structure, activity levels, and even behaviors such as gulping air in fish. Lima and Dill (1990) have argued that such behavioral responses to predators can have profound effects on animal ecology, for almost every animal is a potential prey item (see also Lima 1992). Potential hosts are at least as ubiquitous as potential prey, and parasite avoidance may also profoundly affect animal ecology in ways that are only beginning to be recognized.

Avoidance behaviors differ depending on whether the threat is from a parasitoid, an ectoparasite, or an endoparasite, and those distinctions form the outline of this chapter (fig. 4.2). At times, the divisions may seem artificial; to avoid some ectoparasites, for instance, is to avoid the endoparasites they transmit. In general, however, modes of host acquisition differ among these

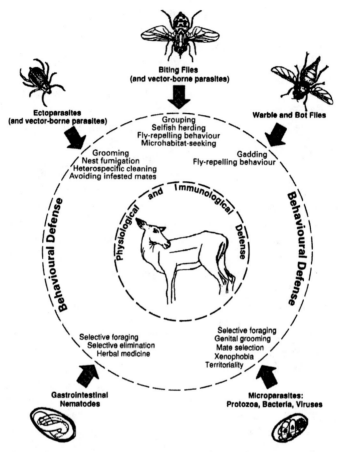

Figure 4.2. Behavior is the first line of defense against a variety of parasitic assaults. Physiological and immunological defenses are only necessary if behavioral defenses are breached. (Hart 1994; reprinted from *Parasitology*, vol. 109, p. 140, with permission of Cambridge University Press.)

groups, and parasite avoidance behaviors reflect that. Parasitoids and ectoparasites often seek their hosts, and hosts respond by moving away from them or discouraging access. With the exception of some actively penetrating forms or parasites that use insect vectors, endoparasites are more likely to be acquired as a result of the host's routine activity (e.g., foraging), which may be modified if parasite acquisition is a threat.

AVOIDANCE OF PARASITOIDS

Parasitoids, which are generally lethal (but see Karban and English-Loeb 1997), are vigorously resisted by potential hosts (e.g., Stamp 1982, 1984; Feener and Brown 1993). In the case of insects, such resistance often takes the form of morphological or habitat modifications (e.g., webs) used in conjunction with defensive behaviors that are stimulated by the approach of a parasitoid (Gross 1993).

Behavioral modifications can also be effective. For instance, potential hosts may alter the timing of activities such as foraging to avoid overlap with parasitoids. Ants may change their foraging cycles in response to phorid flies (*Apocephalus* spp.). *Pheidole titanis* is a desert and thorn forest ant of the southwestern United States and Mexico that eats termites and conducts dry season raids on foraging termite parties early and late in the day. In the wet season, these raids occur at night, probably because the phorid fly is diurnal and increases its activity during the wet season. This restriction of activity on the part of the ant apparently resulted in a (65%) decrease in food for the ants. Thus, fly avoidance takes precedence over foraging on termites and, for that matter, over defense against other ants, even though parasitoid prevalence is low (perhaps because avoidance is so successful? infection so costly?), and the most exposed castes (workers and soldiers) are themselves sterile. Feener (1988) noted that this was an example of a strong ecological effect of a parasitoid despite its apparent low abundance (see also Orr 1992, Folgarait and Gilbert 1999).

The phorid parasitoid *Apocephalus attophilus* lurks around the foraging trails of the leaf-cutting ant, *Atta columbica*, on Barro Colorado Island, Panama, awaiting an opportunity to oviposit. Oviposition requires the phorid to land on a leaf fragment being carried by the prospective host ant. Leaf cutters that carried small minim workers on their leaf fragments were much less likely to be parasitized, and this may explain why these "hitchhikers" were carried. Curiously, another defensive strategy, that of simply dropping the leaf fragment—was never observed (Feener and Moss 1990; see also Feener and Brown 1993).

Ants have been the subject of much ecological scrutiny, both as members of foraging guilds and as organisms of interest in their own right. If parasitoids can shift ant foraging activity from day to night and impinge on other aspects of what we usually regard as basic natural history, they are a force to be reckoned with in ecological investigations.

Likewise, although much has been written about parasites and mate choice, the possibility that activities surrounding courtship and mating might attract parasites has been given little attention. In the case of field crickets (*Gryllus integer*), however, the existence of acoustically orienting tachinid flies (*Euphasiopteryx ochracea*) that are attracted by calling male crickets may mean that satellite males (nearby males that call infrequently) are far more common than such males in congeners (*G. veletis, G. pennsylvanicus*) and in *Teleogryllus africanus* (Cade and Wyatt 1984; but see Allen 1998). These flies will even larviposit on speakers emitting cricket song (Cade 1975).

Parasitoids are not always the explanation for satellite males; silence may also be an alternative strategy for obtaining mates in areas of high cricket density. Hawaiian field crickets (*Teleogryllus oceanicus*) differed from congeners elsewhere in the parameters of their song and in its abrupt initiation and end. The Hawaiian crickets had to contend with an acoustically orienting tachinid, *Ormia ochracea*, and their congeners in Moorea and Australia did not. Silent satellite males occurred in all three locations, indicating that silence was not uniquely related to parasitoid presence. Instead, selection pressure imposed by the parasitoid may influence the abrupt end and beginning of singing that is peculiar to the Hawaiian population (Zuk et al. 1993). This may be opposed by female preference for song patterns that also appear to attract flies (Zuk et al. 1998a). To add complexity, a male harboring a parasitoid may also be less likely to call (Zuk et al. 1995).

Biological control has fostered great interest in parasitoid traits. Oddly, host defensive characteristics, including behavioral ones, have not been studied with the same thoroughness. If host defenses are effective, they may have a large impact on the success of a control program.

AVOIDANCE OF ARTHROPOD VECTORS AND ECTOPARASITES

While we have little trouble acknowledging parasitoids as lethal, many people (especially those in temperate climates) tend to think of ectoparasites as annoyances rather than life-threatening agents. Witness our word "gadfly"—the usual connotation is one of an irritant, not necessarily a deadly adversary. ("Gadding" refers to the galloping behavior of cattle when they are harrassed by flies.) Although recently shaken by the spread of Lyme disease, this cavalier attitude is surprisingly persistent, and may be an anthropocentric view fostered by our relatively large size and our stubbornly hopeful faith in miracle drugs. Tashiro and Schwardt (1953) estimated that tabanids (gadflies) can take more than 300 cc of blood per day from cattle. This is nontrivial; for smaller animals, an ectoparasite may be quite costly, and for all of us, ectoparasites can transmit life-threatening diseases. Avoidance of ectoparasites is therefore a major part of the behavioral repertoire of many animals, and can save blood, reduce defensive energy expenditure, and minimize exposure to disease (see table 4.1, p. 220).

Moving Away

Of course, the most obvious thing to do in the face of ectoparasites might simply be to move away from them. Large mammals may literally run away from biting flies; this is the response of mule deer (*Odocoileus hemionus hemionus*) and elk (*Cervus elaphus nelsoni*) to horseflies and of cattle to mosquitoes. This is not a generalized response to all flying insects: *Tabanus punctifer* was more disturbing than *Hybomitra opaca* (horseflies), and deer and elk paid no mind to white-faced hornets (*Vespula maculata*), predators of horseflies (Collins and Urness 1982; Ralley et al. 1993).

Warble flies (*Hypoderma tarandi*) may be at least partially responsible for the postcalving migrations of reindeer (*Rangifer*). Migrations are often attributed to food acquisition or predator avoidance, but it seems that these flies,

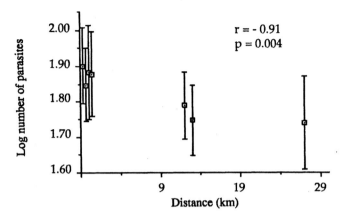

Figure 4.3. Warble flies decrease as reindeer migration distance increases. (Folstad et al 1991; reprinted with permission from *Canadian Journal of Zoology*, vol. 269, p. 2426.)

which live in the skin and can be lethal at high intensities, are not as numerous in migratory herds as they are in herds that remain near the calving grounds (Folstad et al. 1991; fig. 4.3). Of course, migration can also expose a host to new parasites; perhaps these are preferable to warble flies.

Failing to move away has consequences. The adoption of a sedentary, agricultural lifestyle provided the mosquitoes that originally fed on birds and rodents with a predictable resource: humans. Edman (1988) suggested that one way to decrease human–vector contact would be to find behavioral ways for humans to decrease this predictability.

Many species of birds avoid or desert nests and even young infested with ectoparasites (Duffy 1983; Emlen 1986; Barclay 1988; Loye and Carroll 1991; C. R. Brown and Brown 1996; table 4.1). For instance, wintering great tits (*Parus major*) roost in holes. Christe and co-workers (1994) found that these birds clearly preferred parasite-free nests to those infested with hematophagous hen fleas (*Ceratophyllus gallinae*), and would avoid infested nests even if they were the only nests in their territories. And no wonder: In the breeding season, hatching success and subsequent brood size were smaller in infested nests (Oppliger et al. 1994). One might ask whether the birds were really responding to parasites or whether these infested nests shared some other trait that the birds found offensive. Because of the randomized design of the study, this latter possibility is unlikely: Christe and co-workers (1994) generated parasite-free nests by microwaving infested ones and infesting others with known numbers of parasites.

In general, nest parasites wreak havoc with reproductive success of birds. It occurred to Møller (1989) that this might have disturbing implications for avian field studies of birds using nest boxes. Because it has been routine for workers to clean out the boxes between breeding seasons, the ectoparasites that would normally greet new occupants are greatly reduced in such studies. This should be taken into account when interpreting data from nest-box studies. In great tits, lay date, desertion, and hatching success were significantly affected by hen fleas; these data support Møller's concerns (Oppliger et al. 1994; but see Allander 1998).

Shifting Habitats

Putting distance between a host and would-be parasites, whether by stampeding or abandoning nests, is one way to reduce the number of available hosts. However, if parasites are associated with certain habitats, moving away from them may entail not so much geographic separation as selection of a different habitat—one that is less likely to contain parasites (fig. 4.3; but see de Garine-Wichatitsky et al. 1999). Again, this seems to be a strategy directed in large part against ectoparasites.

Hippopotamus habitat choice is drastically influenced by tabanids. When the flies are abundant, hippopotamuses avoid foraging on land and remain almost totally submerged in the water. If the flies persist, the hippopotamus can wriggle its ears in circles, splashing the top of its head and preventing flies from alighting. In the winter and on overcast summer days, with less solar insolation and no flies, hippopotamuses are far more likely to be ashore. Tinley (1964) observed a hippopotamus that was startled by a boat and attempted to leave the water; after encountering tabanids, it apparently decided to brave the presence of the boat.

If ectoparasites transmit disease, avoidance becomes more critical. Avian malaria was introduced to the Hawaiian Islands and probably began claiming the lives of birds in the early 1900s. The mosquito vector is nocturnal and restricted to lower altitudes. Some native birds apparently avoid exposure to malaria by roosting at higher elevations and foraging in malarious areas only during the day (van Riper et al. 1986). This is a variation on the caribou fly-avoidance strategy of seasonal altitudinal migration (Downes et al. 1986).

There are few experimental tests of the power of parasites to alter habitat selection, but stickleback fish (*Gasterosteus* spp.) provide one example (Poulin and FitzGerald 1989a). Juvenile three-spined and black-spotted sticklebacks altered habitat preference when at risk of ectoparasitism by the hematophagous branchiuran *Argulus canadensis*. The parasite tends to be near the bottom, in vegetated areas, and stickleback parasitism was greater among fish that inhabited vegetation than among those that were in the open. In experimental tanks that did not contain the parasite, the sticklebacks swam near the bottom in and near the vegetation. In tanks with *A. canadensis*, the fish spent more time near the surface and in open habitats (fig. 4.4). The black-spotted stickleback was most noticeable in its response and shifted its distribution to include equal use of open and vegetated microhabitats. (The parasite did not shift its distribution in response to the presence of fish, however.)

The sticklebacks did not avoid the parasite-rich area altogether, and Poulin and FitzGerald (1989a) recognized that many influences may determine where

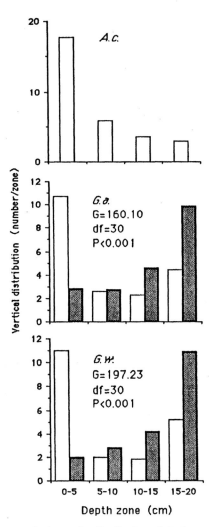

Figure 4.4. The top graph shows the distribution of the hematophagous branchiuran *Argulus canadensis*. The lower two graphs show the distributions of the sticklebacks *Gasterosteus aculeatus* and *G. wheatlandi* in the presence (filled bars) and absence (open bars) of the parasite. (Poulin and Fitzgerald 1989a, reprinted with permission from *Canadian Journal of Zoology*, vol. 67, 16.)

fish spend time. For instance, fish may be safer from predators when away from the surface and in vegetation; in such a case, parasite avoidance and predator avoidance would support conflicting strategies. This hypothesis remains to be tested, but *A. canadensis*, although not a predator, does reduce survival among juvenile sticklebacks and may transmit diseases as well.

Shifting habitats, flying up and down mountainsides, braving possible predators—these are costly behaviors and indicate perhaps as clearly as more quantitative measures such as blood loss, for instance, that ectoparasites are a threat as well as a nuisance to their hosts.

Adjusting Posture

Habitat shifting is not always necessary to avoid ectoparasites. Sometimes even a small movement will help. Red deer (*Cervus elaphus*) simply reclined when harassed by head flies (*Hydrotaea irritans*). They spent twice as much time lying down on days of heavy harassment, and avoidance reactions were halved after reclining, perhaps because both exposure and attractants were reduced. Of course, unlike experiments with fumigated nests, such field observations may be difficult to interpret, for variables such as weather can influence both host behavior and insect density independently, and the behavior may be insect- or host-specific (Espmark and Langvatn 1979). By changing posture, birds may also be protected from exposure to biting flies and the malaria they transmit. For instance, when sleeping, captive Hawaiian birds protect their heads and legs by sheltering them in plumage (van Riper et al. 1986).

Insects will go to great lengths to overcome host avoidance behavior; after all, host seeking is a matter of insect survival. A nose bot fly, *Cephenemyia jacksoni*, stalks deer (*Odocoileus hemionus*) carefully. It tones down its usual buzz and hovers in blind spots, such as under the abdomen or below and anterior to the nose. It has to do this because the deer protects its nose when confronted by nose bot flies. The fly's congener, *C. apicata*, takes advantage of the natural curiosity of naive deer and larviposits when the deer investigate it. (Apparently, experience with the nose bot fly dampens this curiosity considerably.) The deer's avoidance behavior is so successful that larviposition appears to occur on the run, with the fly squirting larvae into the deer's nose as she flies by (Anderson 1975).

Joining a Group

Parasite avoidance can influence social behavior (table 4.2). Group formation can have many benefits, ranging from information exchange to predator and parasite avoidance (Gross 1993; McCauley 1994; but see Lima 1995). In some cases, social grouping is increased in response to parasitism, although, ironically, parasite transmission is also seen as a major disadvantage of social behavior (Alexander 1974; Wilson 1975; Hoogland and Sherman 1976; Hoogland 1979; Brown and Brown 1986; Moore et al. 1988; Møller et al. 1993). The mode of transmission itself might favor or discourage some social structures; for instance, feline immunodeficiency virus is transmitted by biting, whereas feline leukemia virus is also transmitted by other, less aggressive salivary exchanges. The prevalence of these two viruses reflects various elements of cat behavior (Fromont et al. 1997). Møller and co-workers (1993) reviewed many ways that parasites might influence social behavior and offered a series of thought-provoking questions. Might mixed-species flocks offset the cost of parasitism because the diversity of species might impede some transmission? Are dominance interactions and territoriality favored by disease avoidance? Are social species more fastidious in their avoidance of parasites?

In a thought-provoking paper, that begins by asking us to "Imagine a circular lily pond," Hamilton (1971) explored the notion of the selfish herd. Predators, parasitoids, and biting flies are all more likely to attack herd members on the edge of the aggregation than those in the center, and this may motivate herd formation in the face of such threats. In addition, there may be an encounter-dilution effect: in general, the probability of discovery of a group is not directly proportional to its size, and a large group dilutes the effect of a predator or parasite on a per capita basis if larger groups are not discovered more readily (Foster and Treherne 1981; Mooring and Hart 1992; Krause 1994).

The efficacy of these responses depends on the nature of the threat (Mooring and Hart 1992; Hart 1994). As mentioned, group formation is a complex response to parasitism in that the risk of parasitism increases with increasing group size in the case of some parasties and with decreasing group size in the case of others. Bunching in Holsteins increases with intensity of face flies *(Musca autumnalis)*; the cattle stand in a formation much like a rosette, and central individuals are relatively protected from flies (Schmidtmann and Valla 1982; fig. 4.5). The behavior of horses on Assateague Island, Maryland, con-

Figure 4.5. Cattle frequently bunch in response to biting fly attack. (Photo courtesy of Terry Galloway.)

trasts with that of these cattle. Tabanids did not seem to motivate herd formation or enlargement, although distance between individuals was decreased
as a result, and the horses were generally irritated (Rutberg 1987; see also
Duncan and Vigne 1979; fig. 4.6).

In cattle, horseflies cause the most obvious reaction, bunching (which may
cause thermal stress), whereas mosquitoes tend to result in an increase in in-

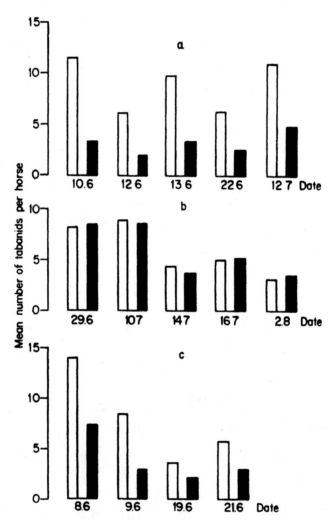

Figure 4.6. Open bars represent the mean number of tabanids that settle on horses
in small groups; filled bars represent the mean number of tabanids that settle on
horses in large groups. (a) Large and small groups are separate; (b) small and large
groups associated; (c) large and small groups separated but in comparable habitats
chosen by investigator. (Reprinted from Duncan and Vigne 1979, with permission of
Academic Press.)

dividual defensive movements such as ear flicks. Stable flies attack lower on the body and do not cause bunching; individual defenses are apparently quite effective (Ralley et al. 1993). Cattle run from warble flies, which oviposit rather than engorge and therefore can use multiple hosts (see above). In this case, group formation would only offer more opportunities for oviposition (Hart 1994).

Selfish herds thus seem to be in evidence. Are there then selfish roosts (Hart 1997a)? This question has not been explored. Across bird species, however, both the number of species and the intensity of ectoparasites in nests were positively related to host sociality, even when host size and phylogeny were taken into account (Møller et al. 1993). Poulin (1991a) found that contact-transmitted parasites (e.g., feather mites) were more prevalent in group-living passerine species, but there was no difference between group-living birds and solitary birds in the prevalence of mobile parasites such as hippoboscid flies. Instead, other factors such as migratory behavior and body weight seemed more influential (Poiani 1992).

In the same vein, Côté and Poulin (1995) classified parasites as either contagious (and relatively immobile) or mobile (and actively host seeking). Their meta-analysis of studies on a variety of host taxa showed that the intensity of mobile parasites increased with decreasing group size, whereas increases in contagious parasite intensity were correlated with increased group size. Thus, these parasites influence group size, but in ways that may oppose one another, depending on transmission mode (see also Rubenstein and Hohmann 1989; Poulin 1999). Of course, correlation does not indicate causation, but Côté and Poulin pointed out that in several studies, temporal increases in group size follow increased abundance of biting flies, and that such changes in fly abundance were probably not resulting from changes in host group size.

Much of the evidence for group formation in the face of fly attack is indeed correlational. Field studies often do not offer clear views of the relationship between sociality and exposure to infection. An experimental approach is no less crucial to understanding the avoidance of parasitism than to any other area of parasite-induced behavioral alterations. Schmidtmann and Valla (1982) used this when they sprayed some Holstein heifers with fly repellent and others with water. The repellent-sprayed heifers did not form groups as much as the water-sprayed control heifers did.

The impacts of sex and social behavior can sometimes be confounded if one influences the other. These, too, can be teased apart through experiment and careful observation. For instance, male reindeer had heavier warble fly infestations than females did (Folstad et al. 1989). The fact that females, calves, and young males formed aggregations earlier in the fly season than did males was consistent with the notion that such social interaction impeded warble fly parasitism. In addition, castrated males had abundances similar to those of females, and calves had relatively high levels. This suggests an immunological influence rather than a social one, with immunological depression associated with testosterone (intact males) and youth (calves) taking its toll.

Laboratory experiments also support the linkage between parasitism and social behavior. Individual juvenile three-spined and black-spotted stickle-backs *(Gasterosteus aculeatus, G. wheatlandi)* in larger shoals experienced a decrease in average number of attacks by a potentially lethal branchiuran ecto-parasite, *Argulus canadensis*, and formed larger shoals in its presence (Poulin and Fitzgerald 1989b). In the field (the St. Lawrence estuary, a predator-free zone for juvenile sticklebacks), shoals begin to form when the fish are less than 2 weeks old, indicating that predation is not the immediate motivation for this social behavior. Attacking three-spined sticklebacks in large shoals is risky for the parasite; the chances of being captured and eaten by an intended host are greater in larger shoals (Poulin and FitzGerald 1989b). Although three-spined sticklebacks avoided schools of conspecifics parasitized with *A. canadensis*, probably based on their erratic swimming (Dugatkin et al. 1994; fig. 4.7), not all fish are so suspicious. Guppies *(Poecilia reticulata)* are even attracted to the abnormal swimming behavior of conspecifics infected by the ectoparasitic monogenean *Gyrodactylus bullatarudis* and thus expose them-selves to the parasite (Scott 1985).

Perhaps the most far-flung experimental evidence for selfish herds comes from lycaenid butterfly larvae *(Jalmenus evagoros)* attended by ants

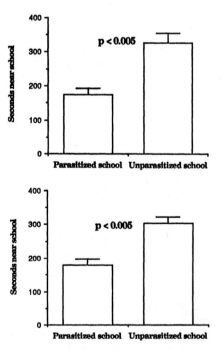

Figure 4.7. Sticklebacks consistently avoided schools of conspecifics parasitized with the ectoparasite *Argulus canadensis*. Parasitized fish were in different sides of the chamber in these two experiments. (Reprinted from L. A. Dugatkin et al. 1994, with permission from Kluwer Academic Publishers.)

(*Iridomyrmex* spp.). The larvae produce food droplets for the ants, which in turn protect them from parasitoids and predators. Axén and Pierce (1998) experimentally manipulated larval group size and showed that the larger the group, the lower the investment in ant reward by each group member.

In summary, increased sociality may influence parasite transmission and resulting parasite intensity. By similar reasoning, does it influence parasite species richness? Do more social hosts have more kinds of parasites? Looking at "contagious" ectoparasites (copepods and monogeneans), Poulin (1991b) determined that this prediction was not supported for 60 species of Canadian freshwater fishes. Host age, size, and range were also unrelated to parasite richness. As for terrestrial environments, I have always been curious about the parasites of ground squirrels: The host exhibits a wide range of sizes and sociality. How do these attributes affect parasite communities in a clade of social mammals?

Polyspecific groups can confer the advantages of increased group size while evading some of the risk of increased exposure, at least where species-specific parasites are concerned (Møller et al. 1993). Primates do this, and the size of their sleeping groups may also be influenced by fly avoidance (Freeland 1976, 1977).

In contrast, individual and sleeping group size could account for most of the interspecific variation of *Plasmodium brasilianum* prevalence in Amazonian monkeys; species characterized by large individuals and large sleeping groups exhibited disproportionately higher prevalence. Davies and co-workers (1991) pointed out that studies of vertebrate group size and hematophagous flies have been generally inconclusive. Perhaps selfish herd principles are less applicable to ectoparasites when the herd is asleep.

Swatting, Biting, Slapping. . . .

For some hosts, the alternative to running away from would-be parasites is to stand and fight, using a wide range of manuevers: ear twitching, tail swishing, foot stamping, head tossing, and the like (Hart and Hart 1994). An elephant's use of branches to swat flies simply incorporates tool use into these opposition behaviors, which increase with ectoparasite density (Hart 1994). Such efforts on the part of the host have a notable impact on the ectoparasite and can cause great injury to it. One of the best documented cases involves biting flies (Edman and Scott 1987). While multiple attributes influence host choice (e.g., availability, time of day, habitat, host size and age, and odor, to name a few; Day and Edman 1984a,b; Edman and Scott 1987), immobilization—that is, the interruption of defensive behavior—is a strong predictor of hosts on which engorgement takes place (Edman et al. 1974; Vale 1977; Waage and Nondo 1982; Coleman and Edman 1987). Indeed, there is ample evidence suggesting that in many cases, defensive behavior (or lack of it) is the primary determinant of a preferred host (Warnes and Finlayson 1987; but see Edman and Spielman 1988). This is not surprising, as host defense is a major source of biting fly mortality.

Because of the deterrent value of host defensive behavior, interspecific differences in behavior can be influential in host species selection by some ectoparasites. For instance, small animals tend to mount the most vigorous defense against blood-feeding flies, perhaps because they are in danger of losing a greater proportion of their blood than a larger animal would (Walker and Edman 1986; Edman and Scott 1987; fig. 4.8). In addition, species with higher natural activity levels are more formidable hosts (Edman and Kale 1971; Edman et al. 1972; Webber and Edman 1972; Edman et al. 1984; Cully et al. 1991). Behavioral differences might be profitably subjected to artificial selection to create behaviorally resistant lines of domestic animals, much as we select for physiologically resistant lineages (Barnard 1989; Norval 1992).

In addition to interspecific differences in defensive behavior, hosts differ in activity rhythms that can influence exposure to parasites. Circadian differences in defensiveness can affect mosquito feeding success, which is highest when hosts are inactive (Edman and Spielman 1988). Mosquitoes that attempted to feed during high host activity experienced great mortality (Day and Edman 1984a). When day feeders do feed on diurnally active hosts, such hosts are often large, and not as defensive (Walker and Edman 1985a). For instance, *Aedes albopictus* prefers humans over dogs in the dark; during the day, human defensive behavior is more effective, and the mosquito switches to canine hosts (Konishi 1989).

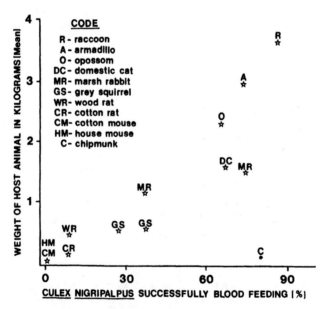

Figure 4.8. Successful blood feeding occurs less frequently on smaller hosts, which tend to be more defensive. (Reprinted from Edman and Scott 1987, with permission of the International Center of Insect Physiology and Ecology Press, Nairobi, Kenya.)

In all these cases, we assume that the relative costs of being fed upon by flies exceed those of defensive behavior, or the latter would not persist, but these costs are rarely compared (e.g., Warnes and Finlayson 1987). In the case of flies that transmit disease, the cost of blood loss must be combined with the probability of incurring bloodborne parasites before total costs of biting flies can be assessed. Unlike grooming, however, for which we have some salivary cost estimates and comparative studies of time (chapter 5), few cost analyses have been performed on parasite avoidance. One exception is a study of howler monkeys (*Alouatta palliata*) on Barro Colorado Island, Panama. Dudley and Mitton (1990) calculated that the monkeys performed >1500 slaps or avoidance movements per 12-hour resting period, spending up to 24% of their metabolic budgets (less basal metabolism) combating flies.

We have seen that across host species, defensiveness can be influenced by attributes such as size, activity level and rhythm, and general behavior. There is also within-species variation in defensiveness, especially in highly defensive species (Kale et al. 1972). Some of this intraspecific variation in host defensiveness is due to age or dominance status (Edman and Scott 1987). Young hosts are not usually as effective in defending themselves as older hosts (Kale et al. 1972; Waage and Nondo 1982).

There are physiological host responses to harrassment by ectoparasites. Under adverse conditions, many animals increase their endogenous opioid-mediated analgesia. In mice, these physiological effects of exposure to biting flies may be disseminated from mouse to mouse via bedding (Colwell and Kavaliers 1992; see also Colwell and Kavaliers 1990) and are also subject to learning (Kavaliers et al. 1999). The results are intriguing, but because endogenous opioids mediate functions ranging from growth and development to immune responses, fitness correlates of this analgesia are not clear.

In summary, the risk of being bitten by a fly is not spread homogeneously across potential hosts. Size, behavior, and activity patterns are influential among host species. Within a species, age, previous experience, and other individual traits can matter a great deal (Scott et al. 1988).

What are the consequences of host defensiveness for the enterprising ectoparasite? In a word, serious. Both fecundity and life itself are at risk. Mosquito populations may incur density-dependent regulation in part because individual mosquito foraging success decreases with increasing mosquito density, as increasingly annoyed hosts become increasingly defensive; because successful foraging is needed for reproduction, such density-dependent host reactions may partially regulate mosquito populations (Edman et al. 1972, 1985; Waage and Nondo 1982; Walker and Edman 1986).

Host defensiveness not only reduces parasite fecundity, it can be lethal for parasites (Poulin and Fitzgerald 1989b). Laboratory strains of *Culex quinquefasciatus* were much more vulnerable to chick-induced mortality than were less domesticated mosquitoes, suggesting that natural selection maintains risk-aversive foraging behaviors (Day and Edman 1984a). The risk from host-induced mortality may be at least partially responsible for lengthy between-meal

intervals in flies such as the tsetse (Randolph et al. 1992) and for the reduction in persistence seen in mosquitoes attempting to forage on defensive hosts
(e.g., Walker and Edman 1985a). Thus, the difficulty that many infected vectors have in successfully acquiring blood meals (e.g., increased probing; chapter 3) may involve extended fitness costs.

Of course, if the ectoparasite is ingested by the host, it is the end of the
ectoparasite. It may not be the end of the life cycle for some parasites they
can contain, however. Malaria, for instance, can be transmitted with an ingested vector (Edman et al. 1985). Thus, risky vector behaviors may not be
harmful to all parasites. Do such parasites depress risk sensitivity in their
vectors?

In general, vectors present a puzzle for many epidemiologists. Frequently,
small proportions of vectors are observed to be infected, and vector survival
is known to be low. In order for a bloodborne parasite to be transmitted, the
vector must become infected and survive to feed on the second host, to which
the parasite is transmitted.

Because host defensiveness influences the likelihood of ectoparasitism, it
can have consequences for the transmission of these vector-borne parasites.
As early as 1974, Edman and co-workers noted that "attraction" was not the
sole determinant of host choice and that the preference for tolerant hosts could
have epidemiological consequences. As we have seen, one expected consequence is that tolerant species should be more involved in disease transmission than intolerant ones. Another less obvious consequence has to do with a
rather pedestrian behavioral alteration expressed by some parasitized hosts:
malaise. Sick animals may not be as defensive as healthy ones. This reduction in energy-intensive defensive behavior can be part of a suite of adaptive,
energy-conserving changes that occur when hosts are combatting parasites
(Hart 1988; chapter 5). Mahon and Gibbs (1982) did not measure defensive
behavior in hens; they found no differences between infected and uninfected
animals when they checked such variables as body temperature, CO_2, and water vapor output, but they did find that traps near birds infected with Sindbis
alphavirus caught more *Culex anulirostris* than traps near uninfected birds,
and caged, infected birds were fed upon more often. This means that all hens
were not identical in their epidemiological effects, and workers should take
this into account when performing such studies (but see Scott et al. 1988, Scott
and Edman 1991). Similar results have been found with *Culex pipiens* engorging on lambs infected with Rift Valley virus (Turell et al. 1984) and tsetse
flies (*Glossina pallidipes*) feeding on cattle infected with *Trypanosoma congolense* (Baylis and Nambiro 1993).

Not only are sick animals often more likely to provide meals for ectoparasites, but if their pathogens are vector borne, they may be especially transmissible to the vector at the time that defenses are most subdued, primarily
because of reduced defensive behavior on the part of the ailing host (chapter
5). Even if acquiring the parasite causes trouble for the ectoparasite later, initially it enjoys a relatively risk-free blood meal (Rossignol et al. 1985).

Finally, *Ae. triseriatus*, for instance, is less persistent on a defensive host than on a host that is not so defensive (Walker and Edman 1985a). If blood meals are defensively interrupted, then host defensiveness may have consequences similar to those of increased probing as the ectoparasites seek to feed again. [In at least some mosquitoes, blood meals must reach a minimum size before host-seeking behavior is inhibited (Klowden and Lea 1978, 1979).]

Interactions between parasitism and the likelihood of being bitten by a vector can have profound consequences for epidemiology. Most models have assumed that vectors bite hosts randomly with respect to infection. If this is not true, then both the persistence of the parasite (its reproductive ratio) and the capacity of the vectors will be greater than what is predicted under assumptions of homogeneity, and control may well be more difficult to achieve. Prevalence of the infection will be higher (Dye and Hasibeder 1986; Edman and Scott 1987; Kingsolver 1987). Moreover, some groups within host species (e.g., young animals) will be more important as reservoirs than more defensive conspecifics (Scott et al. 1988; Scott and Edman 1991).

In summary, not only do ectoparasites encourage the expression of defensive behavior, which in turn varies between and within species in ways that affect the probability of successful engorgement, but the parasites that they transmit may in some cases alter the expression of defensiveness, increasing the likelihood that infected hosts are bitten. Again, this is not universally true; some arboviruses reduce avian host defensiveness, but others do not (Scott et al. 1988), and ectoparasites do not always prefer infected hosts (Freier and Friedman 1976; but see Mahon and Gibbs 1982).

Of course, this cautionary note about generalities should not be applied too generally! In Papua New Guinea, malarial and microfilarial infections are endemic and asymptomatic. There is no consistent pattern of infection and anopheline feeding. Burkot and co-workers (1989) suggested that asymptomatic human malaria is common and that concerns about non-random feeding might be more appropriate in symptomatic, epidemic situations. Do uninfected hosts defend themselves more vigorously against vectors in areas where vector borne diseases are more severe? Are infected hosts less defensive?

Using Camouflage

Host defensiveness can involve more than hostile movements. For instance, hosts can find ways to go unnoticed. Small or patterned targets, including stripes less than a certain width, are not attractive to biting flies, at least not from a distance. Thus, although striped patterns may be useful against predators, or in thermoregulation or social interactions, tsetse flies may be another selective force favoring zebra stripes. Waage (1979, 1981) presented this hypothesis and marshalled a variety of arguments supporting it, ranging from biogeographic co-occurrences of tsetse and boldly patterned zebras to comparative studies with other equiids. Although other cues are important, if everything else is equal, a striped host may not be as attractive as a similar solid-colored host. Indeed, zebras are not a preferred host for tsetse flies compared to other ungulates, including domestic horses. In addition, most biting flies are attracted to dark, moving objects (see Moore 1993 for review). Might such preferences bias tests of parasitism, coloration, and sexual selection? There is no evidence for such an effect (Yezerinac and Weatherhead 1995), but this does not minimize the potential biases that can be introduced by nonrandom host selection by vectors.

In fact, pelage itself can deter insect feeding and in turn influence ecto-parasite behavior. When presented with anesthetized chipmunks, *Ae. trise-riatus* fed primarily on eyelids, ears, nose—places where hair was sparse (Walker and Edman 1985b). Tabanid mouthpart lengths reflect the depth of hair in preferred feeding areas on cattle (Mullens and Gerhardt 1979).

The business of avoiding ectoparasites can consume a significant portion of host time and energy. Much like parasite-induced behavioral alterations re-viewed in chapter 3, these attempts can result in altered habitat choice, altered social behaviors, and a variety of special movements. Additional alterations, such as foraging adjustments, are more likely to be aimed at avoiding en-doparasites.

AVOIDANCE OF PROPAGULES AND INTERMEDIATE HOSTS

Social Influences

The avoidance of ectoparasites by group formation stands in contrast to the risks of parasite transmission imposed by social living (Alexander 1974). The behaviors accompanying fever (anorexia, failure to groom, depression) or, for that matter, exhaustion, conserve energy and benefit the host as it mounts a defense against pathogens, but they also serve to isolate the host socially, thus defending it from exposure to yet other pathogens. This is compounded by the fact that many animals avoid conspecifics that behave in this way (Loehle 1995).

Although the effects of endoparasites are not as likely to be mitigated by grouping of hosts as are those of ectoparasites, many endoparasites and ecto-parasites can be transmitted more effectively when interhost distances are small (e.g., C. R. Brown and Brown 1986; Loehle 1995). Thus, the upper limits in group size frequently may be influenced by increased likelihood of contact transmission and the lower ones by avoidance of biting flies.

There is no single strategy that confers minimal risk of exposure to all par-asites. This is because of the diversity of transmission modes among parasites. Some require small interhost distances for successful transmission, while oth-ers negotiate distances with great success. Moreover, the transmission of par-asites with indirect life cycles is largely independent of host group size (Moore and Simberloff 1990; Côté and Poulin 1995; Loehle 1995).

Such biases may affect the study of parasite communities. One often thinks of such studies as divorced from the host–parasite interaction, at least within age and sex groups; of course, host defense may alter parasite community composition, but so little is known about that at a specific level in most wild animals that it is often treated as a black box. In a study of helminth parasite communities of bobwhite quail that I conducted with Dan Simberloff and Mike Freehling (Moore and Simberloff 1990), we wondered what effect covey size might have on parasite community composition. Coveys are formed in the nonbreeding season and their average size may vary by a factor of two. Most of the parasites in the community were transmitted when the bobwhites consumed intermediate hosts; in addition, one had a relatively lengthy life cycle that might incorporate a facultative intermediate host, and one (*Trichostrongylus tenuis*) had a rapid life cycle with no intermediate host. It seemed reasonable to expect that if exposure to any of the parasites was affected by host social group size, it would be exposure to *T. tenuis*. Indeed, we found the intensity of *T. tenuis* to be most consistently related to covey size (Moore et al. 1988; see also Côté and Poulin 1995). Although we know that exposure to all possible parasite fauna is not identical (all diet items are not equally preferred, for one thing), more subtle influences such as host social behavior can also play a role in parasite community structure, emphasizing the importance of exposure (or avoidance) itself in such patterns (Simberloff and Moore 1997). [NB: *Trichostrongylus tenuis* in bobwhite quail does not seem to be as pathogenic as it is in grouse, and is probably a different species (chapter 2; Durette-Desset et al. 1993; Freehling and Moore 1993).]

In fact, it is well known that some communities of symbionts may actually confer protection against invading parasites (e.g. Hutcheson et al. 1991). In this vein, Lombardo and co-workers (1999) suggested that female birds may adjust copulatory frequency to increase their exposure to such beneficial symbionts.

Xenophobia in primates may function in part as a kind of quarantine that limits exchange between groups and thus limits exposure to new parasites (Freeland 1976; see also Møller et al. 1993). Physical contact between primate social groups is uncommon; home ranges are maintained and mating occurs within the group, with females choosing vigorous, high-ranking, presumably healthy males. Strangers are put through a stressful and prolonged period of admission that may reveal an otherwise subclinical infection (Freeland 1976). Congruent with this pattern, Freeland (1979) found that larger *Cercocebus albigena* groups had more species of intestinal protists, and that all groups of primates that he studied, with the exception of savannah baboons (*Papio anubis*), exhibited strong intergroup differences in the composition of intestinal protistan faunas. Savannah baboons have fewer parasites, on average, and they tend to have larger groups; they may be less xenophobic, as well. Freeland (1979) suggested that social barriers could impede the transmission of contact-spread diseases, but would not limit vector-borne transmission, which must be confronted by other means (e.g., grooming and avoidance). Although the fact that many forms of territorial defense involve noncontact displays such as vocalization and visual signals have traditionally been attributed to injury avoidance, these displays, in place of physical contact, would also limit exchange of pathogens (Loehle 1995).

Keeping strangers on the periphery may also result in an initial low-dose exposure to novel parasites that could serve an immunizing function (Hart 1990). Animals may also avail themselves of a kind of vaccination if their behavior includes brief exposures to antigens, be it from introducing young to prey carcasses (Hart 1990) or conspecifics (Freeland 1976). Some parasitologists have even wondered if the ritual of baptism may not have originated as a form of schistosome immunization.

Within a group, ailing individuals may be assisted [perhaps a form of kin-selection (Rasa 1983], but they may also be shunned or attacked. Burgett and co-workers (1990) appropriated the word "lycurgan" to describe the possible removal of mite-infested bee larvae from the hive, after Lycurgus, a Spartan who advocated infanticide as a disease control measure.

Insect-to-insect contagion may be responsible for differing levels of resistance to pathogens in solitary and gregarious species. Although Hochberg (1991) presented this hypothesis with many caveats, it it nonetheless worth more consideration, and could have serious consequences for biological control programs.

Thus, several questions emerge from the relationship between social behavior and parasitism: Can subdivision of groups have meaningful epidemiological effects, and does this happen in nature in the face of a challenge from directly transmitted parasites (e.g., Freeland 1979)? To what extent are group-formation behaviors, like some parasites, inherited from ancestors (Côté and Poulin 1995)? To what extent is host aggregation an essential part of some parasites' life cycles? The relationship between social group size and parasite transmission is part of an epidemiological axiom: many communicable diseases have population thresholds below which they cannot persist, having exhausted the supply of new, susceptible hosts.

A related possibility has received scant attention: Given the influence of social group size on parasite transmission, how might parasites manipulate social interaction and thus enhance transmission? For instance, we might expect parasites to increase the aggregation tendencies of hosts from which direct transmission is possible. In fact, we would expect increases in social interactions of most types. There is little evidence for this, in part because there have been few studies of the social behavior of hosts infected with directly transmitted parasites. Might parasitized animals be unusually attractive? Studies of guppies with monogeneans and sticklebacks with parasitic crustaceans yield conflicting results (see above; Scott 1985; Dugatkin et al. 1994).

Returning to the notion that exposure is the antipode of avoidance, the history of human disease may be seen in part as the continued effect of increasing population growth and cultural changes on the disintegration of pathogen avoidance mechanisms. I have already mentioned the probable effect of a shift to agriculture, which made humans a spatially dependable resource for a variety of blood-feeding flies (Edman 1988) and perhaps for schistosomes (Combes 1990). Irrigation practices may have compounded this effect, as well as created exposure to more parasites (McNeill 1979). Early hunter-gatherers were probably hosts only for those parasites that were easily transmitted and conveyed little immunity (Dobson and Carper 1996). Successful agricultural practices allowed the growth of cities and the impetus for intergroup travel, which eventually created population levels that could sustain endemic measles and smallpox, both of which may have been acquired from domestic animals, as well as typhoid. The addition of trade routes and the transport of small mammals along those routes probably contributed to the Black Plague in the fourteenth century (Dobson 1992).

There are several books that detail this history of human culture and parasitism (Ashburn 1947; Zinsser 1963; Sigerist 1965; McNeil 1976, 1979; Anderson and May 1992; Desowitz 1997; Diamond 1997). The important feature is that at every turn, as transmission increases, as susceptibles increase, the predictions of the Basic Model are again at the center of events. *The Coming Plague* (Garrett 1994) is perhaps the most gripping account of the process as it continues today.

Habitat and Dietary Influences

Animals may avoid some endoparasites by judicious choice of diet items. Although contented cows may appear to have no worries as they graze through a pasture, young cattle did not forage randomly in a paddock heavily infected with the lungworm *Dictyocaulus viviparus* (Michel 1955). When samples from grazed areas were compared to samples taken at random, the grazed areas contained fewer nematode larvae. [Recall that *D. viviparus* has its own counterstrategy, using mushroom sporangia as a catapult to leave the vicinity of the fecal deposit (Robinson 1962; chapter 3).] In general, calves avoid grazing near fresh cow pats (Gruner and Sauve 1982). Such avoidance may or may not contribute to good health, for often fecal deposits will degenerate before the development of the parasites they carry, leaving little visible warning (Christie 1963). As with studies of most microscopic parasite propagules, quantitative experiments can be challenging. In the words of Whitlock et al. (1972, p. 418), "It would involve a lot of work to disperse sheep over a pasture in such a manner as to force them to defaecate at random."

Horses are thought to be even more selective, defecating in certain areas and feeding in others (Ödberg and Francis-Smith 1977; Rubenstein and Hohmann 1989), whereas in cattle "there [is] no trace of instinct for the division of the field into dining room and lavatory" (Taylor 1954, p. 61). Carnivores can be more choosey yet. If you enjoy the company of a cat or dog in your house, chances are that you have taken advantage of their fastidious nature in the housebreaking process (Hart 1990; fig. 4.9).

Such hygienic behavior makes the occurrence of coprophagy even more remarkable. As a dietary strategy, it clearly has some advantages, and fecal deposits are routinely consumed by a variety of organisms, large and small. However, feces are also a major transmission route for pathogens, and as such, the benefit of a relatively easy meal for the consumer must be offset by the risk and cost of intersecting that route.

Human hookworm infection is closely linked to the human attitudes and

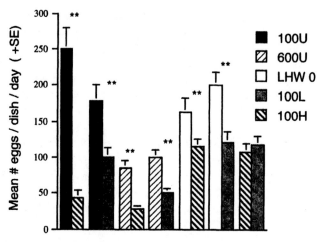

Figure 4.9. Even mosquitoes avoid parasites, which thereby influence mosquito choice of oviposition site. Ovipositing *Aedes aegypti* avoided water containing heavily parasitized mosquito larvae in favor of sites with fewer or no parasitized mosquito larvae. The parasite, *Plagiorchis elegans,* a trematode, could not be passed from mosquito to mosquito, but if some mosquitoes at a site are parasitized, this indicates the presence of an infected snail that is probably shedding cercariae. U = unparasitized larvae, L = lightly parasitized larvae, H = heavily parasitized larvae. (Reprinted with permission from Lowenberger and Rau 1994b, p. 1212).

practices that surround feces. At first glance, one would assume that cultures that found feces especially repugnant would experience a low level of hookworm infection; in fact, because human feces are considered unattractive in most cultures, it is surprising that human hookworm persists. Persist it does, as do several fecally transmitted helminths, in well more than 1 billion humans (Crompton 1999, citing Chan et al. 1994), and all in spite of human antipathy toward feces. Several human helminths are transmitted in excrement, and Nelson (1990) noted that if some groups had exhortations against fecal contamination of the environment that were as serious as their regard for dietary taboos, many parasites would be far less common. For instance, some Hindu and Muslim societies add religious proscriptions about feces to more generalized aversion; hookworms may nonetheless thrive because latrines are also avoided and not maintained. In such societies, elaborate rituals may surround the act of defecation, to the point of considering the left hand defiled (if used for ablution after defecation), and unsuitable for food preparation. Some human cultures do not favor special locations for defecation, so avoidance is difficult, especially if the potential host does not understand the transmission dangers. Even if these are known, fecal deposits (and places where deposits have been) are not always easily identified. In yet other cultures where there are no religious taboos about feces, and where it is well-regarded as fertilizer (e.g., rural China), hookworm exposure is more likely (Nelson 1990).

Avoiding areas with feces certainly gives the appearance of parasite evasion. Timing of the avoidance is also important, however, because many parasites are not immediately infective. For instance, sheep defecate in particular areas, usually at night, and then do not graze there for 5–8 days. On first inspection, this could be taken as evidence of parasite avoidance, but in fact, the trichostrongyle nematodes that live in sheep must develop outside the host for approximately a week before they are infective. They gain entry when they are eaten with forage by the sheep that have returned to graze the contaminated area at just the time that most larvae are infective. What of the selective defecation? This may simply coincide with the tendency for sheep to keep to one area of the pasture at night (Crofton 1958). Similarly, Ugandan mangabey (*Cercocebus albigena*) movement patterns are consistent with a hypothesis of parasite avoidance, but again, other factors may be influential (Freeland 1976, 1980).

Revulsion at encountering the odors of rotting food may also play a part in parasite avoidance. The odors that occur with this sort of bacterial action interfere with ingestion, and thus may discourage dietary acquisition of pathogens (Janzen 1977; Williams and Nesse 1991). Prophylactic attention to what we eat may have influenced regional cuisine, which may reflect a response to parasite exposure levels. The secondary compounds in many spices have antimicrobial properties. Billing and Sherman (1998) compared spice use in meat-based traditional recipes from 36 countries; it correlated well with mean annual temperature (i.e., food spoilage). The hypothesis of spice prophylaxis fit their data better than competing hypotheses did. Billing and Sherman noted that spices have always been prized elements of trade, and that spice appreciation is usually acquired, after parental encouragement. Although some spices do have medicinal value when taken in large doses, the routine use of spice in food may signal an antimicrobial function before consumption (Sherman and Billing 1999).

Pathogen avoidance may even influence flower choice and pollination ecology. Shared flowers can spread the trypanosome *Crithidea bombi* across *Bombus* species. The probability of infection is related to the complexity of the inflorescence, with visits to accessible flowers (e.g., *Rubus caesius*) resulting in fewer infections than visits to flowers with narrow corollas and spiral inflorescences (Durrer and Schmid-Hempel 1994).

What of predators that acquire parasites by consuming infected intermediate hosts? Often those infected prey may behave in ways that increase exposure to predators (chapter 3). Can predators avoid such prey? What happens when the reward for taking an easily captured prey item is tempered by the cost of acquiring the parasite in that prey? I asked this question when I chose the starling–isopod–*Plagiorhynchus* system for study (chapter 3). Unlike other acanthocephalans that altered intermediate host behavior, *Plagiorhynchus* had been reported to be pathogenic for its final host, and allegations of its lethality had entered textbooks. I wanted to find out not only if this acanthocephalan, like others, would change intermediate host behavior, but also if starlings could avoid these parasitized isopods, no matter how easy they are to eat (Moore 1983a). As I have indicated, starlings do eat unusually high numbers of parasitized isopods; they do not avoid them. Although infected birds exhibit reduced standard metabolic rates and infected males increase their energy intake/excretion (Connors and Nickol 1991), death from *Plagiorhynchus* is probably rare (Moore and Bell 1983); the reports cited in textbooks were anecdotal, based on dead birds infected with *Plagiorhynchus* (e.g., Cheng 1973, 1986) and did not take into account all the living, seemingly healthy birds with the parasite. I concluded that the minimal cost of parasitism for a starling and the increased encounter with parasitized isopods meant that starlings should not be expected to avoid eating infected isopods. The isopods, in contrast, had much to lose if they became infected. Because of behavioral alterations induced by the parasite, they were more exposed to avian predation, and in addition, ovaries did not develop in infected females. In fairly artificial tests (the female isopods chose between starling feces without worm eggs and feces to which an aqueous suspension of worm eggs had been applied), female isopods nonetheless showed a tendency to avoid "parasitized" feces (Moore 1983a). This is by no means a general rule for potential intermediate hosts. Recall the possible attraction of *Tribolium* to cestode-infested rat feces (chapter 3; Evans et al. 1992; Pappas et al. 1995; Shostak and Smyth 1998).

Lafferty (1992) created a model that balanced the energetic value of intermediate host prey items with the cost of parasitism for a predator. He found that if prey capture is enhanced by parasites that impose only moderate costs on the predator, predators should not avoid parasitized prey (see also Lozano 1991, Minchella and Scott 1991).

Traditionally, ecologists have held that predators frequently select sick prey, either because they are more noticeable or because they are easier to catch. Milinski and Löwenstein (1980) decided to test this using *Daphnia* that moved oddly. The answer was complex and depended on the amount of training. Milinski and Löwenstein speculated that the importance of prey disability may vary with the predator's strategy; a predator that invests a lot of energy in the capture of a few prey may pay more attention to prey disability than a predator such as a stickleback does, which captures many prey, fairly successfully. This echoes the Holmes and Bethel's (1972) discussion of efficiency among predator/final hosts and its implication for behavioral alteration of intermediate hosts.

In this vein, there is limited, but tantalizing, anecdotal evidence for potential hosts avoiding ingestion of parasites in intermediate hosts. When the clam *Macoma balthica* is infected with larval stages (sporocysts) of the trematode *Parvatremis affinis*, the parasites are visible as white cysts (600 × 350 μm) in the clam when the valve is removed. Clams with this parasite do not burrow normally but crawl just beneath the surface of the sand, making conspicuous tracks that assist in their discovery by oystercatchers (chapter 3). At least one oystercatcher was observed to reject heavily parasitized molluscs, possibly assessing the infection. The nature of the cue that then prompts rejection (the visible sporocyst or some other abnormality sensed by the bird) remains a matter of conjecture. This would tell us whether all infected *M. balthica* were at risk of rejection, or only those with a certain distribution or density of sporocysts. In addition, we know virtually nothing about how this bird acquired its ability to discriminate among clams, or how widespread this ability is. Although other oystercatchers have been observed to reject *Macoma* in the field, the infection status could not be determined. Despite the small sample size ($N = 1$), this is the first example of refusal of parasitized prey, and to date, few others have been described (Hulscher 1982; see Goss-Custard 1984).

In contrast, a study of sticklebacks, copepods, and the cestode *Schistocephalus solidus* produced no evidence for the ability of a host to avoid acquiring a debilitating parasite (Wedekind and Milinski 1996). Sticklebacks are the second intermediate host for the cestode. All but lightly infected fish fail to spawn (McPhail and Peacock 1983), and infected fish may be less fearful of predators (Giles 1983, 1987a; Milinski 1985). Selection against consuming infected copepods, which are more active than uninfected ones, should be strong, but there is no evidence for it. Sticklebacks attacked infected copepods preferentially as well as the more active uninfected copepods. They did not switch to *Daphnia* when given the opportunity, however, even though *Daphnia* were more active than copepods. Moreover, the infection history of the stickleback source population did not influence the tendency to forage on copepods; sticklebacks that came from populations with high risk of infection were just as eager to consume these copepods as fish from uninfected populations. Stickleback discriminatory ability is not likely to be the problem; they can tell the difference between copepods and *Daphnia*, and they can discriminate against parasitized mates (Milinski and Bakker 1990). Although we might speculate about constraints that prevent them from avoiding parasitized copepods, there is no clear explanation for their willingness to consume such expensive prey (Wedekind and Milinski 1996).

AVOIDANCE OF PARASITES THROUGH MATE SELECTION

Judicious selection of mates may contribute to avoidance of both ectoparasites and endoparasites. Freeland (1976, 1981) suggested that competition for mates could reveal relative parasitemia among competitors. The choice of a healthy mate could increase fitness directly and indirectly, by both reducing exposure to disease and enhancing the genetic predisposition for disease resistance among offspring of the union.

Hamilton and Zuk (1982) explicitly addressed the second possibility: that animals should eschew parasitized mates to minimize the transmission of genetic susceptibility to offspring. They suggested that some secondary sexual characteristics were especially good indicators of health—hence their importance in courtship displays. This possibility generated a good deal of interest among behaviorists and evolutionary biologists, and the resulting body of work has been the subject of several reviews (e.g., Clayton 1990, 1991; Read 1990;

McLennan and Brooks 1991; Clayton et al. 1992; Møller 1992, 1994; Zuk 1992; Andersson 1994; Møller and Saino 1994; Wedekind 1994; Able 1996; Shykoff and Widmer 1996; Hamilton and Poulin 1997; John 1997).

Although information exists for some host–parasite associations (Wakelin and Apanius 1997), we do not know whether there is a genetic basis for resistance to most parasites, a void that, along with absence of experimental infections, has compromised some of the tests of the Hamilton-Zuk hypothesis. With or without genetically based resistance, an animal may still benefit from choosing a parasite-free mate if, in so doing, it avoids contagion. Thus, parasite-mediated sexual selection could occur if

1. it favors mates with genetically-based resistance to parasites (a "good genes" model);
2. it favors healthy mates and avoidance of transmission (a self-protection model);
3. it favors mates that are healthy enough to help rear young (a good parent model); or
4. if the preferred traits coincide with low parasite infestations (a coincidental preference model).

Only the first instance requires a genetic component (Clayton 1991; Andersson 1994). These conditions are not mutually exclusive, and given the heterogeneity of parasite communities, could occur in combination (Møller 1994).

Moreover, if parasites impede the ability of an animal to compete successfully for mates, this could give the appearance of parasite-mediated sexual selection (Freeland 1976; Howard and Minchella 1990). For instance, male fence lizards with malaria did not display as frequently as uninfected lizards and relinquished courtship activities rapidly with the advent of a noninfected competitor (Schall and Dearing 1987; Schall 1996). If the parasite is sexually transmitted, then the parasite may benefit from reduced detection [i.e., decreased virulence (Knell 1999)] and may actually increase mating success (e.g., McLachlan 1999).

Another consideration is the fact that parasites can cause hosts to shift habitat preferences (chapter 3). In this case, infected hosts may be spatially segregated from uninfected conspecifics. Such separation limits mate choices and can result in apparent "preferences" for infected or uninfected mates (Zohar 1993; F. Thomas et al. 1995, 1996a,b; Zohar and Holmes 1998).

Much of the work on mate selection and parasitism has been purely behavioral in nature, providing a record of mate preferences with respect to parasitism. In some systems, a more mechanistic approach has provided additional insight. For instance, anogenital odors of male meadow voles (*Microtus pennsylvanicus*) that had recently and vigorously groomed themselves were more attractive to female voles than those of males that groomed less. A lower parasite load associated with higher grooming could be one of several attributes advertised in this way (Ferkin et al. 1996), although, again, an animal that grooms more may signal initial heavier parasite intensities.

Female mice preferred the scent of uninfected males to that of males subclinically infected with *Eimeria vermiformis*; when parasitized, males were disinterested in estrous females unless the infection was patent. These responses are mediated in part by changes in endogenous opioids, which in turn can be related to endocrinological and neuroimmune phenomena. Similar results were obtained with nematode infections (*Heligmosomoides polygyrus*; Kavaliers and Colwell 1992a,b, 1993a, 1995b,c; Kavaliers et al. 1997b, 1998a; but see Zuk et al. 1997). The nematode infection also decreases dominance in male mice (Freeland 1981).

This work brings up another question, all but ignored by researchers who seek associations among visual appearance (or even auditory signals), parasitism, and mate choice: What about chemical signals? Many diseases are associated with specific odors, and chemical evidence for health status is more abundant than we usually consider (Penn and Potts 1998a). Moreover, in both humans and mice, mate choice is influenced by odor cues that reflect the genetic makeup of the major histocompatibility complex; in both species, females prefer males with an MHC that differs from their own (Wedekind et al. 1995; Penn and Potts 1998b; see Penn and Potts 1999 for review). Moreover, although viruses (Penn et al. 1998), protistans (Kavaliers et al. 1997b), and nematodes (Klein et al. 1999) have been shown to affect odor preferences, in some cases such preferences may not influence partner preference (Klein et al. 1999).

Oddly, until recently, behavioral ecologists did not extend their interest in sexual signaling and parasites to immunological influences on both. It has long been known that the immune and endocrine systems are linked by the action of steroids (reviewed by Grossman 1985; see also Barrow 1955). For instance, because testosterone can have opposite effects on secondary sexual characteristics and immunocompetence, enhancing one and depressing the other, the interplay of immune and endocrine systems may be of special relevance to understanding parasite-mediated sexual selection (Folstad and Karter 1992; Wedekind and Folstad 1994; Poiani et al. 2000; but see Hews and Moore 1997; Braude et al. 1999). Much more investigation is needed on a wide range of hormones using physiological levels of these compounds under natural conditions (Hillgarth and Wingfield 1997a,b). In fact, it is possible that the type of parasitic onslaught (and host response) might be revealed by the effect on different aspects of sexual ornamentation (Wedekind 1992). For instance, cartenoid-based coloration pales in male guppies (*Poecilia reticulata*) parasitized with a monogenean, and they become relatively unattractive to females (Houde and Torio 1992; chapter 3). Beware, this is not a general rule. Schall (1990) summarized two studies in which lizards with blood parasites were more brightly colored (i.e., attractive to females) than unparasitized conspecifics (see also Ressel and Schall 1989; Schall 1982). In addition, much of the work just mentioned pertains to vertebrates. Although physiological systems do not operate in a vacuum, and we expect interplay, immunological and endocrinological functions differ substantially across the animal kingdom. The nature and degree of their impact on parasite-mediated sexual selection probably will be equally variable.

In addition to mate choice, parasites have been proposed as influential in almost every area of mating behavior. Mating within stable groups may reduce exposure to novel, external pathogens. Even postcopulatory guarding may serve a protective function, reducing the risk of pathogen transmission from subsequent mates (Freeland 1976). Parasites have been suggested to play a role in the evolution and maintenance of environmentally hostile female reproductive tracts, with a variety of implications for sperm competition (Sheldon 1993). Mating systems may reflect parasite challenges, even among humans (Hamilton 1990; Low 1990; Read 1991). The risks of pathogen transmission may make monogamy especially advantageous for long-lived organisms, and Loehle (1995) suggested this influence could be tested by examining animals such as birds and canids. He also noted that nonpromiscuous polygamy (e.g., horses) or the combination of within-group promiscuity and xenophobia could achieve low transmission levels similar to those characterizing monogamy. Polygyny may also be limited by the negative immunological effects of androgens, which are higher in polygynous organisms (Møller et al. 1993). Ultimately, parasites are likely to have favored the evolution of sex itself (Hamilton 1980; Lively 1987, 1992, 1996; Hamilton et al. 1990; Lively et al. 1990; Clay 1991; Johnson 1992; Howard and Lively 1994; Dybdahl and Lively 1996; Ebert and Hamilton 1996).

As Møller and co-workers (1993) point out, parasites offer a double-edged sword to both social and sexual interaction. Disease transmission may be facilitated within a group, yet groups can protect individuals from predators and dilute the effect of vectors. Diseases may be sexually transmitted, but sexual reproduction can generate resistant genotypes.

If we think back on some of the ways that animals avoid parasites—migration, habitat selection, group formation, mate choice—we see that in most of the natural world, the avoidance of parasites is a nontrival task, one that is costly and, by implication, worth the cost. The success with which organisms avoid parasites is the extent to which they are behaviorally not susceptible to them and ultimately has an impact on parasite reproductive ratio.

5

Behavioral Alterations and the Fitness and Longevity of Infected Hosts

Despite an impressive repertoire of parasite avoidance techniques, animals still acquire parasites. If behavioral alterations in parasitized animals were all the result of natural selection favoring increased parasite transmission and therefore increased R_0 for parasites, the world I am describing here would be uncomplicated. Of course, most of the mechanisms by which parasites alter behavior remain a mystery, and the evolutionary events that enabled them to seize upon these mechanisms are equally unknown to us. These questions certainly lead into complicated territory. Nonetheless, the general result, as far as basic reproductive rate goes, would be fairly straightforward. R_0 simply would be larger than what one would anticipate based on random encounters between propagules and susceptibles, prey and predators.

This is not always the case. In contrast to what one might expect from polymorphids in amphipods sorting out altered habitat preference by final host (Bethel and Holmes 1973), and digeneans increasing mosquito vulnerability to voles and not to guppies (Webber et al. 1987a), not every parasite has a happy ending—not even those that alter host behavior.

As noted earlier, contact with susceptible animals has a large impact on microparasite R_0, as does the proportion of immune hosts. For macroparasites, the number of successful female offspring can be an indicator of R_0. Altered behavior that enhances transmission and parasite fitness essentially increases contact with susceptible animals and increases numbers of successful female offspring. The host, however, is also responding to selective pressures to minimize the negative effects of parasitism and thus maximize its reproductive rate (see Hart 1990, 1994; Clark 1991; Lehmann 1993 for reviews of some costs of parasitism).

Parasitism is an ancient way of life—perhaps as old as the origin of eukaryotes or older—and the happy circumstance of enjoying resources accrued by someone else is not recent. In response, the ability to distinguish between self and non-self is pervasive among living things. One of the end results of all of evolution is that organisms have quite an arsenal with which to protect themselves if accosted by parasites. In so doing, they can severely limit parasite reproduction and life span, parameters that, like transmission, greatly affect R_0 (table 5.1, p. 226).

BEHAVIORS THAT REDUCE PARASITE SURVIVAL

When we consider this arsenal, the immune system and related physiological responses usually spring to mind. Indeed, these are among those phenomena that can heavily influence the probability of encountering susceptible individuals, for instance, and ultimately determine if R_0 exceeds 1. Invertebrates as well as vertebrates have elaborate physiological defense mechanisms, topics that are fascinating, but well beyond what can be covered in this book. In turn, there are remarkable adaptations on the parts of parasites to limit the influence of host responses or even use those responses to benefit the parasite (Hayunga 1979; Parker and Schneider 1981; Damian 1987; Loker 1994).

There are also behavioral defenses that can be strikingly effective—behaviors that can engender or accompany fever or chills, that can purge external or internal parasites, that can protect kin (Hart 1988, 1990, 1994). These are behavioral responses that minimize the success of parasites, perhaps as effectively as physiological defenses do.

Hart (1990, 1994) suggested two requirements for attributing defense against parasites to a given behavior. First, the parasite should reduce host fitness; there should be a cost associated with having the parasite. Second, the behavior should limit or eliminate the parasite. The cost of the behavior would be offset by the benefit of reducing the cost of the parasite. As discussed in chapter 4, such criteria are often difficult to satisfy. Much like behaviors that are alleged to enhance parasite transmission, behaviors that are alleged to enhance host defense frequently lack rigorous experimental examination. This chapter covers aspects of behavior that do not necessarily always meet Hart's criteria; some are even anecdotal. I include them because they illustrate the broad scope of the possible, indeed perhaps even the likely, if we only knew more about these interactions. One would do well, however, to keep Hart's criteria in mind.

In considering these defensive behaviors, I limit them to behaviors that are prompted by the successful transmission/invasion of a parasite (i.e., the behavior of the infected host) and that thus can truncate parasite fecundity and/or life span within or upon the host. Behaviors that result in parasite avoidance (thus reducing successful transmission itself) are discussed in chapter 4.

Many defensive behaviors have been recognized only recently as important contributors to animal health, and I suspect more will be discovered in the near future. Wakelin (1997) has suggested that the dangers of pathology associated with immune response may result in moderate immunological responses of limited effect. If this is the case, then behavioral avoidance and resistance may be even more important in host defense than previously acknowledged.

Fever

Some defensive behaviors can be confused easily with behaviors that enhance transmission. As we have seen, parasitized animals often crawl up on objects in the environment, move into well-lighted areas, and/or choose substrates that differ from the ones unparasitized conspecifics would choose. In addition, many become hyperactive (tables 3.2, 3.6, 3.7, 3.12). These and other changes customarily have been viewed as beneficial for the parasite, aiding in dispersal, whether by spreading propagules to the surrounding area or increasing vulnerability to a subsequent predatory host.

But what if in order to survive a fungal infection, an insect had to spend hours at 40°C? Might it not become hyperactive, or bask at the tops of plants or on warm substrate? In fact, such high temperatures for 8 hours during the first 3 days of infection allow houseflies *(Musca domestica)* to survive an otherwise lethal infection with *E. muscae*. The timing of this fever is important; survival rates dropped below 20% if high temperatures were not available until after day 3. When allowed a choice of temperatures in the laboratory, infected flies opted for warm temperatures, and did so for long enough for the heat to be therapeutic (Watson et al. 1993). This is consistent with field observations that show that flies allowed to bask suffered lower fungal mortality (Watson et al. 1993, citing Oleson 1986). Again, the timing of such a choice clearly is critical, and flies were not seen to generate fever on days 4 or 5 (nor, strangely, on day 1, when it would have been most effective). In addition, warm temperatures on day 4 or 5 or brief exposures to high temperatures early in the infection, while insufficient to rid the fly of the infection, could inhibit sporulation or parasite development, respectively. (In a reversal of defensive preferences, the cool temperatures that favor sporulation are chosen by dying flies.) Warm temperatures late in infection may also cause the failure of dead flies to exhibit the extended appendages and distended abdomen that are typical of this infection and that are favorable for fungus dissemination (Watson et al. 1993). The deleterious effect of warm temperatures on the fungus is probably not unrelated to the low prevalence of the fungus in southern California during the summertime and its much higher prevalence (over 45%) in colder months (Mullens et al. 1987). Fever can also influence insect oogenesis, which can occur more rapidly than fungal development does at temperatures exceeding 22°C (Mullens 1990). Thus, parasites induce altered host behavior by prompting the host to behave in ways that promote host defense while giving the appearance of dissemination (e. g., elevation seeking; table 3.6), or vice versa (Horton and Moore 1993).

OVERVIEW   Fever has been defined as "a change in the thermoregulatory set point" (Kluger 1991, p. 94). It is complex and can be physiological or behavioral—and is usually accompanied by a constellation of defensive responses, depending on the host–parasite taxa, that can include neutrophilia, increased plasma proteins, and anorexia. Fever has some notable beneficial effects in the presence of pathogens, including enhanced interferon and white blood cell activity, neutrophil migration, depletion of plasma iron necessary for bacterial growth, and T-cell proliferation. These and other febrile events occur in response to a variety of insults and are mediated by endogenous pyrogens (Mitchell et al. 1990; Kluger 1991). Fever is metabolically costly, and although it is not inevitably beneficial, it has been shown to enhance overall survival of infection (Kluger 1986). We are just now beginning to understand the influence of fever and other physiological defenses on the nervous system and behavior (Maier and Watkins 1998, 1999).

Fever is widespread in the animal kingdom and probably is phylogenetically ancient (Kluger 1979, 1991; Cabanac 1990). Given the energetic cost of fever, Kluger argued that its persistence can be taken as evidence of its utility. Indeed, the wisdom of interfering with the expression of at least moderate fever, as well as some other manifestations of disease, has been challenged for years (Dixon 1976; G. C. Williams and Nesse 1991; Nesse and Williams 1994), and the treatment of neurosyphilis with malaria resulted in a Nobel Prize (Hart 1988). On the other hand, clear associations between body temperature and mortality from disease have been difficult to establish (Mitchell et al. 1990), and we are aware of some cases in which fever has proven to be harmful or in which fever that is effective against one pathogen may benefit another. After all, fever is a departure from a temperature that is usually the most favorable for the host when not infected (Heinrich 1993). In general, although many aspects of defense against disease may be maladaptive in certain circumstances, their widespread benefits may outweigh these more peculiar costs (Kluger 1991). As I demonstrate below, the results of fever for parasite and for host can be anything but straightforward.

What are the mechanisms by which high temperature affects the course of infection? Our understanding of invertebrate fever remains murky (Boorstein and Ewald 1987; Mitchell et al. 1990; Heinrich 1993). Elevated temperature alone is not necessarily lethal to the parasite, but may interact with other responses (Heinrich 1993). Hart's (1988) overview, while addressing mammalian fever, is probably generally applicable; fever can enhance the immune response and alters what may have initially been a beneficial thermal environment for the pathogen. Microparasites can be suppressed by fever and, in the case of metazoan parasites especially, development may be impaired (see below).

As we have seen, fever can have life-or-death consequences for insects (as well as their parasites), yet pathogen-related defensive responses of ectotherms have not received much attention until the last two decades or so. In the past, the absence of mammalian-type immunoglobins in invertebrates led immunologists to assume that little of interest was occurring in the so-called "lower" taxa; we are rapidly becoming disabused of that idea. In like fashion, the absence of metabolically based thermal regulation was, for many zoologists, mistakenly seen as synonymous with the absence of all thermal regulation.

Scientists working with fish had noticed that fish infected with prokaryotic and eukaryotic parasites often seek warmer water (see Smith and Kramer 1987), but because pyrexia (the high body temperature associated with fever) had been understood in terms of thermoregulatory set point, ectotherms were not considered to have fever. That began to change when Vaughn and coworkers (1974) injected desert iguanas (*Dipsosaurus dorsalis*) with killed, pathogenic gram-negative bacteria, causing a 2°C rise in body temperature that was behaviorally induced.

BEHAVIORAL FEVER   The occurrence of fever in invertebrates is especially tantalizing given their roles as intermediate hosts and vectors. Temperature is paramount among environmental influences on microbial infection of insects, affecting susceptibility and parasite multiplication (reviewed by Watanabe 1987). Marikovsky (1962) noticed that some fungus-infected caterpillars and grasshoppers that climbed high were able to rid themselves of the disease, and speculated that solar radiation was influential. The first evidence that insects could become feverish came from the hissing Madagascar cockroach, *Gromphadorhina portentosa*, injected with an *E. coli* suspension or an endotoxin; in both cases, treated cockroaches preferred temperatures that were higher than those of controls, and the response to *E. coli* injection was dose dependent (Bronstein and Conner 1984).

This and other work over the past 20 years tells us that a febrile response is not only possible among ectotherms, but is widespread, although certainly not inevitable (McClain et al. 1988; Mitchell et al. 1990). Thus, we should consider fever as a possible part of the behavioral repertoire of a parasitized animal. Indeed, the linkage of neurohormones and hemocyte behavior in cockroaches (Baines et al. 1992) suggests one mechanism whereby behavior could be coordinated with other defensive reactions. What is of more concern to most evolutionary ecologists is the nature of such behaviors in the face of infection with parasites themselves and the effect of these behaviors on host and/or parasite fitness. Although interest in behavioral fever stems from clinical conditions in homeotherms in which the set point for thermoregulation is increased, any deviation from the preferred temperature of an uninfected animal, colder or warmer, could have fitness effects for parasite or host. As we shall see, examples of both exist.

Temperature choices can have measurable fitness consequences for infected insects and their parasites. When grasshoppers (*Melanoplus sanguinipes*) were infected with *Nosema acridophagus* and placed on a gradient, they preferred temperatures 6°C higher than those chosen by uninfected grasshoppers (Boorstein and Ewald 1987). When uninfected grasshoppers were kept at these higher temperatures, growth and fecundity were reduced compared to that of uninfected grasshoppers at normally preferred temperatures. Infected grasshoppers, however, did better at those febrile temperatures than did their infected counterparts at the temperatures preferred by uninfected animals; they grew more, produced more offspring, and experienced increased survival. Moreover, while all infected grasshoppers were less fecund than uninfected grasshoppers at lower temperatures, fecundity was similar for both classes of grasshoppers at febrile temperatures. Thus, behavioral fever imposes a cost that would be maladaptive for uninfected animals, but for infected animals, it can be a case of making the best of a bad situation, with a significant difference in fitness (see also Bordat et al. 1984).

When Mediterranean crickets (*Gryllus bimaculatus*) were injected with a lethal intracellular parasite (*Rickettsiella grylli*) and allowed to select a temperature in a gradient, they fared well compared to similar crickets kept at a variety of fixed temperatures (Louis et al. 1986). The crickets kept at 28°C or less experienced the highest mortality: Only 50% survived after 20 days, and all were infected. In comparison, 75% of crickets kept at temperatures of at least 30°C survived, with infections progressing slowly and abnormally. The results were most dramatic in the temperature gradient, where crickets themselves were allowed to choose their ambient temperatures: 90% of the crickets survived, and only 20% developed the infection. Animals in the gradient occupied the hottest part of the cage, and they chose temperatures of 33°C compared to controls at 26.6°C. The timing of the choice was also significant: Infected crickets began to choose hot temperatures about 10 days postinfection, when the parasite multiplied. Until then, differences in thermal preference were not apparent. In addition, crickets did not remain in hot areas, but moved around. Parasites in these crickets showed the same deformities that those kept at 32°C did, thus indicating that constant fever was not necessary to combat the infection.

These results lead to some puzzles. Crickets that were able to choose their own temperatures fared even better than those that were kept consistently at the hottest temperatures. Behavioral fever, although clearly effective, may not have to be continuous and thus may be unpredictable in its manifestation (Inglis et al. 1996). Moreover, a host that responds with fever to one parasite may not respond that way to another (Adamo 1998a).

Of course, selecting different thermal regimes may pose other fitness costs for the host. In some viral and fungal infections, fever may be especially risky. Infected insects are often less heat tolerant and may succumb to heat more quickly than uninfected ones (Wantanabe 1987; see also Vaughan and Coble 1975, for fish and ectoparasites). Basking can increase predation risk, as can other behavioral corollaries of a parasite-induced shift in the thermal preference. For instance, bullfrog tadpoles (*Rana catesbeiana*) injected with dead bacteria (thus stimulating an acute phase response; e.g., Stadnyk and Gauldie 1991) reduced their activity levels and did not seek shelter. In the presence of

a predator (salamander), control tadpoles lowered their temperature preference, and in salamander-conditioned water, they reduced activity levels; injected animals did not. When refuge vegetation was available, salamanders
consumed more injected than control tadpoles (see also Lefcort and Blaustein
1995). Thus, a routine defensive response against pathogens may also reduce
defenses against predators. This leads to a paradox, for if predation is important in the parasite life cycle, immune recognition may serve the parasite well
(Lefcort and Eiger 1993).

Given such preferences on the part of feverish hosts, what of animals that
act as vectors for the parasites they contain? What if these vectors, once infected, are drawn to warm temperatures? Fleas, for instance, disseminate
plague bacillus (*Yersinia pestis*) more readily when blocked, that is, at temperatures below 28°C (chapter 3).

We might expect that preference for high temperatures could enhance host
location—a host for both the bacillus and the flea. Behavioral fever might
characterize *Y. pestis*–flea associations, and might have been co-opted by the
bacterium if it favored transmission. Once the flea is blocked by feeding on
an infected host, however, it must leave that host to transmit the plague organisms to a new individual. This would require greater tolerance for cold
temperatures. R. E. Thomas and co-workers (1993) found that infected fleas
(*Xenopsylla cheopis*) moved farther than uninfected fleas on temperature gra-
dients from the warm end to the cooler end, and thus finished at cooler temperatures (1.5–1.6°C cooler); this was not because of increased activity on the
part of infected fleas, because the two groups did not differ in the distances
they moved in the absence of a gradient. The lack of relationship between infection rates and distances that fleas moved led to the conclusion that *Y. pestis*
probably did not induce the behavior. However, altered behavior in parasitized
animals is not inevitably dose dependent. Clearly, fleas and plague provide one tantalizing example of the possible complexities associated with
temperature-mediated effects on transmission and fitness. If the flea chooses
a warmer location, it will perhaps dissolve the blockage that inhibits its own
feeding and that favors parasite transmission. At the same time, for both the
flea and its parasite, warmer temperatures may signal suitable hosts. A preference for cooler temperatures, while maintaining the blockage, may not result in immediate host location for either flea or plague, but may be a necessary first step in leaving an infected host and finding a new host.

Meanwhile, although fever is widespread (including annelids, arthropods,
fishes, amphibians and reptiles; Cabanac 1990; Kluger 1991), it is not ubiquitous, and in some instances where it has been demonstrated, there may be
methodological problems. For example, fever has been induced in some studies with injections of pyrogenic substances rather than true infectious agents.
This can be criticized as an unnatural situation, but it might also illuminate
aspects of fever more clearly than perhaps a more variable response to living
parasites would.

The absence of fever is equally fraught. It does not always occur when we expect it to. Fever was absent in female *Drosophila* spp. parasitized by the nematode *Howardula aoronymphium*, even though the flies are sterile and more likely to die at temperatures below 28°C than at 29°C, where they can reproduce (Ballabeni et al. 1995, using a 20°–30°C gradient). The absence of fever may be attributed to a variety of conditions and is not necessarily a sign of the inability to produce fever on the part of the host taxon (Cabanac 1989; Kluger 1991). Do large parasites escape the immune system by virtue of their size? Are some examples of laboratory-induced fever artifactual (Mitchell et al. 1990)?

To date, macroparasites have not been experimentally shown to cause arthropod fever (see Karban 1998), and once we have more information, this may be relevant to an explanation of the absence of fever in such infections. Perhaps high temperatures enhance host defenses against some parasites, while other defenses require cooler conditions (see below).

The possibility of behavioral fever adds to the ambiguity encountered when one tries to extrapolate fitness corollaries of behavioral alterations. Is a positive response to light—or a negative response to gravity—a manipulation that will end in parasite transmission or host suicide? Is it an attempt to kill a host or a parasite?

BEHAVIORAL CHILLS   Fever is not the only thermal change that can harm a parasite. Bumblebee workers (*Bombus terrestris*) normally return to the nest at night. Those that were parasitized by a conopid fly, however, remained outside the nest, in colder temperatures (Müller and Schmid-Hempel 1993). When given a choice of temperatures, parasitized animals spent more time in cold areas than unparasitized ones did. At lower temperatures, workers lived longer and fewer parasitoids completed development.

Similarly, snails (*Biomphalaria glabrata*) infected with *Schistosoma mansoni* selected cooler temperatures than uninfected conspecifics did. The effects of this preference on snail and parasite were not documented, but Lefcort and Bayne (1991) pointed out that such behavior would be consistent with other literature that reports a negative effect on schistosomes at both temperature extremes (see Blankespoor et al. 1989 for a review) and with a hypothesis of parasite damage (e.g., Stirewalt 1954). Some other work with snails is inconclusive, however (Rothschild 1940; Vernberg and Vernberg 1963, 1967, 1968; Chernin 1967; Vernberg 1969; Cabanac and Rossetti 1987).

The cockroach and its acanthocephalan, *Moniliformis moniliformis*, seem like good candidates for behavioral fever (fig. 5.1). In the cockroach, the acanthocephalan can only develop between the temperatures of 20°–37°C, with notable abnormalities occurring below 24°C and above 32°C, respectively (fig. 5.2). If infection is initiated at 37°C and shifted to 28°C, all parasites are melanized; the reverse sequence (28°C to 37°C) results in some development, though many are abnormal (Lackie 1972).

When offered a choice of temperatures, however, cockroaches infected with

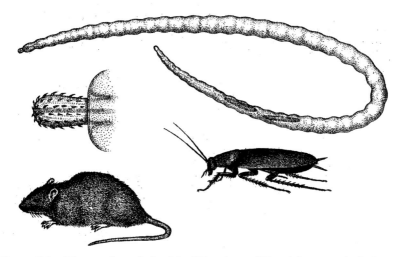

Figure 5.1. The acanthocephalan *Moniliformis moniliformis* has a particularly ver-
minous life cycle, developing to a cystacanth in a cockroach and spending adult life
in the intestine of a rat. (Moore 1984a; reprinted with permission from *Scientific
American.*)

M. moniliformis did not seek out hot surroundings, but spent enough time in
cool temperatures to severely retard the development of the parasite. This effect was much more pronounced in the small cockroach, *Supella longipalpa*,
than in large *Periplaneta americana* (Moore and Freehling submitted). Are
developmentally retarded acanthocephalans able to influence host behavior?
Do cool temperatures also exact a reproductive price; is gamete production
retarded as well?

It is clear that some parasites engender fever and others, chills. Do these
divergent choices accompany different suites of defenses in the host? Might
some host defenses work better at low temperatures, and others, at high ones?
We do not know.

PARASITE DEVELOPMENT AND HOST THERMAL PREFERENCE   Although reports of
the defensive use of thermal preference are still relatively uncommon, studies
that link parasite development to ambient temperature are more numerous, and
provide us with ample reason to expect alterations in thermal preferences on
the parts of parasitized animals, be they for the benefit of parasite or host. Although there is abundant evidence that heat treatment can promote host survival and damage some parasites (Horton and Moore 1993), changes in temperature do not inevitably benefit hosts. There appear to be at least two possible
outcomes of these interactions for parasites if thermal preference is subject to
alteration by parasitism: parasites might manipulate host thermal preference
to enhance their own development, or hosts might select a thermal regime that
would injure the parasite (Horton and Moore 1993).

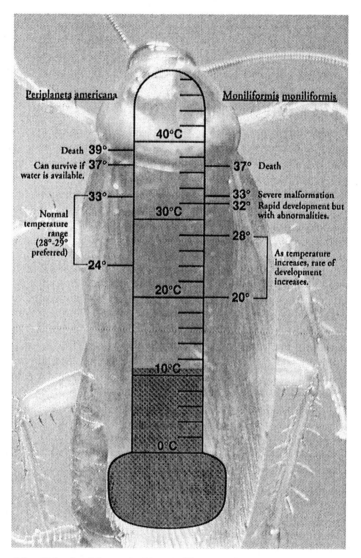

Figure 5.2. Behavioral fever would seem like a good strategy for a cockroach (e.g., *Periplaneta americana*) infected with *M. moniliformis*, which shows developmental abnormalities at temperatures above 32°C; roaches can survive much higher temperatures. Experiments did not confirm this expectation. (Modified photo courtesy of Marie Timmerman.)

Many parasites are extremely susceptible to thermal insult. Developmental times of parasites as diverse as viruses, fungi, and flies can be dramatically affected by ambient temperature, and developmental times can have profound fitness effects (DeGiusti 1949; Stirewalt 1954; Novak 1979; Halvorsen and Skorping 1982; Takaoka et al. 1982; Novak et al. 1986a,b; Watanabe 1987; Turell and Lundstrom 1990; Patel et al. 1991; Tocque and Tinsley 1991; Müller and Schmid-Hempel 1993). As discussed above, altered thermal preference, hot or cold, can harm parasites. There are also examples of altered thermal preferences benefitting parasites. One of these, the sandfly *Lutzomyia vexator*, transmits *Plasmodium mexicanum* to lizards in northern California. When infected, these flies seek temperatures 2° higher than the preferred temperatures of fed, uninfected flies. This increase in ambient temperature does not affect the gonotrophic cycle in the sandfly, but it does mean that the parasite is infective to the next lizard one day sooner than it would have been at the host's usual temperature preference. This is important for transmission, for few sandflies live to take a third blood meal; with this behavioral modification, *P. mexicanum* can be transmitted with the second blood meal. Oddly, the infected lizard does not exhibit behavioral fever, possibly because this parasite can tolerate a wide range of temperatures in the lizard and its growth rate is not temperature dependent between 20° and 35°C (Fialho and Schall 1995).

In addition, variation in thermal regime can have unexpected effects on parasite development. *Plasmodium berghei* requires 3 days at 28°C in early stages of reproduction within its mosquito vector; at a different time in parasite development, such temperatures would be damaging (Vanderberg and Yoeli 1966). Such idiosyncracy in the relationship between temperature and development means that even if parasites were all able to manipulate host thermal preference in ways that contributed to advantageous parasite development schedules, or even if hosts were able to choose thermal regimes that were unhealthy for parasites, predicting those choices would not be simple and uniform.

Not only can developmental changes influence individual host–parasite associations, they can also have population-level consequences. In fact, developmental delays were one of the factors that May and Anderson (1978) found to have a destabilizing influence on host–parasite population interactions. Moreover, because temperature changes affect the course of infections in insects, ambient temperature may be one of the factors limiting the geographic distribution of some parasites (Carruthers et al. 1992; see also Takaoka et al. 1982; Jaenike 1995; Blanford et al. 1998).

The ramifications of thermal limitations and influences on parasites are widespread. If they are to be effective, biological control agents should be disseminated under conditions that favor their survival (e.g., Ignoffo 1981). Success of control agents may vary considerably, depending upon environmental conditions. Given the brief introduction to temperature considerations here, it is not unreasonable to expect that global warming, for instance, would profoundly affect a wide range of symbiotic interactions.

The interaction of development and temperature reinforces the observation that timing of temperature change may be critical. We have seen the potential importance of the timing of temperature changes for parasite development and survival in *E. muscae*, *R. grylli*, and *P. berghei* (Vanderberg and Yoelli 1966;

Louis et al. 1986; Watson et al. 1993). If temperature is used as a host defense or parasite enhancement, then it may be used in ways that are peculiar to each host–parasite association, depending on the developmental biology of the parasite and the tolerances of host and parasite. Opportunities exist for behaviors that exploit thermal environments in ways that may benefit parasite or host, and these behaviors may be inconsistent both across host–parasite associations and within the course of a single infection. Witness the flies that choose warm, defensive temperatures early in fungal infection, but choose cool, fungus-friendly temperatures later.

METABOLIC FEVER AND BEHAVIOR   I have emphasized behavioral fever, given the focus of this book, but metabolic fever, while not generated behaviorally, often has behavioral corollaries (e.g., lethargy, anorexia) that can be as much a part of a fever repertoire as other behaviors might be of, say, hibernation (Hart 1988, 1990; Kent et al. 1992; Maier and Watkins 1999). In the past, these behaviors have been seen as debilitation, but given the expense of fever (13% increase in metabolism per 1°C), they may well be adaptive and energetically frugal (Hart 1988). Animals with metabolic fever may be at increased risk of predation, even as basking animals may be more conspicuous (Carruthers et al. 1992). Ectothermic vertebrates may also exhibit reductions in activity during behavioral (nonmetabolic) fever (e.g., Lefcort and Blaustein 1995). These reductions in eating and other activities are not seen as debilitation, but are responses that routinely accompany fever, caused by the cytokines that cause fever (Kent et al. 1992).

Anorexia deserves special mention (table 3.9). Parasite-induced anorexia is not always associated with fever. When it occurs, it may do so at specific times during parasite development (Crompton et al. 1981). Kyriazakis and co-workers (1998) considered several hypotheses to explain parasite-related anorexia: (1) it benefits the parasite, (2) it starves parasites, (3) it results from decreased energetic efficiency, (4) it assists immune response, and (5) it increases diet selectivity. They found the most support, albeit indirect, for the last two hypotheses. The function of anorexia in parasitism needs more investigation; it is possible that, like fever, it should be seen as a beneficial host response, to be managed during a course of infection, and not overcome (Kyriazakis et al. 1998). Moreover, parasite-induced reductions in food intake do not necessarily lead to a state of deprivation; lambs parasitized by the nematode *Trichostrongylus colubriformis* reduced overall food intake, but did so by shifting to a higher-protein diet (Kyriazakis et al. 1994).

In the case of bloodborne parasites, host malaise can interfere with defensive behaviors that would otherwise prevent the approach and engorgement of potential vectors. Moreover, hyperthermia is not always a requirement for such lethargy. Mice infected with *P. berghei*, *P. chabaudi*, or St. Louis encephalitis virus (SLE) experienced intermittent hypothermia, yet they also exhibited reduced activity at specific times during the infections, perhaps a corollary of defensive energy conservation or perhaps true debilitation. At these times of reduced activity, they were more subject to mosquito feeding (Day and Edman 1983, 1984a; fig. 5.3). In addition, there is evidence that the inhibitory influence of fever in *P. falciparum* infections may synchronize parasite growth and keep the infection at sublethal levels (Gravenor and Kwiatkowski 1998).

In contrast, *Plasmodium yoelii* does not affect mice much, and like some other hitchhiking parasites (chapter 3), it may depend for its transmission on the presence of a second, more severe parasite that reduces defensive behavior (Day and Edman 1983). This happened with an admittedly artificial combination of *P. yoelii*, *Leishmania mexicana amazonensis*, and mice subjected to *Ae. aegypti* feeding. Mice infected solely with *L. mexicana* experienced no effects on activity, whereas mice with *P. yoelii* were somewhat inactive. The most inactive mice, and the ones with the greatest hypothermia, were ones infected with both parasites. Mosquitoes could not feed on mice infected with *L. mexicana* alone, and they had little success with sole infections of *P. yoelii*, but they were quite successful with concurrent infections, especially during times of high parasitemia. Coleman and co-workers (1988) concluded that anything that negatively affected host defensive behavior would probably have a positive influence on vector feeding.

The relationship between symptomatic infection and host choice by vectors may have significance for human malarial and microfilarial epidemiology. Some of these infections do not produce fever and have been considered asymptomatic in Papua New Guinea, where *Anopheles punctulatus* feeding was not consistently related to infection with either parasite. Burkot and co-workers (1989) argued that models based on nonhomogeneous host selection (e.g., Dye and Hasibeder 1986; Kingsolver 1987) may not be applicable to the majority of human malarial areas, where the parasite is largely asymptomatic. In this study, however, fever was viewed as the primary indicator of symptoms; other human behaviors that might distinguish parasitized from unparasitized individuals were not considered.

Fever may have profound effects on other aspects of behavior. As mentioned, many of these are energy saving in nature and may be responses to the metabolic expenses associated with fever. In addition, interleukins and cytokines promote anorexia and sleep (Shoham et al. 1987; Spriggs et al. 1988). It is not uncommon for febrile animals to be anorexic, sleepy, and disinterested in grooming. Thus, in the absence of grooming, as the host musters its defenses against internal parasites, ectoparasite loads frequently increase (Murray 1961; Hart 1988; Clayton 1991).

Grooming

Grooming might be loosely seen as the ectoparasite analog of behavioral fever—that is, a behavioral defense against those parasites that have succeeded in contacting a host. Grooming is often classed with other avoidance behaviors that hosts use in an effort to reduce transmission (e.g., Hart 1990), and when successful, they certainly do truncate host–parasite interaction. Here I choose to consider grooming as a "defensive" measure (postcontact) rather than as an "avoidance" measure (prevent contact; chapter 4).

When we see a pet cat languidly smoothing its fur, or a bird preening, it

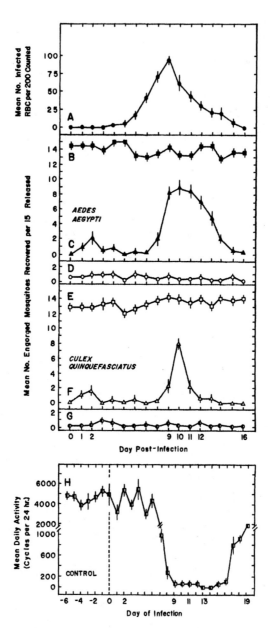

Figure 5.3. Mouse red blood cells infected with *Plasmodium chabaudi* peak simultaneously with feeding success of mosquitoes and at the time during the infection that mouse activity decreases. (Day and Edman 1983; reprinted with permission of *Journal of Parasitology*.)

may not spring to mind that such activities can have measurable fitness consequences (e.g., Clayton 1990). Grooming can occupy a substantial part of an animal's waking activity and time budget (e.g., 30% in the case of rats; Bolles 1960), yet it is rarely mentioned in behavior texts (Hart et al. 1992; Cotgreave and Clayton 1994).

Why assign importance to grooming? Like fever, grooming is costly (Hart 1988, 1997b). For example, costs of grooming must be considerable for antelopes; in maintaining vigilance, territorial males in Kenya groomed half as much as did conspecific males and females (Hart et al. 1992). In moose, the cost may be thermoregulatory. Individuals infested with ticks groomed more than uninfested moose did and, as a result, suffered a premature loss of winter pelage, along with ticks (Samuel 1991). Grooming is even expensive in terms of saliva, equaling one-third of a rat's total evaporative water loss (Ritter and Epstein 1974; see also Murray 1961).

Although grooming is indeed an expensive activity, its efficacy in eliminating ectoparasites is also well documented (reviewed by Hart 1990, 1994). For instance, reducing preening efficiency by placing metal "bits" between the mandibles of rock doves (*Columba livia*) resulted in much higher ectoparasite loads for treated birds than for untreated ones (Clayton 1991). In fact, the absence of grooming may have influenced one of the best-known examples of parasite adaptation to host availability: Hart (1994) suggested that the coordination of flea and rabbit reproduction (Rothschild 1965a,b; see chapter 3) may be especially advantageous for the flea because rabbit mothers do not groom their pups, nor do the pups self-groom.

Ectoparasites are not the only invaders that are subject to the depredations of grooming. In mammals at least, grooming can also limit microorganisms. Male rats deterred from postcopulatory genital grooming were more likely than undeterred rats to be infected with a marked strain of *Staphylococcus aureus* from experimental females (Hart et al. 1987). In this case, the bactericidal effects were minimal, but the very act of washing may have discouraged infection. (*Staphylococcus aureus* is not a genital pathogen of rats, but was used as a marker.)

In contrast, grooming may increase the risk of some infections. At least one cestode (*Dipylidium caninum*) is transmitted when dogs ingest fleas and the larval cestodes they contain; one message here is that dogs groom in ways that cestodes can depend on. (Given the legendary agility of fleas, one can only wonder if parasitized fleas are slow about jumping.) The nematode *Heligmosomoides polygyrus* has a direct life cycle, and infective larvae are ingested when the mouse host grooms itself or others (Hernandez and Sukhdeo 1995). Thus, endoparasites may benefit from grooming behavior.

Social animals groom each other in addition to themselves. The existence of allogrooming is important in a wide range of activities from thermoregulation to the definition and maintenance of social structures, and as such might be argued to be only secondarily related to parasitism. A strong case can be made for a more direct function of allogrooming, however (Hutchins and

Barash 1976; Barton 1985). For instance, in both penguins and primates, solitary individuals have higher ectoparasite loads (Brooke 1985; Hart 1990), indicating the hygenic value of allogrooming.

Among antelope, smaller species (e.g., Thomson's gazelle) groomed more frequently than larger ones (e.g., wildebeest), and did so even in parasite-free environments, thus prompting Hart and co-workers (1992) to propose an endogenous stimulus for grooming in these animals that may be then modified by immediate tick challenges (Mooring 1995; Mooring and Hart 1997; see also Mooring and Samuel 1998a,b). The suggestion that grooming has an endogenous component, combined with the existence of special structures to facilitate grooming and the time that many animals devote to grooming, argues for more attention to grooming than it has been given thus far. In addition, unified studies should address the relative contributions of grooming to social status and to defense from ectoparasites, along with the costs of grooming.

Ectoparasites are not without responses to these insults. Some ectoparasites are fairly mobile and can flee from grooming. Others have morphological adaptations that make them difficult to dislodge. Some parts of host anatomy are more difficult to groom than others, and many ectoparasites prefer those locations either for themselves or their less mobile stages such as eggs (Nelson and Murray 1971; Waage, 1979; Clayton 1991; Hart 1994).

Birds participate in a strange (to humans) behavior—anting—that might be considered a form of grooming. Anting involves intense exposure of birds to ants, with the ants being smeared on the bird's plumage (by the bird) or allowed to crawl there. In one of the few tests of the efficacy of anting against ectoparasites, starling louse and mite loads did not change as a result of anting. This does not address possible effects on parasites other than lice or mites, and other hypotheses (e.g., food preparation, feather maintenance) were not tested (Bennett and Clayton, in preparation, cited in Clayton and Wolfe 1993; see also Judson and Bennett 1992). Grackles and starlings were observed to "ant" using items such as limes or mothballs instead of ants, tangential support for an increasingly likely situation—that at least some passerines have more olfactory awareness that we have given them credit for (Clark and Mason 1987; Clark et al. 1990; Clayton and Vernon 1993).

Perhaps I risk misleading you when I refer to anting as a form of grooming. After all, birds groom in more routine ways (e.g., Clayton 1991). Even though anting defies easy categorization, it is a perfectly good avian behavior in its own right, one that we could learn more about, and one that hints to us of an avian sensory life rich beyond our imagining.

Primates may engage in a similar behavior. In the wet season, capuchin monkeys (*Cebus capucinus*) rub their fur with medicinally active plant material (*Citrus, Clematis, Piper*). This is an energetic process that involves abraiding, pounding, and biting the material. They do this individually or in groups; in the latter instance, monkeys without plant material rub their bodies against animals that have plant material, resulting in a tangled, squirming pile of monkeys, pulp, and leaves. The monkeys are clearly enthusiastic about the process, but its actual function remains speculative (Baker 1996), as does the function of *Trattinnickia aspera* resin in grooming of white-nosed coati (*Nasua narica*) in Panama (Gompper and Hoylman 1993).

In fact, even ants ant. Gaster flagging occurs when an ant (*Solenopsis invicta*) disperses venom by raising and vibrating its gaster, releasing small amounts of venom as an antiseptic on the brood and larger amounts as repellent against heterospecifics (Obin and Vander Meer 1985; see also Oi and Pereira 1993).

To stretch the definition further, perhaps the most far-flung aspect of grooming is that of cleaning symbiosis. Aristotle reported the removal of a leech from a crocodile by a plover, making the relationship between cleaners and clients the first symbiotic association to be recorded in the scientific literature (Sapp 1994). Despite the evocative image of oxpeckers sitting atop African ungulates, silhouetted against a sunset (and making a dinner of engorged ticks), few attempts have been made to understand the evolutionary origins of the relationships between cleaners and their clients, or what currently maintains them (Poulin 1993b; Arnal et al. 2000). Cleaning occurs in several taxa (crustacea, fish and birds) as do clients (fish, reptiles, ungulates; Poulin and Grutter 1996). Why is it so common in marine habitats (Poulin 1993b)? Do clients benefit from cleaning? They certainly seem to solicit cleaning on occasion (Limbaugh 1961; MacFarland and Reeder 1974; Hart 1990). Experimental data are inconclusive, but the time clients devote to cleaning and the number of parasites that may be removed (Hart et al. 1990) can be used as support for a beneficial result (Poulin and Grutter 1996).

Apparent parasites actually may be cleaners. Given all the evidence for grooming, Central American and South American mammals have seemed remarkably indifferent to the amblyopinine beetles that infest them, filled with host blood. In fact, at least some of these beetles feed on engorged ectoparasites; the beetles themselves are not parasitic, and this may explain host tolerance (Durden 1987).

Finally, humans also have a history as clients. In the Middle Ages some humans solicited wound-licking from dogs, perhaps because of their bactericidal saliva (Serpell 1986). Of course, our best friend has also been a source of parasite exposure. While "kissing" dogs is not particularly hygenic, it pales in comparison to traditions in some cultures, where a variety of unsavory canine products, including taeniid tapeworms, are used for alleged medicinal purposes (Nelson 1990).

Self-medication

Janzen (1978) suggested that parasitized animals may use dietary resources differently than unparasitized animals if they are emphasizing items with prophylactic or therapeutic properties, and that some secondary plant compounds may have medicinal value for foragers. Even as foraging behavior may be altered in parasitized animals, hosts and potential hosts may also forage defensively by avoiding exposure and ingesting antiparasitic compounds (Lozano 1991; see Clayton and Wolfe 1993 for a review).

Chimpanzees have yielded much of the evidence for self-medication thus far, especially in the area of ingestion (Huffman 1997). Their ingestion of the leaves, fruit, or pith of more than a dozen plant species that often overlap with those used medicinally by local humans has been hypothesized to be a response to fever, colic, parasites, or other infections. Given the limitations of observation and sample size, the evidence thus far is spare, but intriguing. Among African great apes, geophagy, bitter pith-chewing, and whole leaf-swallowing occur as medicating activities, along with more normal ingestion behaviors (Huffman 1997). In Tanzania, Wrangham and Nishida (1983; see also Wrangham and Goodall 1989) noticed that chimpanzees consumed entire leaves of *Aspilia* (Asteraceae) species without chewing them. *Aspilia* is used therapeutically by humans in Africa. Indeed, Thiarubrine A, which has antifungal and antibiotic properties, has been reported from *Aspilia* spp. It occurs most reliably in the roots of the plant (Rodriguez et al. 1985; Towers et al. 1985; Takasaki and Hunt 1987; Page et al. 1997).

Thiarubrine A is not universally present in these plants, however, and chimpanzees and other apes have now been observed to swallow leaves whole from at least 30 species of plants (Huffman 1997). These leaves are more similar physically than they are chemically: All leaf surfaces are rough and bear trichomes (little hairs), and leaf-swallowing behavior is most frequent at times during the year when parasitic infection is most prevalent. An examination of the dung left by chimpanzees that had swallowed leaves revealed living worms—pathogenic *Oesophagostomum stephanostomum*—often entangled in the leaves. They may be scoured off by the trichomes or may be tangled in the folds of the leaf created by the chimpanzee before swallowing (Huffman et al. 1996a). Swallowing of rough leaves may also combat cestode infections (Wrangham 1995). Across study sites, leaf-swallowing apes select leaves that may have different taxonomic affiliations, but that are structurally quite similar.

Huffman (1997), in speaking with other scientists, found that such diverse vertebrate taxa as bears and geese eat coarse, mostly indigestible grasses that may rid them of gut helminths at critical times such as hibernation or migration. In some ways, leaf swallowing may be analogous to feather eating in great crested grebes (*Podiceps cristatus*). The grebes eat their feathers, especially during the molt, and apparently regulate the amount of indigestible matter in their stomachs in this way. With feathers, and in some cases, fish debris, the grebes form pellets, which are ejected. Across species, the frequency of this habit correlates negatively with parasite intensity, perhaps a testament to its efficacy (Piersma and van Eerden 1989).

In Tanzania, Huffman and Seifu (1989) observed an ailing female chimpanzee chew shoots of *Vernonia amygdalina*, stripping away the poisonous bark and leaves, swallowing only the juice as she discarded the pith. The chimpanzee's symptoms—lethargy, anorexia, and irregularity—matched those of some humans in the region, who used this plant medicinally. This, together with the low frequency of the plant in the normal chimpanzee diet and its bitter taste, led Huffman and Seifu to propose that it might have medicinal properties. One incident of such self-medication was followed by a steep decline in helminth ova in feces (Huffman et al. 1993). Further observations revealed that the *V. amygdalina* used by ailing chimpanzees was rich in vernonioside B, that helminth egg output declined after pith usage, and that the amount that an ill chimpanzee ingested was equivalent to that used in traditional medicinal preparations by local humans (Huffman et al. 1993, 1996b). This plant has a wide spectrum of pharmacological activity, and its use increased in the rainy season when helminth infection also increased (Huffman and Wrangham 1994; Ohigashi et al. 1994).

Work has been done on possibly medicinal geophagy in a number of species. It is particularly informative that in many cases where it has been observed among primates, the consumed soil is rich in clays that have a composition not unlike that of many compounds sold by pharmacies for the purpose of settling one's stomach (Cowen 1990). In both chimpanzees and rhesus monkeys, geophagy has been linked to parasite infection and is thought to act as a therapeutic mitigator of related symptoms such as diarrhea (Mahaney et al. 1996; Knezevich 1998), as well as a source of minerals and antacids (Mahaney et al. 1999).

The study of medicinal plant use by nonhuman primates is difficult in part because the behavior is necessarily observed in field populations and is rare. Nonetheless, the catalog of probable self-medicating behavior in nonhuman primates is expanding, giving rise to questions not only of efficacy and epidemiology, but also of how these animals acquire knowledge. Although neighboring groups use similar plants in similar ways, some similarities transcend species and geographic boundaries. How has this convergence occurred? Do chimpanzees gravitate toward certain plant characteristics when seeking medicine? Although observational learning probably plays a part, it is not clear how the primates initially make the association between eating plants they would normally avoid and gaining relief from ill health.

Ancient humans may have made use of the anthelmintic properties of plants. A survey of 100 samples of desiccated feces from the Desha complex (6800–4800 BC; Utah) revealed a negative correlation between the proportion of *Chenopodium* sp. seed in the diet and the occurrence of parasitic nematodes. *Chenopodium* contains a compound that is toxic for nematodes, and several species of this plant have been cultivated for their anthelminthic properties. Known as "wormweed" in Maryland, it was used by various Native American groups. Of course, intentional use is difficult to demonstrate, but a study of 140 fecal samples at Antelope House in Canyon de Chelly, Arizona, revealed that the only sample that contained large amounts of *Chenopodium* also contained a mass of expelled nematodes (Reinhard et al. 1985; but see Reinhard 1990).

Archaeoparasitology is a fascinating field that gives us much to think about in terms of distribution of disease in historic and prehistoric times and its relationship to human behavior (reviewed by Horne 1985; Herrmann 1988; Kliks 1990; Reinhard 1990). For instance, do discrete layers of *Diphyllobothrium* eggs in the latrines of the Freiburg monastery indicate monastic fasting periods (Herrmann 1988)? Bouchet (1995) connected some of the parasites from the eleventh- to sixteenth-century latrines at the Grand Louvre (Paris) to the fondness of Parisians of the time for eating entrails.

Considering modern humans, Campbell (1977) investigated the notion that ingestion of alcohol might confer some protection against trichinosis. Because the worms are acquired when potential hosts, including humans, ingest meat in which the larvae have encysted, it is reasonable to wonder if concomitant consumption of larvae and liquor might prevent the next phase of larval activity—excystation and maturation in the intestine. High concentrations (12–25%) of ethanol were necessary to irreversibly paralyze the larval nematodes, and Campbell concluded that it was probably unrealistic to imbibe such large amounts of alcohol with dinner no matter how committed to prophylaxis one might be, especially given the diluting nature of other gastric secretions.

Vertebrates are not alone in their medicinal use of dietary items. Insects may also vary diets in response to parasitism, and these choices have serious consequences for host survival. Although parasitoids are often thought to kill hosts (in fact, that is part of some definitions of "parasitoid"), caterpillars have been known to survive tachinid fly parasitism (DeVries 1984). The arctiid caterpillar *Platyprepia virginalis* could survive tachinid parasitism (*Thelaira americana*) and reproduce normally, especially if it had been feeding on poison hemlock (*Conium manculatum;* fig. 5.4). Unparasitized caterpillars did better on lupine (*Lupinus arboreus*). Accordingly, parasitized caterpillars chose

Figure 5.4. The tachinid parasitoid *Thelaira americana* and its host, *P. virginalis*. (Photo courtesy of Richard Karban.)

to forage on hemlock more often than they did on lupine, and unparasitized caterpillars selected lupine. The parasite also benefitted from the hemlock diet and had a greater pupal weight when it came from a host that had been eating hemlock, probably because of the hemlock's felicitous effect on the host (English-Loeb et al. 1993; Karban and English-Loeb 1997).

The implications of the effect of food choice in survival of tachinid parasitism are indeed thought provoking. As Karban and English-Loeb (1997) point out, much parasitoid–insect work is done in the laboratory, and a choice of host plant may not be available. This arctiid–tachinid system would have gone unrecognized had not field cages been used in earlier experiments. The thought that food choice might make a life-or-death difference in host interactions with parasitoids, which are routinely assumed to be lethal, is intriguing and deserves attention from ecologists and behaviorists interested in foraging choices as well as from managers interested in biological control.

Finally, the presence of physiologically active compounds in diet items may not necessarily signal a medicinal role. Despite the focus of behavioral ecologists on clear fitness attributes of foraging choices, some compounds may be consumed because they alter mental states. Joseph Schall (personal communication, 1992) has suggested that the occurrence of some items in the diets of lizards may reflect their psychopharmocologic properties. Hamilton and co-workers (1978, p. 912) studied the diets of chacma baboons and commented, "We identified one additional category, to which we assign no preference level: class 4, euphorics. Euphorics are distinguished by their hallucinogenic properties and their high toxicity to humans and other mammals. . . . We have inadequate information to evaluate the significance of these items to baboon nutrition." Behaviorists, in embracing adaptationism and optimality theory, have not paid as much attention to animal behavior in the pursuit of pleasure that may not be associated in easily predictable ways with fitness consequences. Likewise, some psychoactive chemicals may reduce discomfort associated with disease; such chemical usage, if it occurs in nature, has not been the subject of extensive investigation. [See M. Hart (1990) for a convincing discussion of the beguiling and universal psychological properties of rhythmic sounds and behavior.]

Disinfectants and Repellents

Birds may modify environments to discourage the onslaught of ectoparasites. For example, hole-nesting birds and birds that reuse previous nests sites are more likely than other birds to use green vegetation in the construction of their nests (Wimberger 1984). European starlings are quite selective in festooning nests with green plants—agrimony, wild carrot, fleabane, goldenrod, yarrow, and nettle (fig. 5.5; see also Sengupta 1981). Although these could have several functions, ranging from camouflage, insolation/insulation, and aesthetics to control of humidity or parasites, several lines of evidence support an antiparasitic function. For instance, although we do not usually think of passer-

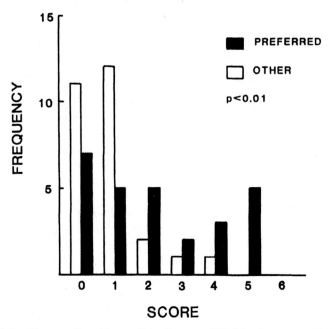

Figure 5.5. Plants preferred by starlings for nests (filled bars) are more likely to have a higher chemical complexity score than other plants (open bars). (Clark and Mason 1985; reprinted with permission from *Oecologia*.)

ines as possessing great olfactory abilities, the olfactory ability of starlings does increase markedly during the breeding season, and this may help in the selection of the plants, which is clearly nonrandom. In addition, experimental manipulation of nests showed that nests without wild carrot (the most preferred plant) had almost a 50-fold increase in hematophagous northern fowl mites and more anemic chicks compared to nests with the plant. In the laboratory, wild carrot and fleabane were unfavorable for mites and prevented molting. Moreover, goldenrod was shown to contain arthropod juvenile hormone analogs. Thus, it is possible that one reason starlings use such vegetation in nests is to discourage mite infestation (Clark and Mason 1985, 1988; Clark 1990, 1991; but see Fauth et al. 1991). In captivity, breeding male starlings responded enthusiastically to greenery, waving it and displaying with it in addition to taking it to nest boxes (Risser 1975). Birds are probably not alone in their attraction to insecticidal plants. Hart (1990) wondered if burrow-dwelling mammals might make similar use of fumigant vegetation.

Combating parasites clearly alters resource use, and Clayton and Wolfe (1993) have pointed out that this may change the way we think about a number of ecological topics. They wondered about the evolutionary consequences of protective secondary chemical compounds in plants being used medicinally by the herbivores they originally defended against, suggesting that choices of medicinal plants may be confusing if foraging theory continues to be seen primarily as an energetic model (see also Lozano 1991). In addition, the ability to recognize and furnish nests with fumigants may figure in mate choice. On a large scale, conservation biologists must consider the importance of preserving communities (with medicinal organisms) along with target species.

Parasites and Inclusive Fitness

Social animals can also care for kin, and in so doing, assist kin in resisting parasites and defend shared genes from parasites. The most dramatic way of helping kin in the face of parasitism was suggested in 1976 by Shapiro, who was intrigued when he found univoltine nymphalid caterpillars (*Chlosyne harrissii*) at a time of year when they should have been butterflies. These caterpillars were rather difficult to miss, sitting atop vegetation in daylight. When collected, they were not particularly interested in food or escape, and each proved to be parasitized by a braconid wasp. Shapiro suggested (in a paper aptly entitled "Beau Geste?") that these caterpillars might have been advertising themselves to predators. If successful, they and the wasp would both perish (the caterpillar is functionally dead already), to the benefit of nearby caterpillars, which, given the reproductive behavior of *C. harrissii*, are probably kin.

Smith Trail (1980) explored the relationship between kin selection and suicidal behavior in more detail. Both Smith Trail and Shapiro saw suicidal host behavior as a potential evolutionary pathway to complex life cycles. If the parasite, fortuitously, could survive in and exploit the predator as a subsequent host, then the suicidal behavior, originally beneficial to suicidal host because it benefitted host kin, would thereafter benefit the parasite.

I doubted whether an evolutionary backdrop of host suicide could explain the behavioral alterations that seem to characterize hosts of some intestinal helminths (Moore 1984b). The ontogeny of some behavioral shifts argues for the parasite as current beneficiary. Moreover, most of the behaviorally altered intermediate hosts of acanthocephalans are not necessarily found among kin, the preinfective period of many acanthocephalans is quite lengthy, and complete castration of intermediate hosts is not universal. All of this means that if kin were not around to benefit, the cost of suicide could be considerably higher than that for a completely castrated or dying host. Kin-selected host suicide may not be highly relevant to acanthocephalan–arthropod associations. Stamp (1981) came to a similar conclusion about some caterpillar–parasitoid associations (see below).

Pea aphids (*Acyrthosiphon pisum*) and their braconid parasitoids (*Aphidius ervi*) may present a different story (McAllister and Roitberg 1987). To say nearby aphids are apt to be close kin is probably an understatement about these parthenogenetic, often sedentary insects. Aphids parasitized before the fourth instar die before reproducing, and the resulting parasitoids may attack close kin. Aphids respond to threats (predators, alarm pheromone) by backing up, running, or dropping off the host plant—behaviors that are both increasingly effective and increasingly immoderate, in terms of energetic costs and risk of desiccation and overheating. When Kamloops aphids and Chilliwack aphids were subjected to alarm pheromone, parasitized Kamloops aphids dropped more often than unparasitized ones did. This is worth noting because Kamloops, British Columbia, is hot and dry and generally inhospitable to aphids that are away from their plants. Parasitism did not affect this response in aphids from Chilliwack, British Columbia, a more benign, coastal locale where the consequences of dropping off the plant are not so dire. McAllister and Roitberg (1987) repeated this experiment with predatory coccinellids that were more or less on a leash (a wire and wooden rod) and guided toward aphids. Once again, parasitized Kamloops aphids dropped off more frequently than unparasitized ones did, whereas parasitism did not affect Chilliwack aphids. This was taken to mean that when escape responses are risky, parasitized aphids are more reckless as they flee.

This conclusion stimulated an exchange of ideas (e.g., Latta 1987; Tomlinson 1987; McAllister and Roitberg 1988) that was followed by a more detailed examination of this system supporting the initial notion of kin-selected suicide. First, cross-infection of aphids showed that the source of parasite (interior or coastal) did not affect the dropping response; parasitism increased the frequency of dropping response to coccinellids in interior aphids no matter what the source of parasitoid, while parasitism did not affect dropping in coastal aphids, regardless of parasitoid source (fig. 5.6). In addition, aphids that were parasitized as second instars (and were therefore incapable of reproducing) dropped more frequently than unparasitized aphids did. In contrast, aphids that were parasitized as fourth instars, thus retaining some reproductive capacity, behaved like unparasitized aphids and took fewer risks in

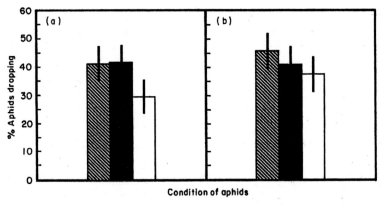

Figure 5.6. When parasitized by the parasitoid *Aphidius ervi*, these pea aphids were more likely than unparasitized aphids to drop off the plant if they were from (a) interior populations than (b) coastal populations, regardless of the source of parasitoid (hatched bars, interior parasitoid; filled bars, coastal parasitoid; open bars, not parasitized). (Reprinted from McAllister et al. 1990, with permission of Academic Press.)

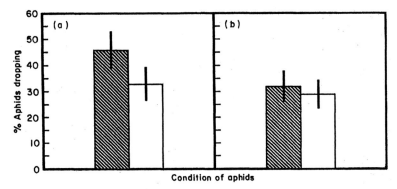

Figure 5.7. Parasitized fourth instar aphids (hatched bars) are more likely to drop off plants than unparasitized aphids (open bars) if they were parasitized (a) at the second instar stage than (b) at the fourth. Animals parasitized in later instars may survive long enough to reproduce. (Reprinted from McAllister et al. 1990, with permission of Academic Press.)

the face of predators, demonstrating further that the behavior was influenced by its cost (McAllister et al. 1990; fig. 5.7).

In the ongoing arms race, parasites frequently adapt to host defenses, and aphid parasitoids may be no exception. If dropping by pea aphids (*Acyrthosiphon pisum*) is harmful to the aphidiid parasitoid *Monoctonus paulensis*, this may partially explain why female wasps chose earlier instar aphid hosts, even though they are smaller. Not only are the smaller insects easier to subdue, but in this species, early instars are less likely to drop than later instars are (Chau and Mackauer 1997). [The fungus *Erynia neoaphidis*, also subject to harm if the pea aphid host drops, may have solved this predicament by reducing the aphid's sensitivity to alarm pheromone (Roy et al. 1999).]

The reluctance of bumblebees parasitized by a conopid fly to return to the nest was initially seen as uncertain in its benefit for the bumblebee colony because parasitized bees do not spend much time in the nest. On the one hand, this means they contribute little to provisioning the colony (a primary measure of bee fitness), and their absence from the nest disperses the larval conopid and protects it from pupating in potentially unhealthy conditions within the nest; on the other hand, the parasitized bees use few provisions and deter the development of future parasitoids (Schmid-Hempel and Müller 1991). Poulin (1992) argued that in this case, the host was benefitting through an increase in inclusive fitness if its time outside the nest resulted in increased risk of predation, superparasitism, or starvation. The inclusive fitness would not encompass the next season's kin, which would be descended from dispersed founding queens, but contemporary kin would benefit from the decoy effect of parasitized workers away from the nest and the reduction in resources used by those workers. In addition, the cost of such loosely defined "suicide" would be negligible, as the worker is not reproductive. Müller and Schmid-Hempel (1992) countered by noting that in bees, workers' reproduction is, essentially, the help they give the colony. Depriving the colony of that assistance over the development of the conopid is a large loss, and may last half the life span of the forager. Moreover, any benefits from diverting predators or parasites are not necessarily directed at the home nest. Thus, they saw the parasite as benefitting from the host's behavior, primarily because it would be dispersed to suitable pupation sites (fig. 5.8).

The realization that these vagrant bees affected parasitoid development by staying away from the nest cast this behavior in a new light. Parasitized bees preferred cooler temperatures than unparasitized bees did, and by so doing, they delayed parasitoid development a great deal, to the point that in about half the hosts, host death truncated the parasitoid life span (Müller and Schmid-Hempel 1993). In warm temperatures, many more parasitoids completed development. By staying away from the nest, especially at night, the bees could extend their working lives while having a negative impact on parasitoids.

The fate of the coral polyps infected with trematodes (Aeby 1991; chapter 3) could offer another view of kin-selected suicide. The coral colony is a single genotype, with polyps being thus profoundly "kin" to one another, and the removal of infected polyps by grazing fish enhances growth of the colony. This is an unusual case of near complete congruence of intermediate host and parasite interests (provided the parasites do not remain and multiply in the vicinity), with the alterations that increase likelihood of predation on infected polyps benefitting both the coral colony and the trematode.

Can hosts simply and directly resist behavioral manipulation by parasites? This is a difficult question to answer because in the extreme, successful re-

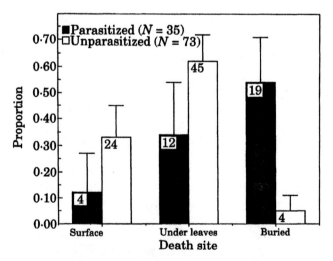

Figure 5.8. If parasitized by a conopid fly, bumblebees are far more likely to bury themselves in soil prior to death compared to uninfected bees. This increases parasitoid survival. (Reprinted from Müller 1994, with permission of Academic Press.)

sistance would go completely unnoticed; the host would behave normally. Indeed, the variability that characterizes behavioral aspects of some host–parasite associations may be the result of such resistance, especially if it is conditional—that is, if the intensity of resistance is a reflection of what might be gained in future host reproduction (Poulin et al. 1994). As is true of many areas, a more thorough knowledge of mechanisms involved in altered behavior might help us characterize potential resistance to such alterations.

David Horton and I (1993) noted that there was a discomfiting amount of ambiguity in the way that many parasitized hosts behaved. Although microhabitat choices and activity levels frequently change as a result of parasitism, the outcome of such changes is not always a foregone conclusion. Altered activity and microhabitat can disseminate or protect parasites, for instance, or they may be signs that the host is resisting parasites. In science as in art, appearance and reality may diverge. Behaviors associated with resistance, such as kin-selected suicide or behavioral fever, may be co-opted over evolutionary time by parasites and used as transmission mechanisms. To the extent that ectoparasites function as intermediate hosts (e.g., fleas and the larval cestodes they may contain), even the act of grooming and destroying some parasites may allow others to complete their life cycles. Some defensive behaviors may be the evolutionary antecedents of behaviors that enhance transmission, and behaviors associated with transmission may have fortuitously given rise to defensive maneuvers. A trait that initially contributed to increased host R_0 may now enhance parasite R_0, and vice versa. What is clear is that even as there is a behavioral subtext for transmission, there are also behaviors that have profound influences on host recovery rate and on the fecundity of parasites.

BEHAVIORS THAT ENHANCE HOST/PARASITE SURVIVAL

So far, we have seen that hosts can engage in a variety of behaviors that will have negative effects on their parasites. This is to be expected if parasites rob hosts of resources or even life. In contrast, some parasites in final hosts simply shed propagules, using the host as an agent of sustenance and dispersal. Most adult worms, for instance, are prodigiously fecund organisms, using final hosts as sources of nutrients and as transportation, while propagules are dispersed over time. Reproductive rate is best enhanced by minimizing harm to this wayfaring larder. In these situations, host survival, one outcome of successful host defense, also is in the best interest of the parasite. In the case of parasitoids (and organisms such as mermithids, nematomorphs, and fungi surely qualify as parasitoids, along with the normative insects), the parasitoid is developing, and its reproductive activity will take place outside the host, or at least will involve a permanent exit. The host therefore is nutrient, and parasitoid reproductive ratio can be best served by enhancing parasitoid development, usually at the expense of the host. When host behaviors are altered in ways that favor parasitoids, the payoffs are more likely to be seen in terms of parasitoid nutrition and survival.

Parasite-induced alterations in behavior that promote host survival or death can be difficult to interpret and are confounded at times by the fact that the host itself may benefit from prolonged life, or host kin may fare better if the host dies (kin-selected suicide). Once again, no single outcome is inevitable. What is clear is that even as host defenses limit parasite reproduction and life span, host survival can influence the survival of the parasite and its reproductive ratio, and these in turn can be influenced greatly by behavioral changes that occur in parasitized hosts. Behaviors associated with the extension or truncation of life—that of the host and the parasite it contains—can greatly affect parasite reproductive ratio.

Prolonged Parasite Survival

Parasites can prolong host life and alter host maturation schedules, thus enhancing parasite development, survival, and dispersal. The "quality" of this life, evolutionarily, may be at times questionable, for parasitic castration is not uncommon, nor is the gigantism that may accompany it. Fecundity may also be reduced as a result of the alteration of nonreproductive behaviors of the hosts; altered habitat choice, for instance, may limit mating encounters.

Indeed, the continued ability of the host to behave, at least until parasitoid emergence, is part of what defines two categories of parasitoids: (1) the koinobionts, which allow hosts to move, defend themselves, and in some cases even pupate, perhaps all to the advantage of the parasitoid, and (2) the idiobionts, which often paralyze hosts before consuming them (Askew and Shaw, 1986). In fact, in some instances, host behavior may benefit koinobiont parasitoids even after emergence. Such is the case with *Pieris brassicae* caterpillars parasitized by the braconid wasp *Cotesia glomerata*. The caterpillar remains with the cocoons of the emerged parasitoids and actually spins a web over them. If disturbed, the caterpillar unleashes aggressive behavior, but does not move away from the web. Thus, host behavior is altered in a way that probably benefits parasitoid survival, and the alteration occurs without continued physical connection between parasitoid and host. The host dies before it pupates, but not before it provides this macabre extended care for *C. glomerata* (Brodeur and Vet 1994). At least two congeners of *C. glomerata* have no such effect on their hosts (Beckage 1997).

Postemergence behavioral alterations such as those of *P. brassicae* are rarely reported; preemergence ones have been the subject of more investigation and were predicted by Fritz (1982), who hypothesized that parasitoids would manipulate host behavior in ways that decreased the likelihood of predation and thus promoted parasitoid survival. This also would promote host survival, at least until the parasitoid emerged. For instance, when the Baltimore checkerspot *(Euphydryas phaeton)* is parasitized by *Apanteles euphydryidis*, it is more likely to crawl high on vegetation. Initially, this might be interpreted as a form of host suicide, in which doomed caterpillars protect their relatives by exposing the parasitoid to predation (e.g., Smith Trail 1980), or perhaps it is an attempt at fever, but Stamp (1981, see above) found that parasitoids in these acrophilic caterpillars had a reduced risk of hyperparasitism and were more likely to encounter mates after emergence. By crawling up on vegetation, the host gave the parasitoid an advantage. Stamp (1981) suggested that host suicide may be less likely among aposematic, distasteful hosts of specialist, multivoltine parasitoids than among cryptic, edible hosts of generalist parasitoids with generation times that are the equivalent of the host's.

When the gregarious braconid *Cotesia congregata* emerges from the tobacco hornworm host (*Manduca sexta*), it would be at risk of being eaten by the host but for the fact that the caterpillar has ceased most locomotion and feeding some 8 hours before the emergence of the parasitoids. Up until this time, the caterpillar behaves normally; indeed, there is no evidence of general debilitation, and the host may live for 2 or more weeks after the wasps emerge. During this time, both living and dead hosts have their dangers for the wasps, which remain for a few days as pupae in cocoons on the caterpillar exterior. A dead host could rot, exposing the attached pupae to pathogens, and an active, living host could eat the pupae. Instead, the caterpillar is lethargic and anorexic, to the benefit of the parasitoid (Adamo et al. 1997; Adamo 1998b). Once again, it is apparent that behavioral changes (e.g., activity, feeding) may have a variety of outcomes.

The speed of host maturation has implications for both behavior and survival. Parasitized animals have been shown to delay maturation in some cases, and advance it in others (Lewis 1960; Welch 1960; Jones et al. 1986). *Cotesia congregata* induces developmental arrest in *M. sexta*, so that the wandering prepupal phase is absent; this occurs in spite of the fact that the parasitized caterpillar may be larger than the size required for metamorphosis. Juvenile hormone titres are elevated [coincident with a reduction in juvenile hormone esterase activity (Alleyne and Beckage, 1997; Beckage et al., 1994; Reed and Beckage 1997]. This probably benefits the parasitoid. If the parasitoids can emerge from a caterpillar rather than from a pupa, they have a better chance of surviving because of purely mechanical influences, if nothing else; the caterpillar exoskeleton provides an easier exit than the tough pupal cuticle does (Alleyne and Beckage 1997). Polydnaviruses (polydisperse DNA viruses) carried by the parasitoid may play a role in this developmental arrest, as well as in the immunosuppression, anorexia, and other changes that accompany parasitization with *C. congregata* (Beckage 1998).

As indicated previously, the beneficiary of a parasite-induced behavioral change can vary across time, even within a given host–parasite association (see Watson et al. 1993). For instance, *Bombus terrestris*, when parasitized by conopid fly endoparasitoids, seeks cool environments, thus delaying the development of the parasitoid (see above; Müller and Schmid-Hempel, 1993). The parasitoid is not killed, however, and toward the end of the infection, the bee digs into the sod, where it dies, providing a protected hibernaculum for the overwintering parasitoid pupa. This behavior contrasts with the death site selected by uninfected bees, which are more likely to die on the surface or under leaves, and quite unlikely to burrow into the soil. In experiments, Müller (1994) showed that parasitoids fared better when their bees were buried than when they died on the surface of the ground.

Beneficiaries can also vary across species. The pea aphid *Acyrthosiphon pisum* apparently resorts to kin-selected suicide as a response to infection with the braconid parasitoid *Aphidius ervi*, and in so doing, increases its inclusive fitness while minimizing parasitoid reproductive ratio (McAllister and Roitberg, 1987; McAllister et al., 1990; see above). In contrast, the potato aphid *Macrosiphum euphorbiae* behaves in ways that protect the overwintering parasitoid *Aphidius nigripes*. Before death, aphids containing diapausing wasps leave the host plant and sequester themselves, thus reducing the risk of hyperparasitism and environmental insult (Brodeur and McNeil, 1989). This is not something the aphid would normally do, for an unparasitized aphid overwinters as an egg. The sequestration results from altered responses to light, color, and touch, with parasitized aphids exhibiting negative phototaxis and thigmokinesis, as well as an attraction to dark substrate (Brodeur and McNeil, 1990). Aphids with nondiapausing parasitoids do not leave the plant, but do move from the lower to upper surfaces of leaves, where they incur reduced hyperparasitism and predation. Brodeur and McNeil (1991, 1992) suggested that perhaps these aphids remain on the plant (rather than exit completely like their counterparts with diapausing wasps) because over a short (nondiapausing) developmental time, ready access to hosts upon emergence may be more important to parasitoids, and biotic effects such as insolation and exposure may have a different impact. A comparison of three aphid hosts and six parasitoids was consistent with this suggestion; parasitoids that took longer to develop also tended to be the ones that prompted aphids to leave their plants when near death (Höller 1991).

There are, however, other ways that aphid parasitoids avoid hyperparasitism. The black bean aphid (*Aphis fabae*) in central Europe is attended by honeydew-collecting ants. These ants protect it from the parasitoid *Trioxys angelicae*, but not from another aphidiid, *Lysiphlebus cardui*. Aphids with the latter parasitoid produced more honeydew than unparasitized aphids did, and the ants reduced hyperparasitism on *L. cardui* (Völkl 1992; see also Müller et al. 1997).

Finally, not all aphid–parasitoid associations are successful at avoiding hyperparasitism. For instance, ant attendance does not always reduce hyperparasitism, and it does not always discourage dispersal from the colony. These associations do exhibit a variety of behavioral changes, but Müller and coworkers (1997) showed that the changes were not necessarily advantageous for the parasitoid. Host–parasitoid associations are behaviorally diverse; behavioral alterations abound, but the fitness consequences are not always obvious.

When helminths and microparasites use the host as a dispersal agent, host survival is often positively linked to parasite fitness. Lefcort and Durden (1996) suggested as much when they noted that the alterations in tick behavior (*Ixodes scapularis*) associated with infection by *Borrelia burgdorferi*, a spirochete that causes Lyme disease, may prolong the life of the adult tick and thus enhance bacterial survival. In such instances, if the parasite is influencing host behavior, we might expect to see behavioral alterations that enhance host survival. This could be responsible for a general pattern seen in complex life cycles: Final hosts, which are the vehicles for propagule dispersal (e.g., eggs) are usually less negatively affected by their parasites than intermediate hosts are.

One might also expect the truly manipulative parasite to minimize the opportunities for its host to be eaten by the wrong predator. One trematode (*Microphallus* sp.) seems to do that with its snail intermediate host. Infected *Potamopyrgus antipodarum* not only forage on top of rocks at a time when the waterfowl definitive host is feeding, they retire to protected places on the bottoms of rocks at times when fish predation is heaviest. In fact, uninfected snails are more likely to be eaten by fish than are snails infected with *Microphallus* (Levri 1998).

The notion that parasites benefit when their dispersal agent continues to function is perhaps one origin of the traditional idea that parasites evolve toward benign coexistence with hosts; after all, in a definitive host, pathogenicity may be considered a stab at the goose that laid the golden egg. The idea of benign coexistence ignores parasite stages that require the death of the intermediate host for their transmission; the precise timing of behavioral alterations seen in some associations reflects the fact that this death, if it is to serve the parasite, cannot be premature. The traditional idea of benign coexistence also does not differentiate between host survival and host fitness, a salient distinction in the case of parasitic castration.

Parasitic Castration and Host Survival

In the case of permanently and completely castrated hosts, increased survival holds little promise for increased individual fitness. Following Rothschild (1941a,b), Baudoin (1975) argued that host castration was a strategy of the parasite that increased parasite fitness in any of several ways—by increasing available energy, host survival, or host growth. Castration diverts host resources while minimizing negative effects on host survival (Obrebski, 1975).

A thorough examination of the reproductive consequences of parasitism is beyond the scope of this chapter (e.g., Hurd 1990, 1993, 1998; Read 1990; Hurd and Webb 1997; table 3.11), but to the extent that host reproductive activities may result in increased risk of predation, not to mention increased energy expenditure (Lima and Dill 1990), parasite survival might be well served by a less fecund or castrated host. To complicate this picture, hosts may reduce (or increase) reproductive effort in an adaptive response to parasitism; such shifts may influence resource allocation to defense, for instance (Forbes 1993; Perrin et al. 1996; Møller 1997). In turn, parasite effects on host fecundity can have severe implications for host populations. These population effects may be greater than the costs we normally associate with parasites, such as debilitation and death (Hudson and Dobson 1997).

Castration is common among parasitized hosts, especially invertebrate hosts, and usually has behavioral consequences. In addition to inducing a reallocation of resources and minimizing predation risk, parasites may co-opt the reproductive behaviors of some castrated hosts, using them in the care and dispersal of the next generation of parasites. Mock oviposition, when hosts engage in oviposition behaviors but only deposit parasites, is usually accompanied by castration, with the parasite filling the reproductive organs, if not the entire abdomen. This is especially common in nematode-infested insects and may involve males as well as females in some instances (see chapter 3).

In addition to mock oviposition, some castrated hosts display what might be called mock parental care. The parasitic barnacle *Sacculina* castrates male and female crabs (e.g., *Carcinus maenas*) and feminizes males through hormonal interference. It is an internal parasite that produces external gonads (the "externa"). When these gonads appear, the parasitized crabs of both sexes begin to behave as if they were reproductive females, going to hide in deeper water and cleaning and fanning the externa as if they were crab eggs. Such behavior is only induced by living externa (Rasmussen 1959; see also Bishop and Cannon 1979).

Parasites can cause their hosts (again, primarily invertebrates) to assume an intersex status or to reverse sex, a phenomenon that is poorly understood (table 3.11). More commonly, castration occurs without such dramatic pirating of reproductive behavior or appearance, and the parasite may experience the dual benefit of redirected resources (otherwise used in reproduction) and decreased risks to the host that would accompany reproductive activity (e.g., Egerter et al. 1986; Egerter and Anderson 1989). Although parasitic castration has been compared to the action of parasitoids [death, in evolutionary terms (Kuris, 1974)], parasitically castrated animals continue to live and participate in ecological interactions. Moreover, parasitic castration may be absolute or partial and may be reversible or permanent. In some host–parasite associations, these outcomes may be more variable than we currently realize. Jaenike (1996) used degree of reduction in fecundity as an indicator of virulence, and cautioned that what we call "virulence" for any given parasite should be assessed across the taxonomic range of its hosts; the effect may not be uniform, and certainly is not in nematode (*Howardula aoronymphium*)–*Drosophila* spp. associations.

Although the traditional argument for the benefits of host castration is based on increased host longevity and energy reallocation for the parasite (see Zelmer and Esch 1998), there are numerous examples of shortened life spans in castrated hosts. *Limnicolaria* sp., the gastropod first-intermediate host of *Dicrocoelium hospes*, which alters ant behavior in such a remarkable way (chapter 3), does not produce eggs when infected with the trematode. In addition, these infected snails have a reduced life span (Frank et al. 1984); if this benefits the parasite, the benefit is not obvious, for the snail releases parasites in slime balls, and it would seem the parasite would be more successfully dispersed by long-lived snails. Moreover, in the case of intermediate hosts that must be eaten for parasite transmission to occur, increased life span (at least beyond the point of infectivity) would not necessarily benefit the parasite, which awaits predation on its host in order to reproduce, nor would reallocation of resources seem to be a great boon to an infective parasite that has completed growth and development. Resources may still be in demand for maintenance, warding off host defenses, or modifying host behavior, but the resource requirements of these activities (in fact, of most parasite activities) are largely unknown. There are nonetheless many examples of such fully or partially castrated intermediate hosts, including several acanthocephalan–arthropod associations (Schmidt 1964; Oetinger and Nickol 1981; Brattey 1983; Moore 1983a; but see Lyndon 1996).

Although we have only a rudimentary understanding of both mechanism and adaptive consequences of parasite-induced fecundity reduction in insects, Hurd (1998) did a masterful and thought-provoking job of synthesizing existing physiological data and possible adaptive scenarios. We have glimpses of linkages among nervous, immune, and endocrine systems in some invertebrates (de Jong-Brink 1995) that may someday make more sense of such apparent accidents.

We are beginning to understand how at least one cestode modifies beetle reproduction, and this surprising story belies the "side effect" of "pathology" so frequently invoked when other explanations for reduced fecundity are lacking. The cestode *Hymenolepis diminuta* reduces fecundity and longevity in *Tribolium confusum* (Maema 1986; Robb and Reid 1996) and *Tenebrio molitor* (Hurd and Arme 1986, 1987). In the latter instance, the parasite inhibits vitellogenin uptake, thus decreasing egg volume and viability. It does this by producing a 10–50 kDa substance that affects vitellogenesis, reducing it by almost half. Metacestodes early in development are especially effective at limiting insect vitellogenesis (Webb and Hurd 1999). Is this pathology? Is it resource redirection?

The fecundity of vectors such as biting flies is often affected by the presence of the parasites they transmit. The continued activity and viability of the vector is in the best interest of the parasite, and if limiting fecundity promotes such a situation, the parasite would benefit. Moreover, if limited blood-feeding success is also beneficial for the parasite, requiring increased probing, for instance, and if this limits fecundity, then the parasite may be doubly blessed. Such a direct effect on fitness could generate selective pressure favoring resistance to the parasite, but then again, this could be said of any parasite that reduced host fitness. Meanwhile, as Rossignol and co-workers (1985) have suggested, if the parasite also reduces the cost of initial blood meals for uninfected flies by interfering with host defensive behavior or modifying characteristics of the blood, then later fitness costs may be balanced by initial benefits (see chapter 3).

In some cases, the effect of parasites on fecundity is not straightforward.

Richner and co-workers (1995) manipulated brood size to show that male great tits (*Parus major*) provisioning enlarged broods were more likely to become infected with malaria than males caring for normal-sized broods (see also Norris et al. 1994; Ilmonen et al. 1999 for similar results with other blood parasites). In parallel fashion, females with larger clutches were more likely to develop malaria (Oppliger et al. 1997; see also Ilmonen et al. 1999). When eggs were removed from nests, females compensated by laying more eggs. These overly fecund females were at greater risk of developing malaria (Oppliger et al. 1996a; see also Richner and Heeb 1995). Thus, the ability to raise current brood is not compromised, but the production of future ones may be.

Complete castration may favor such strong host defenses as kin-selected suicide; suicide costs the castrated host little. Another way to mitigate fecundity reduction involves life-history changes with concomitant alterations in reproductive behavior (Milinski 1990; Schmid-Hempel and Durrer 1991; McCurdy et al. 1999a). Snail–trematode interactions offer examples of such host responses. Castration has been documented for snails as diverse as the marine dogwhelk and terrestrial hosts of *Leucochloridium*. *Biomphalaria glabrata* exposed to *Schistosoma mansoni* increase egg-laying shortly thereafter, at the expense of later reproduction in exposed but uninfected snails. Infected snails cease reproduction altogether, and early reproduction compensates for some later losses (Minchella and LoVerde 1981, 1983). Minchella (1985) saw such fecundity compensation as an alternative to resistance. He suggested that gigantism could also be a strategy of long-lived hosts, a way to sequester resources from parasites in favor of hosts (see also Michalakis and Hochberg 1994; Jokela and Lively 1995; chapter 3.) Indeed, by selecting for low susceptibility, Cooper and co-workers (1994) produced a *B. glabrata* strain that was also low in fecundity (see also Yan et al. 1997).

Fecundity compensation is not a generalized response to parasitism. Female crickets *Acheta domesticus* compensated when infected with the bacterium *Serratia marcescens*, but did not do so when parasitized by the lethal tachinid *Ormia ochracea*. Might this response be linked to the nature of physiological defenses (Adamo 1999)?

Yet another way to mitigate the cost of castration may be gender-specific castration of hermaphrodites. Although *S. mansoni* disrupts egg production in *B. glabrata*, infected snails are nonetheless able to produce sperm and inseminate other snails. Cooper and co-workers (1996) noted that castration of one gender of hermaphrodites might allow the susceptible genotype to persist in the population.

One prevailing image of a parasitized animal is that of an emaciated, stunted creature barely able to wrest sustenance from the malingerers within. Castrated hosts frequently yield a different picture: Trematode-induced castration probably results from something more complex than overall energy drain, even though larval stages such as sporocysts must demand much energy from snails. This is echoed by the partial castration seen in chaetognath hosts of trematode metacercariae, a stage that probably is far less energetically demanding

Figure 5.9. Larvae (plerocercoids) of the cestode *Spirometra mansonoides* in rodents secrete a plerocercoid growth factor that mimics some of the aspects of mammalian growth hormone (Mueller 1980; Phares 1992, 1997). Although this photograph depicts an extreme case, these hosts can become quite fat; it might be argued that they may not survive well and may have trouble escaping predators. It is not clear how important rodents are as intermediate hosts; frogs and snakes most commonly serve that role in nature (Mueller 1980; Phares 1992). (Mueller 1963; reprinted with permission of the *Annals of the New York Academy of Sciences*.)

than sporocysts or rediae in snails (Pearre 1976). Gigantism is common, and although it is perhaps best known in trematode-infected gastropods, it occurs in many hosts (table 3.2; fig. 5.9). Gigantism may result from any of numerous causes: the secretion of a growth factor, overeating, or failure to undergo metamorphosis (Barnard 1990). Gigantism and other altered appearances (table 3.2) are not behavioral shifts, strictly speaking, but they can have ecological and epidemiological consequences. Rothschild (1962) warned that although gigantism and a preference for exposed locations on the part of parasitized snails could render them more vulnerable to predation, it might also bias sampling, yielding inflated estimates of trematode prevalence. Moreover, not all gigantic snails harbor stages that are transmitted by predation.

A mark–recapture study of trematode-infected *Cerithidea californica* revealed no gigantic tendencies, and Sousa (1983) argued that infection would enhance the growth of only sexually mature snails in which reproduction commanded a relatively large portion of the energy budget (i.e., short-lived, semelparous snails). In long-lived, iteroparous snails (characteristically marine), reproductive costs are spread over several seasons, and hence castration and the elimination of such costs do not result in significant changes in energy allocation. Without such changes, gigantism is unlikely. Likewise, prereproductive snails are not investing heavily in reproduction and therefore should not respond to parasitism with increased growth. Sousa predicted that gigantism would therefore occur most commonly in short-lived, semelparous gastropods of reproductive age.

Minchella (1985) came to the opposite conclusion. He saw gigantism as a host strategy that was more consistent with the snail making the best of a bad situation. Given this reasoning, he suggested that gigantism is more likely to occur among long-lived snails, which would actually benefit from it if they could outlive their infection.

Such tidy expectations, even conflicting but tidy ones, are in the process of being disassembled. Fernandez and Esch (1991) discovered that in the field, short-lived, infected snails grew more slowly than uninfected counterparts, possibly because there were insufficient resources to result in gigantism. They suggested that perhaps some cases of gigantism were artifacts of abundant nutrients in the laboratory. At this point, there is no clear understanding of gigantism in trematode-infected snails. Is it adaptive? Is it a reflection of life-history constraints? Nutritional constraints?

Ballabeni (1995) experimentally infected the freshwater hermaphrodite *Lymnaea peregra* with the trematode *Diplostomum phoxini*. He exposed snails from a population that is normally exposed to the parasite and snails that had not had a history with *D. phoxini* to an allopatric lineage of the trematode. The former group of snails did exhibit gigantism, as well as lower mortality, compared to the historically naive snails. Although none of the infected snails reproduced, Ballabeni remained optimistic that had there been larger sample sizes, a smaller degree of infection, or a sympatric parasite, some reproduction might have occurred. He concluded that the expression of gigantism was most likely a host property, rather than a parasite property, although the hypothesis of nonadaptive consequence cannot be eliminated.

De Jong-Brink (1990, 1995) viewed gigantism and castration as beneficial to the trematode parasite in the snails she studied. She reviewed extensive work on the mechanisms that underlie interference in *Lymnaea stagnalis* reproduction by *Trichobilharzia ocellata*, a relative of *S. mansoni*. A host-derived factor called schistosomin, widespread among trematode-snail associations, appears in the snail hemolymph at the time trematode cercariae develop and acts on the parts of the neuroendocrine system that regulate reproduction and growth. Schistosomin may also inhibit internal defenses such as hemocyte activity. In addition, neuropeptides may be released earlier in infection, and these can inhibit gonadal development and defensive responses. Thus, nervous, endocrine, and immune systems are probably linked in these snails (de Jong-Brink et al. 1997, 1999), a troika that is familiar in vertebrates, but that has yet to be demonstrated in most invertebrate hosts (Beckage 1993a,b, 1997). De Jong-Brink raised the intriguing possibility that schistosomes might interfere with both defensive and neuroendocrine systems in ways that are synergistic and, as yet, largely unexplored. Recently, Hoek and co-workers (1997) reported that genes in parasitized snails are not regulated in the same ways those in uninfected snails are. Significantly, this is especially true for genes that express neuropeptides.

Many host–parasitoid interactions, including castration and developmental alterations, may have a hormonal basis, which may be quite complex and involve parasite manipulation of host hormones or parasite production of hormones (Beckage 1985). Solitary parasitoids consume both host hemolymph and tissues, whereas gregarious parasitoids subsist on hemolymph alone. Complex endocrine strategies are more widespread among the latter group, which often inhibit host pupation (Lavine and Beckage 1995). Even more complicated scenarios arise when one considers possibilities such as "hormonal decoys," which may bind to receptors and interfere with host hormone function (Beckage 1991).

Fecundity Reduction as a Result of Altered Behaviors

Failure of a host to reproduce need not have a noticeable anatomical basis (Barnard 1990). Cockroach hosts (*P. americana*) of the acanthocephalan *M. moniliformis* appear normal. Although female reproduction may be somewhat affected, there is no evidence of castration or pathology. Nonetheless, male cockroaches appear to be insensitive to female sex pheromone; under such conditions, they would be "behaviorally castrated" compared to their uninfected competitors (Carmichael et al. 1993). They would be unlikely to devote many resources to reproduction. These are, however, intermediate hosts, and their prolonged survival beyond parasite infectivity is of little benefit to the parasite (see also Hurd and Parry 1991).

Gammarids (*G. lacustris*) that are intermediate hosts to acanthocephalans (*P. paradoxus*) also experienced reduced mating success, in part because of

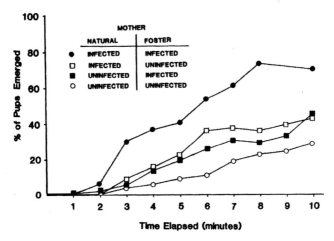

Figure 5.10. Mouse pups born to or raised by *Trichinella spiralis*-infected mothers left the nest before those born to or raised by uninfected mice. (Rau 1985; reprinted with permission of *Journal of Parasitology*.)

their altered habitat preferences. They overwintered in shallower water than uninfected gammarids did, and this placed them in the company of juveniles and small females (Zohar 1993; Zohar and Holmes 1998).

Once mated, infected animals may make poor parents (table 3.11). Although these hosts have invested some energy in reproduction, they have limited investment in parental care, whether evidenced by oviposition choices or nurturing young. In addition to these effects, the young of parasitized parents are often compromised compared to those of healthy parents (examples in Holmes and Zohar 1990; fig. 5.10).

In summary, many behavioral alterations affect host and parasite survival. The survival of parasites and of hosts obviously has implications for parasite and host fitness, but like other aspects of parasite–host interaction, these implications are not necessarily simple or easily predictable. At first glance, both parasite and host should benefit from increased host survival. Indeed, in some cases, especially those in which the parasite has a long developmental period in the host or in which the parasite uses the host for long-term dispersal, the parasite benefits from host survival, which may in turn be enhanced by limited reproductive investment on the part of the host. Parasite transmission may also be favored in cases such as mock oviposition and mock parental care, where host reproductive behaviors can be co-opted and used for parasite dispersal. Although parasite reproduction and survival, along with host survival, have traditionally been thought to increase as a result of the diversion of host resources away from host reproduction, there are numerous examples of intermediate host castration; it is not clear that increased intermediate host survival would benefit the parasite once it develops to a transmissible stage.

There are also examples of decreased longevity in castrated hosts. In addition, immunosuppression, although apparently beneficial to parasite survival, may put the host at risk of increased parasitism and may put the parasite at risk of increased competition and of host debilitation and death. Moreover, if the host population structure is suitable, and if host reproductive options are severely curtailed, decreased host survival (e.g., kin-selected suicide) may benefit the host more than survival would. The behavioral changes that affect host and parasite survival are rarely addressed from a rigorous evolutionary standpoint, yet given the strong relationship between survival and reproductive ratio, these may be pivotal in the evolution of host–parasite interactions.

6

Concluding Remarks

All of them [trematodes] provide splendid material for the still non-existent study of parasite behaviour which, sooner or later, has to come into being. . . .

—Szidat

The study of parasites and host behavior is a field that is overdue some attention. This may seem like a strange comment as we squint through the dust flurries, not yet settled, surrounding Hamilton and Zuk's (1982) suggestion that parasites should influence mate selection. It is true that the proposal of that hypothesis set off something akin to a scientific stampede. The resulting literature was rich and exposed new questions and new ways to look at mate selection.

If we look beyond sexual selection, however, we see that parasites still command relatively little attention from ecologists, epidemiologists, neurobiologists. Given the content of this book, such neglect is remarkable, for parasites alter the behavior of animals in ways that impinge upon most of the areas of interest to ecologists, and as they do so, they violate many of the assumptions of epidemiological models. Moreover, although some of the mechanisms by which parasites alter behavior are pedestrian, in other cases, apparently specific lesions of the nervous system are involved (Adamo 1997; Hatalski and Lipkin 1997; table 3.1); in the majority of altered behaviors, the mechanism is unknown.

I have argued that host behaviors can influence the evolutionary trajectories of host–parasite associations. The more immediate ramifications of altered host behavior can also be farflung (Moore 1987, 1995; Minchella and Scott 1991; Poulin and Thomas 1999). Ecology is the study of the distribution and abundance of organisms, and parasites affect both, on local and global scales (Piersma 1997; Tompkins and Begon 1999). They do so in obvious ways, such as increasing mortality, and in not so obvious ways, such as altering fecundity, foraging choices, and habitat preferences (e.g., Tables 3.6, 3.9, 3.11, 3.12; F. Thomas et al. 1999). Even pollination behavior may reflect the presence of parasites (Schmid-Hempel and Schmid-Hempel 1990; Schmid-Hempel and Stauffer 1998). A snapshot follows.

Parasites transmitted in the food chain (by predation upon intermediate hosts, for instance) are certainly capable of altering encounter rates between predators and prey, as well as ease of capture and dietary breadth (chapter 3). Parasites can also alter encounter rates between study organisms and sampling scientists (Hailey, 1967; Ashraf and Berryman 1970; Curtis and Hurd 1983; Lykouressis and van Emden 1983; Burkot et al. 1990). In fact, there is evidence that some dietary choices, as well as foraging patterns, may reflect self-medication or parasite avoidance (chapters 4, 5). Parasitized hosts may eat more or less than their unparasitized counterparts (table 3.9; but see Russell 1980), and feeding alterations can have a temporal component that distinguishes them further from behaviors expressed by unparasitized animals (Levri and Lively 1996). Resource use in general may be modified. In some cases, parasitized hosts may simply be inefficient (Crowden and Broom 1980). In other instances, they may choose different foods than unparasitized animals do (Boland and Fried 1984; Curtis 1985) and their competitive interactions with conspecifics and other species can be subject to parasite influence (Park 1948; Haldane 1949; Utida 1953; Milinski 1984; Price et al. 1986, 1988; Jaenike 1995; Yan et al. 1998). Parasites may even act as barriers to invasions of colonizing species in some cases (Freeland 1983). Parasites promote genetic variation (Coltman et al. 1999), contribute to mate choice (chapter 4), alter the cost of parental care (Christe et al. 1996a,b; Tripet and Richner 1997) and influence many social interactions (table 4.2). In fact, behavior alone can be used as a reliable indicator of health status in organisms as complex as the great apes (Alados and Huffman 2000). The importance of parasites in conservation biology is only beginning to be realized (Scott 1988; Loye and Carroll 1995; McCallum and Dobson 1995; Dobson and McCallum 1997). Seen in this light, the ways in which parasites alter behavior are not only instrumental in the long vision of host–parasite evolution, they jostle and disarrange our expectations of daily ecological interactions.

How then do we investigate these host-parasite amalgams? Experimental infections are necessary wherever possible, lest the differences we observe between infected and uninfected hosts be the preexisting cause of their infection status and not the result. Such a conflation can yield striking results: In the field, young birds parasitized by acanthocephalans, young reindeer parasitized by nematodes, and lizards with high tick loads were all larger than less parasitized conspecifics. One could interpret this as being evidence for the beneficence of the worms and arthropods in question, but, in fact, the large young birds and young reindeer probably ate more and better, thus increasing their exposure (Moore and Bell 1983; Halvorsen 1986a). The same might be said of the large lizards, which were ensconced in good home ranges, with established tick populations (Bull and Burzacott 1993). By the same token, in field studies, one should not confuse prevalence, or percentage of infected hosts, with susceptibility. Uninfected hosts may be unsusceptible, but without information to the contrary, we must also consider that they may be unexposed. It is therefore all the more unfortunate that in some cases, laboratory strains of hosts have provided different behavioral results than wild strains (e.g., Platt et al. 1997). In reality, some host–parasite systems are not amenable to laboratory investigation; either the parasite, the host, or in some cases, both, resists cultivation. Whenever possible, however, infections should be experimentally induced, and they should be induced in hosts that are not far removed from their wild progenitors.

Of course, I am writing as if an animal is either infested by a given parasite or it is not, for behavioral purposes. In fact, many altered behaviors are intensity dependent, that is, the degree of change is correlated with the level of parasitism. In experimental infections, such correlation is evidence for the parasite causing the alteration; lack of correlation does not refute such a hypothesis, by the way, because a threshold effect may be occurring instead. In natural infections, it is virtually impossible to know if the shift in behavior, even if correlated with intensity, is a result of parasitism. The host may have always behaved differently, thus increasing its exposure to the parasite in question; this can then be compounded by behavioral changes induced by the initial inoculum of parasites (e.g., Anderson et al. 1978; Poulin et al. 1991b).

Intensity-influenced outcomes can be confusing in field studies for other reasons. For instance, because mice with one botfly (*Cuterebra*) larva are not as likely to emigrate as uninfected mice are, they may appear to live longer (i.e., be present on a plot) than uninfected mice do. Mice with more than one larva are so impaired that they are readily caught by predators and give no such impression of longevity. To complicate things, mice that live longer are probably exposed to more botflies (Goertz 1966; Miller and Getz 1969; Hunter et al. 1972). When exposure to parasitism and age and growth/size are confounded, parasitized hosts may experience increased predation because they have a greater intensity of parasites or because they are older and bigger. Parasitic castration may also result in size increases that attract predators (Moore 1983a).

Host sex is another influence on behavioral alterations. For instance, Gotelli and I (1992) found that in *Periplaneta americana* infected with *Moniliformis moniliformis*, the effects of parasitism on activity were gender related; unlike males, infected females moved greater distances than uninfected females (Gotelli and Moore 1992; see also Yan et al. 1994 for beetle/cestode examples). The impact of infection on behavior may also vary with the method of exposure (Draski et al. 1987) or with the age of infection (Hay and Aitken 1984; Dolinsky et al. 1985) or host (Poulin 1993a).

As with many ecological and behavioral investigations, the plethora of experimental and observational methods used by various scientists does not simplify interpretation. My colleagues and I studied the males of seven species of cockroaches infected with *M. moniliformis*; if activity was affected by the infection, it was decreased (Moore and Gotelli 1996 and references therein). The measurements of activity were taken for 15 minutes after a 20-minute acclimation period. When Wilson and Edwards (1986) measured cockroach activity (no acclimation period) for 5 minutes, infected male *P. americana* traveled greater distances at greater speeds than uninfected roaches did (see also Etges 1963; Moore 1983b). Across the host–parasite associations in table 3.7, activity has been assessed with running wheels, open field observations, mazes, stamina chambers, acoustic chambers, tethering, flight tunnels, and continuous electronic monitoring and video tapes, to name a few. The test periods range from 2 minutes to many days. As with most investigations, the duration of observation and the methods used greatly influence our perception of parasite influence. Taking these and similar concerns into account is no small task, and many host–parasite associations may simply disallow addressing all of them.

An examination of alterations in activity can showcase the ambiguities that surround the study of parasite-induced behavior. The seeming ubiquity of activity changes may occur in part because they are noticeable. Many workers, addressing questions that have little to do with either behavior or parasitism, may observe that the parasitized animals in their study are sluggish or immobile. Other forms of behavioral changes, such as microhabitat alterations, may go unnoticed without explicit experiments.

In addition, just because activity change is frequently reported does not mean we understand its significance. Even as increased activity may cause an intermediate host to be more noticeable to a predator (Quinn et al. 1987) or may cause greater dispersal of parasite propagules by other hosts (Goldstein 1929), decreased mobility may likewise lower chances of escaping a predator, whether by impediments to locomotion or reduced familiarity with local refugia (Rau 1983a). In addition, if a host resorts to fever to combat an infection, lethargy may play a supportive role in energy conservation (Hart 1988).

Reports of altered activity also raise questions about how much we can generalize from one example of altered behavior. Parasites may differ greatly in their effects on the same host species. Pinworms, which have no intermediate host, but are directly transmitted, are a case in point. *Syphacia obvelata* causes mice to reduce exploratory activity. A different pinworm, *Aspiculurus tetraptera*, has no effect on activity (McNair and Timmons 1977). Likewise, the nematode *Dispharynx nasuta* increased photophilia in terrestrial isopods (Moore and Lasswell 1986); the acanthocephalan *Plagiorhynchus cylindraceus* did many things to isopods, but an effect on photophilia was not among them (Moore 1983a).

Conversely, the same parasite may affect the activity of different host species in different ways. The tapeworm *H. diminuta* decreased emigration tendencies in one flour beetle intermediate host (*T. castaneum*), but increased that behavior in another (*T. confusum*). These changes were influenced by intensity, host sex, and strain (Yan et al. 1994; see also Yan and Phillips 1996). The acanthocephalan *M. moniliformis* decreased some measures of activity in males of four species of cockroaches; males of three other species were unaffected (Moore and Gotelli 1996 and references therein). Such seeming inconsistencies underscore a serious deficit in our study of host–parasite interactions and one that is not likely to be corrected soon: not only are we ignorant of life cycles for most parasites, but where life cycles are known, we rarely understand which host species are most important in nature, much less what developmental constraints might pertain.

Describing a behavioral change is interesting and can fortify the already strong allegation that parasites do matter to a broad range of investigations, but we should look beyond the description. What are the fitness effects of the alteration on both parasite and host? What is the mechanism that underlies the behavioral change? The answers to these questions play substantial parts in understanding the evolution of parasite takeover or host resistance. While many workers invoke the cost of various aspects of host–parasite relationships, in fact, accurate estimates of such costs (or benefits) are difficult to achieve.

Which hosts are important in natural life cycles? Many species may be susceptible (e.g., Freehling and Moore 1993; Moore and Crompton 1993), but few may be "chosen," to paraphrase, somewhat perversely, Matthew 22:14. If a behavioral alteration is discovered in an experimental host that is not a normal part of the parasite's life cycle, this could be an indicator of some generalized response to parasitism, or of a mechanism that was effective in a broad range of hosts because of shared physiology, for instance. It would not be particularly relevant to the evolution or ecology of the parasite in question, however. Conversely, a parasite with low host specificity, which may invade any of a number of host species, may be constrained in the extent to which it becomes adapted to any particular host (Combes 1997).

Do any of these phenomena have a genetic basis? For the ability to alter behavior (or to have behavior altered) to evolve, it must have a genetic basis and, with the exception of plague, fleas and blocked feeding, that has yet to be demonstrated experimentally. The question of genetics looms over every altered behavior we might speculate about, whether considering evolution or mechanism. There is reason to be encouraged here. Some genetic differences are associated with either changes in parasite behavior or in the behaviors of their hosts (although not with behavioral alterations themselves). Schistosomes provide strong evidence for the role of genetics, in this case, evidence of a genetic basis for rhythmic shedding of propagules (cercariae) that favors transmission. Populations of *Schistosoma mansoni* cercariae in the French West Indies exhibited different chronobiological patterns of shedding, including an early one (1100 hours) where humans were likely final hosts and a late one (1600 hours) where rats were final hosts (Théron 1984, 1985). If these strains are crossed, they produce an F1 that sheds at intermediate times, and F1 crosses produce an F2 that exhibits early, intermediate, and late patterns (Théron and Combes 1988).

In addition, two genetic isolates of the nematode *Trichinella spiralis*, one from a raccoon and one from a coyote, produced different behavioral reactions in the mice *(Peromyscus leucopus)* they infected. The raccoon strain induced decreased exploratory activity, while the coyote strain increased that activity (Minchella et al. 1994).

Within a host–parasite association, how much uniformity can we expect? The consideration of genetics leads to a bugaboo in the study of parasite-induced behavioral alterations: the question of variation. Of course, the life cycle of *Trichinella* (chapter 2) predisposes it to strain formation, and many parasites may occur in a greater degree of genetic isolation than free-living organisms usually do. Even without such influences, we expect to see some geographic variation in both parasites and hosts. We know there is such variation in other attributes of host–parasite interactions (e.g., Schmid-Hempel and Koella 1994), and it is to be expected in behavioral ones. First of all, uninfected hosts may vary in behaviors, both within populations and across geographic ranges. This may limit the extent to which parasites are able to induce behavioral modifications. Second, parasites may vary in the nature and extent of the alterations they induce. The genetic work cited above supports such an expectation, as do many of the examples cited previously in this book. In addition, there is evidence for geographic variation in parasite-induced changes in appearance, be it pigment dystrophy in acanthocephalan-parasitized crustaceans (Seidenberg 1973; Pilecka-Rapacz 1986) or demelanization of sticklebacks infected with *Schistocephalus solidus* (LoBue and Bell 1993). Perhaps many behavioral alterations that occur in response to parasitism will differ greatly across the parasite's range, depending on gene flow and differences in selective regimes experienced by both host and parasite. We do not know.

What are the constraints that parasites and hosts face in this behavioral warfare? It is all well and good for a parasite to persuade a host to go to lighted areas, but what if the (uninfected) host is already highly photophilic? How could an investigator detect such constraints? Perhaps the best example of constraints can be found in the mosquito *Anopheles stephensi* infected with malaria *(Plasmodium bergei)*. Probing behaviors are not altered in this association, in part because salivary apyrase levels are already low in uninfected *An. stephensi*. The parasite probably cannot lower them (Li et al. 1992). If uninfected animals vary in their behaviors, then some behaviors may have reached a limit at which no alteration can be expressed, while others may be more volatile.

Phylogenetic studies may be able to expose some of these pitfalls, and there is good reason for doing them (Poulin 1998). A glance at table 3.1 reveals that the mechanisms that underlie behavioral alterations range broadly and do not allow general predictions for all parasite–host associations (see also Thompson and Kavaliers 1994, Kavaliers et al. 1998c). Parasites seem to have been opportunistic in their alterations of host behavior, using anything from brute force (demolition of organs and muscles) to precise neurochemical meddling. Hosts have responded with an equally diverse arsenal. In any one association at any one time, the resulting behaviors may reflect numerous parasite influences and host responses; there is even the tantalizing possibility, yet to be examined, that at one point in the association, some host defensive behavior may have been co-opted by parasites, or vice versa. The level at which to ask multispecies evolutionary questions about behavioral alterations should be the host or parasite clade. To date, only one experimental study has done so, and found no influence of host phylogeny on the ways that *M. moniliformis* altered cockroach behavior (Moore and Gotelli 1996). The possibilities for comparative study are intriguing. The ants are especially enticing, with their seemingly convergent behavioral responses to parasitism, their complex social interactions, and their constellation of parasites—cestodes, fungi, nematodes, and a suite of related trematodes. Malaria and filarial worms and their vectors also beckon researchers; geographic expansion paralleled shifts into different host species, and this was often accompanied by adaptation to different host activity patterns.

The existing literature in the field of behavioral alterations and parasites hints at many possibilities, mostly unfulfilled. If these alterations are adaptive, how specific are they? To what extent are behavioral alterations tailor made for the life cycle at hand? What are the evolutionary pressures and possibilities that might yield such precision, and what are its limits? There is evidence that a few alterations are precise (e.g., Bethel and Holmes 1977); in the case of others, such fine-tuning seems hopeless. In some respects, this is a subtopic of the evolution of species specificity. Why such precision in one case, and untidiness in another?

If hosts can oppose manipulation by parasites (e.g., Poulin et al. 1994), or, for that matter, even if they cannot, do parasites oppose host manipulation on the part of other parasites? What happens when two manipulator species find themselves in the same host? There is evidence that multiple infections result in different behaviors than monospecific infections do (Curtis 1987; Burkot et al. 1990), but this has not been examined experimentally. Does another kind of warfare erupt, as parasites vie for high manipulation stakes, or do they collaborate? What is the cost of manipulation? In some cases, we might expect it to be significant; in others (e.g., taking up residence in the eye), it could be purely beneficial for the parasite—a good location as well as increased transmission opportunity. What is the cost of behavioral defense and avoidance? Is there a hierarchy of behaviors that we expect to see, depending on the magnitude of the parasitic threat and the cost of the response? How many hosts do not exhibit behavioral alterations and go unrecorded in a literature that emphasizes significant differences (Poulin 2000)?

Can we make use of parasite–host behavior discoveries in our management of agricultural and public health challenges? Norval (1992), impressed with the success of the zebu and sanga cattle in avoiding ticks, suggested that European cattle be bred for behavioral and morphological traits that would limit ectoparasite infestations. Moreover, if a parasite predisposes its host to predation, might it be as effective in controlling that host as one that kills its host outright? How important are group sizes and the availability of medicinal plants in our design of wildlife refuges? Do infected vectors have different habitat preferences than uninfected ones? If so, can we use this information to target vector control? Finally, how can we use insights from the ways animals react to their parasites to design responses to emerging infectious diseases of humans? On the whole, those diseases are not appearing de novo; their R_0 begins to exceed 1 in humans as a result of changes in human demography and, ultimately, behavior.

My hope for this book is that it can be at once an undeniable statement of the pervasiveness of effects of parasites on behavior, their ecological and potential evolutionary significance, and an open door for those who would like to join in their study. The questions that remain are numerous, fundamental, and, quite frankly, great fun. They require knowledge of parasites, hosts, behavioral biology, immunology, ecology—as much diversity as parasites and their hosts have to offer. They are not easily fenced in.

A cautionary word here: The study of parasites and host behavior is not without its challenges. One of them is the persistence of many colleagues in attempts to classify this work according to traditional scientific categories. This can be frustrating, especially when required to fit some pigeonhole or other for the purposes of publication or funding. Such a failed designation can be perversely satisfying, however, for it mirrors the peculiar and wonderful amalgamation that makes host–parasite associations differ from either hosts or parasites.

I became fascinated with parasites and host behavior while taking Clark Read's parasitology class as an undergraduate. Read always left us with more questions than answers, and his admonition to his class was deceptively simple: "Think like a trematode!" Upon consideration, this is enough to give most folks pause, and I admit that I have yet to master that assignment. If we ever approach such a mental state, however, many things would probably become clear, and among them: Animals that are homes to trematodes are not like other animals, and the same can be said for all hosts of parasites. It is not just that they are "sick," whatever that means. They are engaged in a contest of great significance, sometimes life or death, almost always a contest of fitness outcomes. They are engaged from the level of cells and self-recognition to that of community and ecosystem interactions. The rules of engagement are varied, the possible outcomes are numerous, some border on the fantastic, and our knowledge exceeds our understanding.

Appendix: Tables

Table 3.1. Some Proximal Causes of Parasite-Induced Behavioral Alterations

PARASITE/HOST	PROXIMAL CAUSE	BEHAVIOR	SOURCE
	VIRUSES/BACTERIA/PROTISTA/FUNGI		
Dengue 3 virus/*Aedes aegypti*	Salivary gland and nervous system invasion	Prolonged feeding	Platt et al. 1997
Semliki forest virus/*Aedes aegypti*	Salivary gland pathology	Repeated probing	Mims et al. 1966
Rabies virus/many mammals	Invades centers of aggression in CNS, salivary glands	Increased aggression, salivation, impaired swallowing	Steck 1982; Hart 1990; Hatalski and Lipkin 1997
Streptococcal infection/children	Antibody damage to caudate nucleus of basal ganglia of brain	Obsessive-compulsive disorder	Brown 1997
Trypanosoma spp./tsetse fly	Cover labral mechanoreceptors; salivary gland damage	Reduced foraging; increased probing	Jenni et al. 1980; Molyneux and Jenni 1981; Patel et al. 1982; Golder et al. 1987
Trypanosoma brucei gambiense/rats	Changes in levels of neuroactive compounds	Inactivity	Stibbs and Curtis 1987
Plasmodium mexicanum/Western fence lizard	Anemia	Reduced aerobic ability (stamina)	Schall et al. 1982
Plasmodium gallinaceum/mosquito	Salivary gland pathology	Salivary apyrase deficiency; repeated probing	Rossignol et al. 1984, 1986
Plasmodium yoelii/*Anopheles stephensi*	Nutritional demands or tissue damage	Reduced activity	Rowland and Boersma 1988
Toxoplasma gondii/many mammals	Invades brain	Reduced coordination, altered activity	Hutchison et al. 1980a,c; Hay and Hutchison 1983; Webster 1994

Parasite/host	Mechanism	Behavioral effect	Reference
Eimeria vermiformis/mice	Increases endogenous opioids	Reduced predator anxiety, water maze performance; altered responses to estrous female odor	Kavaliers and Colwell 1994, 1995a; Kavaliers et al. 1995, 1997a
Sarcocystis rauschorum/lemmings	Encysts in muscle	Increased activity, decrease grooming	Quinn et al. 1987
Haemogregarina sp./lizard	Anemia	Lower locomotor speed	Opplinger et al. 1996b
Glugea anomala/stickleback	Encysts at base of tail	Propulsion inhibited	Arme and Owen 1967
Entomophthorales/ants	Hyphae in vicinity of nervous system	Elevation seeking	Schmid-Hempel 1998
Entomopathogenic fungi/insects	Use rhizoids to anchor host in elevated position	Elevated location	Marikovsky 1962; Carner 1980; King and Humber 1981

MONOGENEA/TREMATODA/CESTODA

Parasite/host	Mechanism	Behavioral effect	Reference
Himasthla elongata/cockles	Damages foot	Burrowing compromised	Lauckner 1984
Curtuteria australis/cockles	Stunted foot	Failure to burrow	Thomas and Poulin 1998
Gymnophallus fossarum/bivalve	Thickens pallial edge	Impeded valve closure, exposure	Bartoli 1976–1978
Gyrodactylus bullatarudis/guppies	Fin rays stick together	Abnormal swimming	Scott 1985
Diplostomulum spathaceum/fish	Inhabits eyes	Reduced foraging efficiency	Crowden and Broom 1980
Euhaplorchis californiensis/killifish	Invades surface of brain	Erratic motion, surfacing	Lafferty and Morris, 1996
Ornithodiplostomum ptychocheilus/minnows	Inhabits brain	Reduced schooling	Radabaugh 1980a,b

continued

Table 3.1. (*continued*)

PARASITE/HOST	PROXIMAL CAUSE	BEHAVIOR	SOURCE
MONOGENEA/TREMATODA/CESTODA (CONTINUED)			
Dicrocoelium spp./ants	Inhabit brain	Elevation seeking, inactivity	Anokhin 1966; Lucius et al. 1980; Frank et al. 1984
Brachylecithum mosquensis/ants	Inhabits brain	Circling behavior, reduced activity	Carney 1969
Microphallus papillorobustus/amphipods	Inhabits brain	Hyperactivity, negative geotaxis, positive phototaxis	Helluy 1983a, 1984
Maritrema subdolum/amphipod	Anemia	Increased surface activity	Mouritsen and Jensen 1997
Psilostomum ondatrae/perch	Encysts in lateral line	Disorientation	Beaver 1939
Schistosoma mansoni/hamsters	Influences endogenous opioids	Activity changes	Kavaliers and Podesta 1988
Schistosoma mansoni/CF1 mice	Reduces testosterone	Castration	Isserhoff et al. 1986
Trichobilharzia ocellata/snail	Neuroendocrine product (schistosomin)	Castration, gigantism	de Jong-Brink et al. 1999
Ribeiroia sp./Pacific tree frogs	Encysts near limb bud	Limb deformity	Johnson et al. 1999
Nanophyetes salmincola/salmon, trout	Muscle migration	Transient activity depression	Butler and Milleman 1971
Echinococcus granulosus/large ungulates	Invade lungs, interfere with breathing	Lower resistance to predators	Crisler 1956; Mech 1966, 1970
Spirometra mansonoides/rodents	Produces growth hormonelike factor	Increased growth	Mueller 1980; Phares 1992, 1996, 1997; Phares et al. 1990; Phares and Corkum 1974

continued

Ligula intestinalis/rudd	Gonad stimulating hormone eliminated	Reduced gonads; do not gather in shoals	Orr 1966
Schistocephalus solidus/stickleback	Fills body cavity, inhibits foraging	Inefficient feeding, reduced predator avoidance	Milinski 1984, 1990; but see Ness and Foster 1999
NEMATODA/ACANTHOCEPHALA			
Loa loa/biting fly	Invades muscles	Decreased activity	Lavoipierre 1958
Brugia pahangi/mosquito	Invades flight muscles	Decreased activity	Hockmeyer et al. 1975; Berry et al. 1986
Dirofilaria immitis/mosquito	Invades Malpighian tubules	Decrease/increase activity (varies with mosquito species)	Berry et al. 1987a, 1988
Tetrameres americana/grasshopper	Encysts in muscle	Inactivity	Cram 1931
Moniliformis moniliformis/cockroach	Disturbs giant interneuron spiking activity	Delayed escape response	Libersat and Moore 2000
Acanthocephalus spp./crustacean	May compete with host for pigment precursors	Pigment dystrophy	Oetinger and Nickol 1982a,b
Anguillicola crassus/European eel	Lives in/around swim bladder, affects gas concentration	?	Würtz et al. 1996
Trichinella spiralis/mice	Encysts in muscle	Decrease in activity	Rau 1983a; Zohar and Rau 1986;
	Reduces muscle power output	Decrease locomotion	Harwood et al. 1996

Table 3.1. (continued)

PARASITE/HOST	PROXIMAL CAUSE	BEHAVIOR	SOURCE
Trichinella pseudospiralis/mice	Encysts in muscle	Slight hyperactivity	Rau 1984a
Trichinella pseudospiralis/ kestrels	Encysts in muscle	Reduced activity	Saumier et al. 1986
Toxocara canis/mice	Encysts in brain	Reduced activity and exploration, disorientation	Dolinsky et al. 1981, 1985; Burright et al. 1982; Cox 1996
Pneumostrongylus tenuis/reindeer	Encysts in nervous system	Motor dysfunction, stumbling (but nonpathogenic in white-tailed deer)	Anderson 1971
Polymorphus paradoxus/ amphipod	Changes in serotonin levels	Skimming, clinging	Helluy and Holmes 1990; Maynard et al. 1996
ECTOPARASITES/PARASITOIDS			
Ornithomyssus bursa/ *Hirundo rustica*	Anemia	Reduced song	Møller 1991a
Stomoxys calcitrans/mice	Increased endogenous opioids	Decreased spatial learning ability	Kavaliers and Colwell 1995d; Kavaliers et al. 1998b
Sphecid wasp (*Ampulex compressa*)/cockroach	Head sting may stimulate dopamine receptors	Greater stimulus required for walking; continuous grooming	Fouad et al. 1992, 1996; Weisel-Eichler et al. 1999
Cardiochiles nigriceps/*Heliothis virescens*	Poison gland product	Reduced feeding	Guillot and Vinson 1973

Table 3.2. Some Effects of Parasites on Host Appearance

HOST	PARASITE	ALTERATION	SOURCE
	VIRUSES/BACTERIA/FUNGI		
Bollworm (*Heliothis armigera*)	Nuclear polyhedrosis virus	Dark color	Whitlock 1974
	Granulosis virus	Light colored, shiny tegument, increased size	Fowler and Robertson 1972
Lepidoptera (*Wiseana cervinata; Witlesia* sp.)	Iridescent virus	Blue color, iridescence	
Looper (*Ruchiplusia ou*)	Nuclear polyhedrosis virus	Pale yellow	Paschke and Hamm 1961
Agrotid (*Persectania ewingii*)	Granulosis virus	Gray	Lower 1954
Noctuid (*Eupsilia transversa*)	Granulosis virus	Grayish brown; red below	Edland 1965
Cabbage looper (*Trichoplusia ni*)	Polyhedral virus	Pale	Semel 1956; Hostetter and Biever 1970
Pandora moth (*Coloradia pandora*)	Polyhedral virus	Orange brown	Wygant 1941
Terrestrial isopod	*Rickettsiella* sp.	Opaque white masses	El-Aal and Holdich 1987
Caterpillars (several spp.)	*Nomuraea rileyi*	Yellow/brown spots	Ignoffo 1981
Fly (*Musca domestica*)	*Entomophthora muscae*	Swollen abdomen, white conidiophores	Møller 1993
Tiger moth (*Creatonotus gangis*), armyworm (*Pseudaletia unipuncta*)	*Entomophthora creatonotus*	Yellow brown	Yen 1962
Beetle (*Oryctes rhinoceros*)	*Metarrhizium anisopliae*	Creamy white with dark patches	Nirula 1957
Ants (several spp.)	*Myrmicinosporidium durum*	Darker color	Sanchez-Peña et al. 1993
Honey bee (*Apis mellifera*) larva, pupae	*Rhodotorula* sp., *Penicillium cyclopium*	Pink, brownish-green	Prest et al. 1974
Ant (*Formica rufa*)	*Alternaria tennis*	Post-mortem white stripes	Marikovsky 1962

continued

Table 3.2. (*continued*)

HOST	PARASITE	ALTERATION	SOURCE
		PROTISTA	
Ant (*Myrmecia pilosula*)	Gregarine	Brown; soft cuticle	Crosland 1988
Fish (*Gasterosteus aculeatus*)	*Glugea anomala*	White cysts (4 mm diam)	Arme and Owen 1967
Grasshopper (*Melanoplus bivittatus*)	*Nosema locustae*	Pale color; distended abdomen	Henry and Oma 1981
Reduviid (*Rhodnius prolixus*) *Triatoma infestans*	*Trypanosoma rangeli*	Pale color when hemolymph infected	Grewal 1969
	Blastocrithidia triatomae	Swollen abdomen	Schaub and Schnitker 1988
		TREMATODA	
Coral (*Porites compressa*)	*Plagioporus* sp.	Swollen polyp, pink/white	Aeby 1991
Ant (*Camponotus pennsylvanicus, C. herculeanus*)	*Brachylecithum mosquensis*	Obesity, pale intersegmental membranes exposed (striped appearance)	Carney 1969
Chaetognath (*Sagitta friderici* and others)	Hemiurid trematode	Opaque trematode in almost transparent chaetognath; gigantism	Pearre 1976, 1979
Gastropod (*Biomphilaria glabrata*)	*Schistosoma* sp.	Gigantism	Minchella 1985
Gastropods (*Succinea ovalis, Oxyloma retusa*)	*Leucochloridium variae, L. cyanocittae, Neoleucochloridium problematicum*	Brightly colored sporocyst, swollen tentacles, pulsation	Kagan 1951; Lewis 1974, 1977

Many gastropods	Trematodes	Gigantism	Rothschild 1936, 1962; Sousa 1983; Minchella 1985
Bivalve (*Venereupis aurea*)	*Gymnophallus fossarum*	Reversed position, ventral side uncovered; thick pallial edge	Bartoli 1976–1978
Bivalve (*Macoma balthica*)	*Parvatrema affinis*	Large size	Lim and Green 1991
Guppy (*Lebistes reticulatis*)	*Diplostomum spathaceum*	Dark body coloration	Brassard et al. 1982
Three-spined stickleback (*Gasterosteus aculeatus*)	*Diplostomum* spp.	Larger pectoral fins	Brønseth and Folstad 1997
Banded killifish (*Fundulus diaphanus*)	*Crassiphiala bulboglossa*	Metacercariae for black spots	Krause and Godin 1994
	CESTODA		
Amphipod (*Gammarus zaddachi*)	*Diplocotyle* sp.	Increased size (females)	Stark 1965
Ant (*Tetramorium* spp., *Triglyphothrix striatidens*, *Xiphomyrmex* sp., *Pheidole* sp.)	*Raillietina tetragona*	Pale distended abdomen	Nair et al. 1982
Ant (*Myrmica* spp.)	*Raillietina urogalli*	Dark brown cuticle	O'Rourke 1956
Ant (*Leptothorax nylanderi*)	*Anomotaenia brevis*	Yellow, size/shape differences	Plateaux 1972; Plateaux and Péru 1987; Salzemann and Plateaux 1987; Trabalon et al. 2000; see also Stuart and Alloway 1988

continued

Table 3.2. (continued)

HOST	PARASITE	ALTERATION	SOURCE
		CESTODA	
Three-spined stickleback (*Gasterosteus aculeatus*)	*Schistocephalus solidus*	Distension; pale ventral body, larger pectoral fin, depigmentation (integument); hypermelanization (eyes)	Arme and Owen 1967; LoBue and Bell 1993; Brønseth and Folstad 1997; Ness and Foster 1999
	Diphyllobothrium spp.	Larger pectoral fin	Brønseth and Folstad 1997
Rodents	*Spirometra mansonoides*	Obesity	Mueller 1963
Lagomorph	*Taenia multiceps*	Delay in seasonal fur color change	Leiby and Dyer 1971
		NEMATODA/ACANTHOCEPHALA	
Ant (*Pheidole dentata*)	*Mermis* sp.	12 × size of normal worker	Wheeler 1907
Ants	mermithids	Various alterations	Wheeler 1928 (review)
Ant (*Camponotus castaneus*)	*Rabbium paradoxus*	Swollen gaster	Poinar et al. 1989
Amphipod (*Gammarus pulex*)	*Polymorphus minutus*	Orange cystacanth	Hynes and Nicholas 1963; Barrett and Butterworth 1968
	Pomphorhynchus laevis		A. F. Brown and Thompson 1986; Bakker et al. 1997
G. lacustris	*P. minutus*	Depigmented cuticle	Hindsbo 1972
	P. paradoxus, P. marilis	Orange cystacanth	Bethel and Holmes 1977

Host	Parasite	Effect	Reference
Hyalella azteca	*Corynosoma constrictum*	Orange cystacanth	Bethel and Holmes 1977
Isopod (*Asellus intermedius, Lirceus garmani*)	*Acanthocephalus dirus*	Light colored	Seidenberg 1973; Camp and Huizinga 1979; Oetinger and Nickol 1981; Hechtel et al. 1993
Lirceus lineatus	*A. jacksoni*	Light-colored	Muzzall and Rabalais 1975; Oetinger and Nickol 1981
Asellus aquaticus	*Polymorphus contortus*	Heavily pigmented	Pilecka-Rapacz 1984
	Acanthocephalus anguillae	Darker color	Pilecka-Rapacz 1986; Dezfuli et al. 1994; Lyndon 1996
	A. lucii	Darker 2 of 3 years, some colorless remaining year; dark operculae	Brattey 1983; Pilecka-Rapacz 1986; Lyndon 1996
Asellus intermedius	*Acanthocephalus jacksoni*	Depigmentation	Oetinger and Nickol 1982a,b
Armadillidium vulgare	*Plagiorhynchus cylindraceus*	Light substrate preference; slower growth	Moore 1983a
Cockroach (*Periplaneta brunnea*)	*Moniliformis moniliformis*	Increased use of white surface	Carmichael and Moore 1991

continued

Table 3.2. (continued)

HOST	PARASITE	ALTERATION	SOURCE
		PARASITOIDS	
Red-backed cutworm (*Euxoa ochrogaster*)	Encyrtid wasp (*Berecyntus bakeri*)	Large size; supernumerary instar?	King and Atkinson 1928
Potato aphid (*Macrosiphum euphorbiae*)	Aphidiid wasp (*Aphidius nigripes*)	Dark substrate preference	Brodeur and McNeil 1990

Changes in appearance may be difficult to interpret. Although appearance is not behavioral, strictly speaking, it can interact with and complement behavioral alteration. When we see an odd-colored or oversized intermediate host, for instance, we may assume that such an appearance would increase the likelihood of predation and possible parasite transmission. However, color also influences body temperature, and there is no reason that altered coloration in response to parasitism could not be part of a thermal defense strategy on the part of the host in some cases. Other changes such as enhanced growth may also benefit the parasite at times and the host at other times, and may also affect heat load. Finally, there are fascinating examples of the host–parasite association mimicking another organism altogether. Such modifications should not be discounted. When parasitized by a cestode (*Anomotaenia brevis*), the ant (*Leptothorax nylanderi*) differed so much in appearance (color, cuticular markings, size) from its conspecifics that Plateaux (1972) was in the process of new species description when he noticed that every member of the "new species" contained a cestode cysticercoid (fig. 3.12). The yellow color of the parasitized ants is especially intriguing. Is it a result of endocrinological interference? Does it make the afflicted ants easier to see when a woodpecker rips open the nest?

Table 3.3. Some Examples of Behaviorally Enhanced Transmission (Predation)

PARASITE/HOST (PREY)	LAB/FIELD	COMMENTS	SOURCE
		PROTISTA	
Sarcocystis cernae/Microtus arvalis	Field	Kestrel DH; diet /snap trap comparison	Hoogenboom and Dijkstra 1987
*S. dispersa/*mice	Lab	Long-eared owl DH	Voříšek et al. 1998
Frenkelia spp./wild rodents	Field	common buzzard *Buteo buteo* DH; diet /snap trap comparison	Voříšek et al. 1998
		TREMATODA	
Plagiorchus noblei/ *Aedes aegypti*	Lab	Vole DH	Webber et al. 1987a
Microphallus papillorobustus/ *Gammarus* spp.	Lab	Gull DH	Helluy 1983a
Curtureria australis/ *Austrovenus stutchburyi*	Field	Shorebird DH	Thomas and Poulin 1998
Plagioporus sp./*Porites compressa*	Lab, field	Butterfly fish DH	Aeby 1991, 1992
Euhaplorchis californiensis/ *Fundulus parvipennis*	Field	Compared disappearance of infected/ uninfected fish	Lafferty and Morris 1996
		CESTODA	
Schistocephalus solidus/ *Macrocyclops albidus*	Lab	Stickleback Second IH	Wedekind and Milinski 1996
Ligula intestinalis/Leuciscus rutilus	Field	Piscivorous bird DH; compared fishermen's catch to diets	Van Dobben 1952; see also Dence 1958
Hymenolepis diminuta/ *Tenebrio molitor*	Lab	Rat DH	Blankespoor et al. 1997; but see Webster et al. 2000

continued

Table 3.3. (continued)

PARASITE/HOST (PREY)	LAB/FIELD	COMMENTS	SOURCE
Eubothrium salvelini/ Cyclops vernalis	Lab	Increased predation by brook trout tracked increasing behavioral alterations	Poulin et al. 1992
		ACANTHOCEPHALA	
Polymorphus paradoxus/Gammarus lacustris	Lab	Dabbling duck DH	Holmes and Bethel 1972; Bethel and Holmes 1977
P. minutus/G. lacustris	Lab	Dabbling duck DH	Hindsbo 1972
Corynosoma constrictum/ Hyalella azteca	Lab	Dabbling duck DH	Holmes and Bethel 1972; Bethel and Holmes 1977
Plagiorhynchus cylindraceus/ Armadillidium vulgare	Lab, field	Starling DH	Moore 1983a
Acanthocephalus dirus/ Asellus intermedius	Lab	Creek chub DH	Camp and Huizinga 1979
A. lucii/A. aquaticus	Lab	Perch DH	Brattey 1983
Pomphorhynchus laevis/ Gammarus pulex	Lab	Fish DH	Kennedy et al. 1978; A. F. Brown and Thompson 1986; Bakker et al. 1997

DH, definitive host; IH, intermediate host.

Table 3.4. The Timing of Some Altered Behaviors/Appearances

PARASITE/HOST	BEHAVIOR/APPEARANCE	COMMENTS	SOURCE
Polymorphus paradoxus/ *Gammarus lacustris*	Skimming, clinging, reversed phototaxis at infectivity	Enhanced predation demonstrated	Bethel and Holmes 1974
Acanthocephalus dirus/ *Asellus intermedius*	Color change at infectivity	Enhanced predation demonstrated	Seidenberg 1973; but see Hechtel et al. 1993
Acanthocephalus lucii/ *Asellus aquaticus*	Color change; surface seeking	Surface seeking increased with age of infection; enhanced predation	Brattey 1983; Pilecka-Rapacz 1986
Acanthocephalus anguillae/ *Asellus aquaticus*	Color change; surface seeking	57–64 days postinfection; increased with age of infection	Pilecka-Rapacz 1986
Acanthocephalus raneel *Asellus aquaticus*	Color change; surface seeking	Increased with age of infection	Pilecka-Rapacz 1986
Rabbium paradoxus/ *Camponotus castaneus*	Daytime foraging	Usually nocturnal	Poinar et al. 1989
Skrjabinoclava morrisoni/ *Corophium volutator*	Increased surface activity	Diurnal (shorebird definitive host); only involved late-stage larvae	McCurdy et al. 1999b
Microphallus papillorobustus/ *Gammarus* spp.	Attraction to light; hyperactivity when infective	Must be in brain; enhanced predation	Helluy 1983b
Microphallus sp./ *Potamopyrgus antipodarum*	Feeding time overlaps with predator activity (early A.M.)	Only occurs in mature infections	Levri 1999
Eubothrium salvelinil *Cyclops vernalis*	Swim greater distances when infective	Enhanced predation	Poulin et al. 1992

continued

Table 3.4. (continued)

PARASITE/HOST	BEHAVIOR/APPEARANCE	COMMENTS	SOURCE
Schistocephalus solidus/ Cyclops abyssorum	Increased activity when infective	Congener (*C. scutifer*) yielded no evidence of predation consequence	Urdal et al. 1995
Schistocephalus solidus/ Gasterosteus aculeatus	Predator aversion lost Demelanization	In mature infections	Tierney et al. 1993 Ness and Foster 1999
Dicrocoelium hospes/ Camponotus compressiscapus	Elevation seeking	Crepuscular	Lucius et al. 1980
Hymenoleis diminuta/ Tribolium confusum	Reduced velocity when infective	Intensity and age of infection influential	Robb and Reid 1996
H. diminuta/Tenebrio molitor	When infective: decreased activity and photophobia; unresponsive to aggregation pheromone		Hurd and Fogo 1991
Nuclear Polyhedrosis virus/ Noctuid caterpillars	Increased dispersal when shedding virus	Field and laboratory results differ	Goulson 1997
Plasmodium yoelli/Anopheles stephensi nigeriensis	Feeding persistence increases when parasite infective to next host	Otherwise, persistence low	Anderson et al. 1999
Altenaria tennis/Formica rufa	Elevation seeking	Evening elevation promotes fungal growth and dispersal	Marikovsky 1962; see also Rockwood 1950
Manduca sexta/Cotesia congregata	Delayed metamorphosis, large size	May be mediated by polydnavirus	Beckage 1998

Table 3.5. Some Effects of Parasites on Host Antipredator Behavior

HOST	PARASITE	BEHAVIOR	SOURCE
		VIRUSES/BACTERIA/FUNGI	
Red-legged frog tadpole (*Rana aurora*)	*Candida humicola*	Subdued reaction to tadpole-fed newts	Lefcort and Blaustein 1995
Pea aphid (*Acyrthosiphon pisum*)	*Erynia neoaphidis*	Decreased sensitivity to alarm pheromone	Roy et al. 1999
Bullfrog tadpoles (*Rana catesbeiana*)	Alcohol-killed bacteria (*Aeromonas hydrophilia*)	Reduced refuge-seeking in presence of predator	Lefcort and Eiger 1993
Fox and others (*Vulpes vulpes*)	Rabies virus	Loss of shyness	Steck and Wandeler 1980
Badger (*Meles meles*)	Bovine tuberculosis (*Mycobacterium bovis*)	Loss of fear of humans	Cheeseman and Mallinson 1981
		PROTISTA	
Stickleback (*Gasterosteus aculeatus*)	*Glugea anomala*	Dorsal spine not raised in presence of predator	Milinski 1985
Mouse	*Toxoplasma gondii*	Less fearful (defecation measure)	Hutchison et al. 1980a
Rat (*Rattus norvegicus*)		Less neophobic	Webster et al. 1994
		Unafraid of cat odor	Berdoy et al. 2000
Mouse (*Mus musculus*)	*Eimeria vermiformis*	Reduced fear of cat odor	Kavaliers and Colwell 1995a
Varying lemming (*Dicrostonyx richardsoni*)	*Sarcocystis rauschorum*	Caution indicators decreased during exploration	Quinn et al. 1987

continued

181

Table 3.5. (continued)

HOST	PARASITE	BEHAVIOR	SOURCE
TREMATODA			
Upland bully (*Gobiomorphus breviceps*), Common river galaxias (*Galaxias vulgaris*)	*Telogaster opisthorchis*	Antipredator responses decrease	Poulin 1993a
Banded killifish (*Fundulus diaphanus*)	*Crassiphiala bulboglossa*	Shoaling tendency not increased after model predator "attack"	Krause and Godin 1994
CESTODA			
Ant (*Leptothorax nylanderi*)	Dilepidid cestode	Do not flee when disturbed	Plateaux 1972
Copepod (*Macrocyclops albidus*)	*Schistocephalus solidus*	Reduced predator avoidance	Jakobsen and Wedekind 1998
Three-spined Stickleback (*Gasterosteus aculeatus*)	*Schistocephalus solidus*	Recovered more quickly from fright; continued to forage in presence of predator	Giles 1983; Milinski 1985; Godin and Sproul 1988; Tierney et al. 1993; Ness and Foster 1999
Nine-spined stickleback (*Pungitius pungitius*)	*Schistocephalus* sp.	Spent less time away from surface in presence of predator	Smith and Kramer 1987
NEMATODA/ACANTHOCEPHALA			
Mouse	Nematode (*Toxocara canis*)	Reduced caution	Hay et al. 1983; Hay and Aitken 1984; Cox 1996; Cox and Holland 1998
	(*Heligmosomoides polygyrus*)	No fear of weasel odor	Kavaliers et al. 1997a
Reindeer (*Rangifer tarandus*)	Nematode (*Pneumostrongylus tenuis*)	Loss of fear	Anderson 1971
Isopod (*Caecidotea intermedius*)	Acanthocephala (*Acanthocephalus dirus*)	More likely to be near predator	Hechtel et al. 1993

Table 3.6. Some Influences of Parasites on Elevation Seeking

HOST	PARASITE	BEHAVIOR	SOURCE
		VIRUSES/BACTERIA	
Pandora moth caterpillar (*Coloradia pandora*)	Polyhedral virus	Crawl out on terminal branches	Wygant 1941
Noctuid caterpillar (*Eupsilia transversa*)	Granulosis virus	Crawl to tops of trees	Edland 1965
Bollworm (*Heliothis armigera*)	Nuclear polyhedrosis virus, granulosis virus	Crawl to high places	Whitlock 1974
Black-legged tick (*Ixodes scapularis*)	Lyme disease bacterium (*Borrelia burgdorferi*)	Adults: prefer horizontal surface, lower questing height; nymphs: prefer vertical surface	Lefcort and Durden 1996
Southern grass worm (*Laphygma frugiperda*)	Polyhedral virus	Crawl to tips of grass blades	Allen 1921
Nun moth caterpillar (*Lymantria monacha*)	Treetop disease virus	Gather at tops of trees	Smirnoff 1965
Forest tent caterpillar (*Malacosoma disstria*)	Nuclear polyhedrosis virus	Crawl to tops of trees	Stairs 1965
Noctuid (*Mamestra brassicae*)	Nuclear polyhedrosis virus	Crawl to apex of cabbage leaves	Vasconcelos et al. 1996; Goulson 1997
Agrotid caterpillar (*Persectania ewingii*)	Granulosis virus	Crawl to high places	Lower 1954
Looper (*Rachiplusia ou*)	Nuclear polyhedrosis virus	Move to high position	Pascheke and Hamm 1961

continued

Table 3.6. (continued)

HOST	PARASITE	BEHAVIOR	SOURCE
VIRUSES/BACTERIA			
Cabbage looper (*Trichoplusia ni*)	Nuclear polyhedrosis virus	Crawl high on vegetation	Semel 1956; Hostetter and Biever 1970
Porina (*Wiseana* spp.)	Wiseana nuclear polyhedrosis virus	Burrowing caterpillar surfaces to die	Kalmakoff and Moore 1975
Hepialidae (*Wiseana cervinata*)	Iridescent virus	Found higher in soil, near surface (uninfected are >12 in deep)	Fowler and Robertson 1972
FUNGI[a]			
Red locust (*Cyrtacanthacris septemfasciata*)	*Empusa grylli*	Climb high	Skaife 1925
Grasshopper (*Melanoplus* spp.)	*Empusa grylli*	Die at tops of plants	Rockwood 1950
Locusts/grasshoppers	*Entomophthora grylli*	Climb to top of vegetation, clasp with front legs	Roffey 1968
Soldier beetle (*Chauligiognathus pennsylvanicus*)	*Entomophthora lampyridarum*	Climb onto plants, attach with mandibles; postmortem wing spread	Carner 1980
Tiger moth (*Creatonotus gangis*), Armyworm (*Pseudaletia unipuncta*)	*Entomophthora creatonotus*	Seek elevated place	Yen 1962
Loopers (*Rachiplusia ou, Autographa precationis, Trichoplusia ni*)	*Entomophthora sphaerosperma*	Seek elevated place	Yendol and Paschke 1967
Pea aphid (*Macrosiphum pisi*)	*Empusa aphidis*	Fasten to leaves by rhizoids	Rockwood 1950

Host	Pathogen	Behavior	Reference
Sugar-beet root aphid (*Pemphigus betae*)	*Entomophthora aphidis*	Subterranean aphid crawls to crown and leaves of plant	Harper 1958
Ant (*Formica rufa*)	*Alternaria tenuis*	Crawl high on grass near ant trails, mandibles lock	Marikovsky 1962
African Ponerine ant (*Paltothyreus tarsatus*)	*Cordyceps myrmecophila*	Climb on plants, fixed by mandibles to grass	Bünzli and Büttiker 1959; Evans 1982
Ants (*Dolichoderus attelaboides, Cephalotes atratus*)	*Cordyceps* sp.	Hide under leaf litter, bark	Evans and Samson 1982
Ant (*Technomyrmex* spp.)	*Stilbella dolichoderinarum*	Crawl beneath bark	Samson et al. 1981
Ants	*Entomophthora* spp.	Crawl up on vegetation	Loos-Frank and Zimmerman 1976; cited in Romig et al. 1980; Spindler et al. 1986
Onion maggot (*Delia antiqua*)	*Entomophthora muscae*	Adults move to tops of onion leaves, on periphery of field; abdomen raised	Miller and McClanahan 1959; Krasnoff et al. 1995
Fly (*Musca domestica*)	*Entomophthora muscae*	Crawl upward, dying flies spread wings	Mullens 1990; Krasnoff et al. 1995
Carrot fly (*Psila rosae*)		Under leaves on plants surrounding carrot fields	Eilenberg 1987
Yellow dungfly (*Scatophaga stercoraria*)		Climb up, face plant downwind under leaves, spread wings, raise abdomen	Maitland 1994; also Carner 1980; Krasnoff et al. 1995
Fly (*Sarcophaga aldrichi*)	*Entomophthora bullata*	Attach to twigs, tips of leaves	MacLeod et al. 1973
Tadpole, red-legged frog (*Rana aurora*)	Yeast (*Candida humicola*)	Prefer shallow water	Lefcort and Blaustein 1995

continued

Table 3.6. (*continued*)

TREMATODA

HOST	PARASITE	BEHAVIOR	SOURCE
Molluscs: gastropods *Ilyanassa obsoleta*	*Gynaecotyla adunca*	Move high on beaches	Curtis 1987, 1990
Stagnicola elodes	*Plagiorchis elegans*	Move to top of water column	Lowenberger and Rau 1994a
Littorina littorea	*Himasthla leptosoma*	Occur lower on coastal rocks than uninfected	James 1968
Nucella lapillus	*Parorchis acanthus*	Occur in upper part of species' distribution	James 1968
L. littorea, L. littoralis	*Microphallus similis*		
L. saxatilis	*M. similis, M. pygmaeus,* *Parvatrema homoeotecnum,* *Parapronocephalum* *symmetricum*		
L. littorea	Plagiorchid, opecoelid trematodes *Cryptocotyle lingua*	Occur in lower end of range Remain higher on beach	Lambert and Farley 1968
Pseudosuccinea columella	Xiphidiocercaria	Move to top of water lily	Rothschild 1940
Applesnail (*Pomacea dolioides*)	Trematodes (three spp.)	Spend more time on the surface of the water	Bourne 1993
Nassarius obsoletus	Several trematodes, including *Austrobilharzia variglandis*	More likely to be in-shore	Sindermann 1960
Snails	*Cryptocotyle lingua*	Overrepresented in high tide zone	Sindermann and Farrin 1962
Freshwater snail *Potamopyrgus antipodarum*	*Microphallus* sp.	Vary timing of elevation seeking, move to top of rocks in early A.M.	Levri and Lively 1996; Levri 1999

Host	Parasite	Behavioral modification	Reference
Molluscs: bivalves *Venerupis aurea*	*Gymnophallus fossarum*	Ventral side uncovered	Bartoli 1976–1978
Macoma balthica	*Parvatrema affinis*	Occur close to surface; leave tracks, higher in tidal flats	Swennen 1969; Swennen and Ching 1974; Hulscher 1973; Lim and Green 1991; but see Brafield and Newell 1961; Mouritsen 1997
Cockle (*Austrovenus stutchburyi*)	*Curtuteria australis*	Have difficulty burrowing	F. Thomas and Poulin 1998
Cardium edule, C. lamarcki	*Himasthla elongata*	Do not burrow	Lauckner 1984
Ants (*Camponotus compressiscapus*)	*Dicrocoelium hospes*	Crawl to elevated places	Lucius et al. 1980; Romig et al. 1980
Formica sp.	*Dicrocoelium dendriticum*	Crawl to elevated places, lock mandibles at low temperatures	Anokhin 1966
Mosquitoes (*Aedes aegypti*)	*Plagiorchis noblei*	Occur higher in water column if heavily (>3) infected; lower if <3 metacercariae	Webber et al. 1987b
Chaetognaths (*Sagitta elegans*)	Hemiurid trematode	Occur near surface	Pearre 1979
Fish: dace (*Leuciscus leuciscus*)	*Diplostomum spathaceum*	Spend more time near surface	Crowden and Broom 1980
Pejerrey (*Basilichthys*)	*Diplostomum mordax*	Tumble to surface (caused by motionlessness)	Szidat 1969
Minnow (*Pimephales promelas*)	*Ornithodiplostomum ptychocheilus*	Schools of infected fish occur closer to surface	Radabaugh 1980b
Killifish (*Fundulus parvipinnis*)	*Euhaplorchis californiensis*	Dash to surface of water	Lafferty and Morris, 1996; Lafferty 1997

continued

Table 3.6. (continued)

HOST	PARASITE	BEHAVIOR	SOURCE
		TREMATODA	
Cetaceans	*Nasitrema*	Involved in stranding?	Dailey and Walker 1978; Geraci and St. Aubin 1987
		CESTODA	
Bivalve (*Venerupis staminea*)	*Echeneibothrium*	Exposed on gravel beds	Sparks and Chew 1966
Crustacea (*Gammarus zaddachi*)	*Diplocotyle*	Float upward	Stark 1965
Beetle (*Tribolium* spp.)	*Hymenolepis diminuta*	Move to surface	Yan et al. 1994
Beetle (*Tribolium confusum*)	*Raillietina cesticillus*	Occur on top of culture medium	Graham 1966
Three-spined stickleback (*Gasterosteus aculeatus*)	*Schistocephalus solidus*	Occur in shallow water, near surface	Arme and Owen 1967; Lester 1971; LoBue and Bell 1993; but see Ness and Foster 1999
Nine-spined stickleback (*Pungitius pungitius*)	*Schistocephalus solidus*	Occur closer to surface in hypoxic water	R. S. Smith and Kramer 1987
Shiner (*Notropis cornutus*)	*Ligula intestinalis*	Occur in shallow water	Dence 1958
		NEMATODA	
Gastropod (*Theba pisana*)	*Muellerius capillaris*	Express negative geotaxis	Cabaret 1984
Kestrel (*Falco sparverius*)	*Trichinella pseudospiralis*	Reduce perch height	Saumier et al. 1988; Saumier et al. 1991

ACANTHOCEPHALA

Ostracods (*Cypridopsis vidua*, *Physocypria pustulosa*)	Occur high in water	DeMont and Corkum 1982
Isopods (*Asellus intermedius*) *Octospiniferoides chandleri* *Neoechinorhynchus cylindratus*	Crawl up sides of aquarium; more time on top of leaves	Camp and Huizinga 1979; Hechtel et al. 1993
Acanthocephalus dirus	Shift to surface zone	Pilecka-Rapacz 1986
Asellus aquaticus *Acanthocephalus* spp.	Crawl over surface of leaves	Muzzall and Rabalais 1975
Lirceus lineatus *Acanthocephalus jacksoni*		
Amphipods (*Gammarus pulex*) *Pomphorhynchus laevis*	In open water on vegetation, occur more frequently in drift, spiral toward surface	Kennedy et al. 1978; A. F. Brown and Thompson 1986; McCahon et al. 1991a; Bakker et al. 1997
Echinogammarus stammeri	Found more frequently in drift	Maynard et al. 1996
Gammarus lacustris *Polymorphus minutus*	Occur high in water	Hindsbo 1972
P. paradoxus	Skim on surface	Bethel and Holmes 1974
Cockroaches (*Supella longipalpa*) *Moniliformis moniliformis*	Tend to use vertical surfaces more	Moore and Gotelli 1992
Periplaneta americana	Increase use of horizontal surfaces, especially white ones	Edwards 1987; Gotelli and Moore 1992
Blatella germanica	Increase use of horizontal surfaces, especially white ones	Gotelli and Moore 1992
P. australasiae	Reduce use of black horizontal; depends on light regime	Moore et al. 1994
Skink (paratenic) (*Sphenomorphus quoyii*) *Sphaerechinorhynchus rotundocapitatus*	Increase dive length?	Daniels 1985

continued

Table 3.6. (continued)

HOST	PARASITE	BEHAVIOR	SOURCE
ECTOPARASITES			
Atlantic cod (*Gadus morhua*)	Copepod (*Lernaeocera branchialis*)	Occur near surface	Khan 1988
Sticklebacks (*Gasterosteus* spp.)	Branchiuran (*Argulus canadensis*)	Move to surface to avoid parasite	Poulin and Fitzgerald 1989a
PARASITOIDS			
Lepidoptera (*Euphydryas phaeton*)	Braconid wasps (*Apanteles euphydridis*)	Crawl up on vegetation	Stamp 1981
Leucania separata	*A. kariyai*	Crawl up on plant before egression	Sato et al. 1983
Chlosyne harrissii	Braconid wasp	Occur high on vegetation	Shapiro 1976
Pea aphid (*Acynthosiphon pisum*)	Braconid (*Aphidius ervi*)	May be more likely to drop off plant	McAllister and Roitberg 1987; McAllister et al. 1990
Potato aphid (*Macrosiphum euphorbiae*)	Braconid (*Aphidius nigripes*)	Move to upper surfaces of leaves (if diapausing, seek concealed location)	Brodeur and McNeil 1989, 1991

Note: The ecological consequences of elevation seeking can be difficult to interpret. In a terrestrial environment, animals that chose elevated microhabitats may be increasing their conspicuousness to a visual predator, and increasing their risk of dessication (Rockwood 1950), but they may also be decreasing encounters with more earthbound predators. They may be disseminating parasite propagules, or they may be basking, producing a therapeutic fever. In the water, seeking the surface may signal oxygen constraints, a change in thermal preference, or may simply be the default state of a mobility-impaired animal that rarely moves. Elevation change is especially common among parasitized gastropods (James 1968). Some of these differences could result from variations in final host densities; impaired mobility (differential movements) may also contribute to differences in elevation. Seasonal migration is inhibited in a number of gastropod–trematode associations, and this means that infected and uninfected snails are distributed differently in the intertidal (see also Curtis and Hurd 1983). The role of gastropods in trematode transmission is variable, depending on whether the gastropod contains metacercariae, which must be eaten for transmission to occur, or cercariae, which are liberated to seek hosts on their own. Transmission has not been measured for many gastropod-trematode associations.

The mechanisms behind molluscan elevational change in response to parasitism are largely obscure. Nonetheless, zonation in aquatic and marine systems has long been an object of intensive ecological study, and molluscs have figured prominently in that literature. The role that their parasites may have played in these studies is largely unacknowledged and unexamined (Lauckner 1984, 1987). The bivalve *Mytilus edulis*, for instance, does not produce as many byssus threads when infected with metacercariae of *Himasthla elongata* (Lauckner 1984). Does this alter its role as a space occupier in intertidal communities?

[a]Reviewed by Evans (1989).

Table 3.7. Some Effects of Parasites on Host Activity

HOST	PARASITE	BEHAVIOR	SOURCE
	VIRUSES/BACTERIA		
Swaine jack-pine sawfly (*Neodiprion swainei*)	*Borrelinavirus swainei*	Abnormal rearing movements; wander away from food	Smirnoff 1965
Noctuid (*Eupsilia transversa*)	Granulosis virus	Sluggish movement	Edland 1965
Looper (*Rachiplusia ou*)	Nuclear polyhedrosis virus	Sluggish movement	Paschke and Hamm 1961
Velvetbean caterpillar (*Anticarsia gemmatalis*)	Nuclear polyhedrosis virus	Lethargy	Young and Yearian 1989
Noctuid (*Mamestra brassicae*)	Nuclear polyhedrosis virus	Increased activity; high dispersal rate when shedding virus	Evans and Allaway 1983; Vasconcelos et al. 1996; Goulson 1997
Bee (*Apis cervana*)	Apis iridescent virus	Inactivity, crawl on ground	Bailey and Ball 1978
Ticks (*Dermacentor marginatus, Ixodes persulcatus*)	Tick-borne encephalitis virus	Increased activity, faster movement	Alekseev 1991
Mosquito (*Aedes trivittatus*)	Trivittatus virus	No effect	Berry et al. 1987b
Mouse (*Peromyscus maniculatus*)	Venezuelan equine encephalitis	Reduced torpor	Yuill 1987
Bullfrog tadpole (*Rana catesbeiana*)	Alcohol-killed bacteria (*Aeromonas hydrophila*)	Reduced mobility	Lefcort and Eiger 1993
Badger (*Meles meles*)	Bovine tuberculosis (*Mycobacterium bovis*)	Contorted gait; limited movement	Cheeseman and Mallinson 1981
Black-legged tick (*Ixodes scapularis*)	Lyme disease bacterium (*Borrelia burgdorferi*)	Less active, crossed obstacles fewer times (adult)	Lefcort and Durden 1996

continued

Table 3.7. (continued)

HOST	PARASITE	BEHAVIOR	SOURCE
		FUNGI	
Tigermoth (*Creatonotus gangis*), armyworm (*Pseudaletia unipuncta*)	*Entomophthora creatonotus*	Restlessness	Yen 1962
Eastern spruce budworm	*Entomophaga aulicae*	Decreased activity	Tyrrell 1990
Ants	*Cordyceps* sp.	Restlessness	Evans 1982
Ant (*Cephalotes atratus*)	*Cordyceps* sp.	Erratic movement	Evans and Samson 1982
Fly (*Musca domestica*)	*Entomophthora muscae*	Lethargy	Watson et al. 1993
Locust (*Cicada septendecium*, *Magiccicada septendecim*)	*Massospora cicadina*	Easily caught by hand; do not fly as far; no change in speed/endurance	Speare 1921; Goldstein 1929; White et al. 1983
		PROTISTS	
Fire ant (*Solenopsis invicta*)	Microsporidian (*Thelohania*)	Diminished defensive behavior and vigor	Allen and Buren 1974
Mosquito (*Anopheles stephensi*)	*Plasmodium yoelli*	Flight activity reduced when oocysts reach maximum size	Rowland and Boersma 1988
	Plasmodium cynomolgi	Reduced distance, speed and time	Schiefer et al. 1977
Tsetse fly (*Glossina longipalpis*)	*Trypanosoma* sp.	Increased activity?	Ryan 1984
Monarch butterfly (*Danaus plexippus*)	*Ophryocystis elektroscirrha*	Diminished activity in heavily infected females	Altizer and Oberhauser 1999
European smelt (*Osmerus eperianus*)	*Pleistophora ladogensis*	Reduced swimming speed	Sprengel and Lüchtenberg 1991
Stickleback (*Gasterosteus aculeatus*)	*Glugea anomala*	Locomotion impeded	Arme and Owen 1967

Host	Parasite	Behavioral effect	Reference
Rat (*Ratus norvegicus*)	*Toxoplasma gondii*	Increased activity	Webster 1994
Mouse	*T. gondii*	Increased general movement; decreased rearing, digging; falls increased	Hutchison et al. 1980a,c; Hay and Hutchison 1983
Varying lemming (*Dicrostonyx richardsoni*)	*Sarcocystis rauschorum*	Increased exploration, decreased grooming	Quinn et al. 1987
Vole (*Microtus montanus*)	*Trypanosoma gambiense*	Activity rhythms altered	Seed and Khalili 1971
Grasshopper (*Melanoplus* spp.)	*Nosema locustae*	Lethargy	Henry 1972; Henry and Oma 1981; Bomar et al. 1993
Water strider (*Gerris odontogaster*)	Trypanosomatid (*Blastocrithidia gerridis*)	Skating endurance reduced (male)	Amqvist and Mäki 1990
Reduviid bugs (*Rhodnius prolixus*)	Trypanosomatid (*Trypanosoma rangeli*)	Sluggishness	Grewal 1969
Triatoma infestans	Trypanosomatid (*Blastocrithidia triatomae*)	Sluggishness	Schaub and Schnitker 1988
Rat	*Trypanosoma brucei gambiense*	Inactivity	Stibbs and Curtis 1987
Lizard (*Lacerta vivipara*)	*Haemogregarina* sp.	Reduced speed	Oppliger et al. 1996b
Mouse	*Plasmodium* spp.	Reduced defensive behavior	Day and Edman 1983
Lizard (*Sceloporus occidentalis*)	Malaria (*Plasmodium mexicanum*)	Decreased stamina; no effect on sprint	Schall et al. 1982; Schall and Sarni 1987
MONOGENEA			
Guppy (*Poecilia reticulata*)	*Gyrodactylus bullatarudis*	Lethargy; abnormal swimming	Scott 1985
Largemouth bass (*Micropterus salmoides*)	*Dactylogyrus* sp.	Sluggishness	Herting and Witt 1967

continued

Table 3.7. (continued)

HOST	PARASITE	BEHAVIOR	SOURCE
		TREMATODA	
Snail 1st IH (*Limicolaria*)	*Dicrocoelium hospes*	Sluggishness, move less	Lucius et al. 1980; Frank et al. 1984
Snail 1st IH (*Succinea*)	*Leucochloridium*	Sluggishness	Wesenberg-Lund 1931; Kagan 1952
Snail (*Nucella lapillus*)	*Parorchis acanthus*	Late in joining winter aggregations	Feare 1971
Snail 1st IH (*Nassarius obsoleta*)	several trematodes	Off-shore movement inhibited in autumn	Sindermann 1960
Snail 1st IH (*Stagnicola elodes*)	*Plagiorchis elegans*	Reduced activity	Lowenberger and Rau 1994a
Snail 1st IH (*Oxyloma retusa, Quickella* sp.)	*Neoleucochloridium problematicum*	Reduced activity	Kagan 1952
Amphipod (*Corophium* sp.)	*Gynaecotyla aduncta*	Increased diurnal activity	McCurdy et al. 1999a
Corophium volutator	*Maritrema subdolum*	Increased surface activity	Mouritsen and Jensen 1997
Gammarus spp.	*Microphallus papillorobustus*	Hyperactive	Helluy 1983a,b, 1984
Sponge (*Porites* sp.)	*Plagioporus*	Inability to retract polyp	Aeby 1991
Snail 1st IH (*Nassarius reticulatus*)	Three spp. of trematodes	Reduced activity, no migration to deeper water	Tallmark and Norrgren 1976
Common periwinkle (*Littorina littorea*)	*Cryptocotyle lingua*	Little movement; do not migrate in winter; infecteds occur higher in intertidal	Sindermann and Farrin 1962; Lambert and Farley 1968
	C. lingua or *Renicola roscovita*	Move less	I. C. Williams and Ellis 1975

Host	Parasite	Behavior change	Reference
Perch (*Perca flavescens*)	*Psilotrema spiculigerum*	Disorientation	Beaver 1939
Coho salmon (*Oncorhynchus kisutch*) Steelhead trout (*Salmo gairdneri*)	*Nanophyetus salmincola*	Slower, less active time (transient)	Butler and Millemann 1971
Sheepshead minnow (*Cyprinodon variegatus*)	*Ascocotyle pachycystis*	Swimming performance decreased	Coleman 1993
Snail (*Nassarius obsoletus*)	Several trematode spp.	Moved less	Stambaugh and McDermott 1969
Ilyanassa obsoleta	*Gynaecotyla adunca*	Repeated migration to higher level	Curtis 1990
Mosquito larvae (*Aedes aegypti*)	*Plagiorchis noblei*	Increased/decreased activity	Webber et al. 1987b
Guppy (*Lebistes reticulatis*)	*Diplostomum spathaceum*	Sluggishness; intensity dependent	Brassard et al. 1982
Snail 1st IH (*Biomphalaria glabrata*)	*Schistosoma mansoni*	Sluggishness	Chernin 1967; Lefcort and Bayne 1991
Ant (*Camponotus* spp.)	*Dicrocoelium hospes; D. dendriticum*	Motionless when elevated	Anokhin 1966; Lucius et al. 1980; Frank et al. 1984
	Brachylecithum mosquensis	Sluggishness, circling behavior	Carney 1969
Killifish (*Fundulus parvipennis*)	*Euhaplorchis californiensis*	Erratic movement	Lafferty and Morris 1996; Lafferty 1997
Fish (perch and others) (*Perca flavescens*)	*Psilostomum ondatrae*	Disoriented	Beaver 1939
Pejerrey (*Basilichthys* spp.)	*Diplostomum mordax*	Motionless, float to surface	Szidat 1969

CESTODA

Host	Parasite	Behavior change	Reference
Shiner (*Notropis cornutus*)	*Ligula intestinalis*	Sluggishness, easy to capture	Dence 1958
Sockeye salmon (*Oncorhynchus nerka*)	*Eubothrium salvelini*	Decreased swimming performance; slower; increased fatigue	Boyce 1979
	E. salvelini, Diphyllobothrium dendriticum, Proteocephalus sp.	Disoriented migration	Garnick and Margolis 1990

continued

Table 3.7. (*continued*)

HOST	PARASITE	BEHAVIOR	SOURCE
		CESTODA	
Stickleback (*Gasterosteus aculeatus*)	*Schistocephalus solidus*	Slow movement	Arme and Owen 1967; Lester 1971; Meakins and Walkey 1975; Ness and Foster 1999
Sheep, lagomorphs	*Taenia multiceps*	Cerebral infection causes gid; hindered movement	Leiby and Dyer 1971
Amphipod (*Gammarus zaddachi*)	*Diplocotyle* sp.	Sluggishness	Stark 1965
Crustacea (*Cyclops strenuus*)	*Diphyllobothrium* spp.	Reduced motility; impaired escape	Pasternak et al. 1995
Cyclops vernalis	*Eubothrium salvelini*	Increased activity	Poulin et al. 1992
Ant (*Leptothorax nylanderi*)	*Anomotaenia brevis*	Inactivity	Plateaux 1972; Gabrion et al. 1976
Ants (several genera)	*Raillietina tetragona*	Sluggishness	Nair et al. 1982
Beetle (*Tenebrio molitor*)	*Hymenolepis diminuta*	Decreased activity	Hurd and Fogo 1991
Flour beetle (*Tribolium confusum*)	*H. diminuta*	Slow movement; low velocity	Robb and Reid 1996; but see Yan et al. 1994
		NEMATODA	
Amphipod (*Corophium volutator*)	*Skrjabinoclava morrisoni*	Increased diurnal crawling	MuCurdy et al. 1999b
Grasshopper (*Melanoplus* spp.)	*Tetrameres americana*	Decreased activity	Cram 1931
Beetle (*Scolytus ventralis*)	*Sulphuretylenchus elongatus*	Limited flight; delayed emergence	Ashraf and Berryman 1970
Bark beetles	Nematodes	Decreased activity	Thong and Webster 1975 (review) but see Forsse 1987

Host	Parasite	Effect	Reference
Douglas fir beetle (*Dendroctonus pseudotsugae*)	Nematodes	Decreased initial flight time	Atkins 1961
Bumble bee (*Bombus* spp.)	*Sphaerularia bombi*	Flight continued until late summer; do not initiate new colonies; unsteady, near-continuous flight	Poinar and van der Laan 1972; Lundberg and Svensson 1975
May fly (*Ameletus similior*)	Mermithid	Increased activity	Benton and Pritchard 1990
Mosquito (*Aedes communis*)	*Hydromermis churchillensis*	Inactivity (larva)	Welch 1960
Aedes aegypti	*Brugia pahangi*, *Dirofilaria immitis*	Reduced activity	Townson 1970; Hockmeyer et al 1975; Rowland and Lindsay 1986; Berry et al. 1986, 1987a
Culex pipiens	*Wuchereria bancrofti*	Flight inhibited	Hawking and Worms 1961, citing Ormori 1957
Aedes trivittatus	*Dirofilaria immitis*	Hyperactivity	Berry et al. 1988
Fly (*Chrysops silacea*)	*Loa loa*	Reduced activity	Lavoipierre 1958
Mite (*Liponyssus bacoti*)	*Litomosoides carinii*	Reduced movement	Kershaw and Storey 1976
Sockeye salmon (*Oncorhynchus merka*)	*Philonema oncorhynchi*	Disoriented migration	Garnick and Margolis 1990
European smelt (*Osmerus eperianus*)	*Pseudoterranova decipiens*	Reduced swimming speed	Sprengel and Lüchtenberg 1991
European eel (*Anguilla anguilla*)	*Anguillicola crassius*	Reduced swimming speed	Sprengel and Lüchtenberg 1991
Kestrel (*Falco sparverius*)	*Trichinella pseudospiralis*	Reduced flight; increased walking; decreased grooming, preening	Saumier et al. 1988

continued

Table 3.7. (continued)

HOST	PARASITE	BEHAVIOR	SOURCE
	NEMATODA		
Mouse	Trichinella spiralis	Ambulation and exploration decreased; inactivity increased Speed and distance reduced	Rau 1982, 1983a; Edwards and Barnard 1987 Rau and Putter 1984; see also Minchella et al. 1994
Mouse	Trichinella pseudospiralis	Milder effect than T. spiralis (slight decrease); no effect on exploration	Rau 1984a
Deer mouse (Peromyscus maniculatus)	T. pseudospiralis T. nativa	No activity change Decreased activity	Poirier et al. 1995
Mouse	Toxocara canis	Reduced activity; impaired orientation	Dolinsky et al. 1981, 1985; Donovick et al 1981; Hay et al. 1983; Yuhl et al. 1985
		Decrease varied with exposure Reduced exploration	Draski et al. 1987 Burright et al. 1982; Cox 1996; but see Hay and Aitken 1984
Mouse	Syphacia obvelata, Aspicularis tetraptera	Exploration reduced	McNair and Timmons 1977
Rat	Trichinella spiralis	Decreased running ability	von Brand et al. 1954
Golden hamster (Mesocricetus auratus)	Trichinella spiralis	Decreased activity	Bernard 1959; Goodchild and Frankenberg 1962
Reindeer (Rangifer tarandus)	Pneumostrongylus tenuis	Circling, stumbling	Anderson 1971

ACANTHOCEPHALA

Cockroach (*Periplaneta americana*)	*Moniliformis moniliformis*	Greater distances and velocity	Moore 1983b; Wilson and Edwards 1986; Edwards 1987
		Increased activity in wheels; sex influence	Moore 1983b; Gotelli and Moore 1992
P. australasiae		Reduced distances and velocity	Moore et al. 1994
Blattella germanica		Reduced distances and velocity	Gotelli and Moore 1992
Blatta orientalis		Reduced distances and velocity	Moore et al. 1994
Isopod (*Lirceus lineatus*)	*Acanthocephalus jacksoni*	More time wandering	Muzzall and Rabalais 1975
Isopod (*Asellus aquaticus*)	*Acanthocephalus* sp	Increased activity	Pilecka-Rapacz 1986
	A. lucii, A. anguillae	Impaired righting response; attraction to disturbance	Lyndon 1996
Isopod (*Asellus intermedius*)	*Acanthocephalus dirus*	Hyperactivity, crawl up sides of aquarium	Camp and Huizinga 1979; Hechtel et al. 1993
Isopod (*Armadillidium vulgare*)	*Plagiorhynchus cylindraceus*	Increased activity, distances (females)	Moore 1983a
Amphipod (*Gammarus lacustris*)	*Polymorphus paradoxus*	Evasive response altered; clinging	Bethel and Holmes 1973
Amphipod (*Gammarus pulex*)	*Pomphorhynchus laevis*	Spiraling swimming pattern	Kennedy et al. 1978
Echinogammarus stammeri		Increased activity	Bellettato 1979; Maynard et al. 1998
Amphipod (*Hyalella azteca*)	*Corynosoma constrictum*	Sluggishness	Bethel and Holmes 1973
Crab (*Hemigrapsus crenulatus*)	*Profilicollis antarcticus*	Increased activity	Haye and Ojeda 1998; but see Pulgar et al. 1995

continued

Table 3.7. (continued)

HOST	PARASITE	BEHAVIOR	SOURCE
		ECTOPARASITES	
Atlantic cod (*Gadus morhua*)	Copepod (*Lernaeocera branchialis*)	Hyperactive; erratic	Khan 1988
Brook trout (*Salvelinus fontinalis*)	*Salmincola edwardsii*	Increased distance	Poulin et al. 1991a,b
Common lizard (*Lacerta vivipara*)	Lealapid mites	Gender-specific effects on dispersal	Sorci et al. 1994
Cliff swallow (*Hirundo pyrrhonota*)	Flea (*Ceratophyllus celsus*), swallow bug (*Oeciacus vicarius*)	Dispersal	C. R. Brown and Brown 1992
		PARASITOIDS	
Common armyworm (*Leucania separata*)	Braconid (*Apanteles kariyai*)	Little movement during egression; leaves cocoon	Sato et al. 1983
Spruce budworm (*Choristoneura fumiferana*)	Braconid (*Apanteles fumiferanae*)	No dispersal; late emergence from hibernacula	Lewis 1960
Tobacco hornworm (*Manduca sexta*)	Braconid (*Cotesia congregata*)	Reduced locomotion	Adamo et al. 1997
Cockroach (*Periplaneta americana*)	Sphecid (*Ampulex compressa*)	Paralysis; lethargy	Piek et al. 1984; Fouad et al. 1992, 1994
Mouse (*Peromyscus leucopus*)	Botfly (*Cuterebra fontinella*)	Lethargy	Hirth 1959
Mouse (*P. leucopus*)	Botfly (*Cuterebra* sp.)	Reduced emmigration	
P. maniculatus		Reduced time active	D. H. Smith 1978a
P. boylii		Active, not awkward	Brown 1965

IH, intermediate host.

Note: In aquatic systems, activity alteration—even nothing more than failure to move or swim—may result in other altered behaviors, such as elevational change, as an animal floats to the surface (Stark 1965; Webber et al. 1987a,b). In contrast, fishermen have long known that fish may actively choose to be near the surface or in shallow water if they are seeking warmer temperatures (fever?) to assist recovery from injury or disease (Gunter and Ward 1961). In other cases, such choices may result from neuropathology if parasites encyst in elements of the nervous system (Radabaugh 1980a,b; Lafferty and Morris 1996).

Table 3.8. Some Examples of Parasite-Enhanced Predation by Predators Other Than the Next Host

HOST	PARASITE	EFFECT	SOURCE
Dogwhelk (*Nucella lapillus*)	Digenean (*Parorchis acanthus*)	Belated winter aggregation; oystercatcher predation	Feare 1971
Littorina littorea	*Cryptocotyle lingua*	Occur higher in the intertidal, but inshore fishes are 2nd IH	Sindermann and Farrin, 1962; Lambert and Farley 1986; Stambaugh and McDermott 1969; but see E. E. Williams 1964
Damselflies	Mites (*Arrenurus* spp.)	Increased hand capture	Forbes and Baker 1991
Cicada (*Magicicada septendecim*)	Fungus (*Massospora cicadina*)	Easily captured by hand	White et al. 1983
Chironomus hyperboreus	Mermithid	False oviposition increases predation by fish	Rempel 1940
Grasshopper (*Melanoplus* spp.)	Protistan (*Nosema locustae*)	Lethargy, bird predation	Bomar et al. 1993
Largemouth bass (*Micropterus salmoides*)	Monogenean	More vulnerable to predation	Herting and Witt 1967
Bluegill (*Lepomis macrochirus*)	Columnaris disease		
Menhaden (*Breevortia* sp.)	Isopod (*Olencira praegustator*)	More easily captured by trawl	Guthrie and Kroger 1974
Roach (*Rutilus rutilus*)	Cestode (*Ligula intestinalis*)	Increased pike predation	Sweeting 1976
Common shiner (*Notropis cornuta*)	Cestode (*L. intestinalis*)	Easily caught in net	Dence 1958

continued

Table 3.8. (*continued*)

HOST	PARASITE	EFFECT	SOURCE
Stickleback (*Gasterosteus aculeatus*)	Cestode (*Schistocephalus solidus*)	Increased predation by salmon	Jakobsen et al. 1988
Guppies (*Lebistes reticulatis*)	Digenean (*Diplostomum spathaceum*)	Increased predation by trout	Brassard et al. 1982
Red-legged frog (*Rana aurora*)	Yeast (*Candida humicola*)	Reduced predator perception; increased predation	Lefcort and Blaustein 1995
Red grouse (*Lagopus lagopus*)	Nematode (*Trichostrongylus tenuis*)	Conspicuous odor; increased predation	Hudson et al. 1992
Mouse (*Peromyscus maniculatus*)	Botfly (*Cuterebra* sp.)	>1 *Cuterebra* increased shortail weasel predation; decreased arboreal locomotion	D. H. Smith 1978b

IH, intermediate host.

Table 3.9. Some Effects of Parasites on Host Foraging Behavior

HOST	PARASITE	BEHAVIOR	SOURCE
		VIRUSES/BACTERIA/FUNGI	
Spruce budworm (*Choristoneura fumiferana*)	*Bacillus thuringiensis*	Feeding inhibition	Retnakaran et al. 1983
Looper (*Rachiplusia ou*) (*Trichoplusia ni*)	Nuclear polyhedrosis virus	Reduced feeding	Paschke and Hamm 1961 Semel 1956
Bollworm (*Heliothis armigera*)	Nuclear polyhedrosis virus Granulosis virus	Feeding continued Feeding cessation	Whitlock 1974
Noctuid (*Eupsilia transversa*)	Granulosis virus	Reduced feeding	Edland 1965
Yponomeutid caterpillar (*Plutella xylostella*)	Fungus (*Zoophthora radicans*)	Reduced food intake	Reddy et al. 1998
Flea (*Xenopsylla cheopis*)	Plague (*Yersinia pestis*)	Difficulty when engorging	Bacot and Martin 1914; Cavanaugh 1971
Mosquito (*Aedes triseriatus*)	LaCrosse virus	Increased probing; decreased engorging	Grimstad et al. 1980; Patrican et al. 1985
Aedes aegypti	Semliki Forest virus Dengue 3 virus	Difficulty in feeding Prolonged feeding	Mims et al. 1966 Platt et al. 1997
Culex pipiens	Rift Valley fever virus	Reduced feeding success	Turell et al. 1985
Desert locust (*Schistocerca gregaria*)	*Metarhizium flavoviride*	Reduced food intake	Moore et al. 1992
Ants	*Cordyceps* sp.	Decreased feeding	Evans 1982
Artiodactyls	Rinderpest	Increased thirst	Plowright 1982
Fox (*Vulpes vulpes*)	Rabies virus	Refusal to eat	Steck and Wandeler 1980

continued

Table 3.9. (continued)

HOST	PARASITE	BEHAVIOR	SOURCE
	PROTISTS		
Grasshopper (*Melanoplus* spp.)	*Nosema cuneatum, N. locustae*	Reduced consumption	Henry 1971; Henry and Oma 1981
Grasshoppers (*Melanoplus differentialis*)	Microsporidea (*Nosema locustae*)	Reduced consumption (females)	Oma and Hewitt 1984
M. sanguinipes, nymphs		Supressed feeding	Johnson and Pavlikova 1986
Flour beetle (*Tribolium castaneum*)	Microsporidea (*Nosema whitei*)	Increased consumption	Armstrong 1979
Honeybee (*Apis mellifera*)	Microsporidea (*Nosema apis*)	Consumed more sucrose Reduced attendance on queen; forage at earlier age	Moffett and Lawson 1975 Wang and Moeller 1970
Triatomid (*Rhodnius prolixus*)	Trypanosome (*Trypanosoma rangeli*)	Increased probing	Anez and East 1984
Sandfly (*Lutzomyia longipalpis*)	Trypanosome (*Leishmania mexicana*)	Increased probing	Killick-Kendrick et al. 1977
Sandfly (*Phlebotomus dubosqi*)	Trypanosome (*Leishmania major*)	Increased probing	Beach et al. 1985
Tsetse fly (*Glossina* spp.)	Trypanosome (*Trypanosoma* spp.)	More frequent feeding; increased probing and engorgement time	Jenni et al. 1980; Molyneux and Jenni 1981; Roberts 1981
Mosquito (*Aedes sierrensis*)	Ciliophora (*Lambornella clarki*)	Blood-feeding inhibited Reduced host seeking and feeding	Egerter and Anderson 1989 Egerter et al. 1986
Aedes aegypti	Microsporidea (*Edhazardia aedis*)	Decreased feeding if horizontally transmitted	Koella and Agnew 1997
Ae. aegypti	Apicomplexa (*Plasmodium gallinaceum*)	Increased time probing; smaller meals	Freier and Friedman 1976; Rossignol et al. 1984
Anopheles stephensi	*Plasmodium yoelli nigeriensis*	Increased persistence when patent	Anderson et al. 1999

Host	Parasite	Effect	Reference
Anopheles spp.	P. falciparum	Increased probing, more time probing	Wekesa et al. 1992
An. punctulatus	Plasmodium spp.	Increased tenacity	Koella and Packer 1996
Ae. gambiae	Plasmodium falciparum	Full engorgement and multiple hosts more likely	Koella et al. 1998a
Three-spined stickleback (Gasterosteus aculeatus)	Microsporidea (Glugea anomala)	More profitable prey not chosen	Milinski 1984
Rainbow trout (Oncorhynchus mykiss)	Hemoflagellate (Cryptobia salmositica)	Decreased food consumption	Lowe-Jinde and Zimmerman 1991

TREMATODA

Host	Parasite	Effect	Reference
Snail (Ilyanassa obsoleta)	Several species	Response to carrion influenced	Curtis 1985
		Infection and empty gut positively correlated	Curtis and Hurd 1981
Limicolaria sp.	Dicrocoelium hospes	Reduced consumption	Frank et al. 1984
Potamopyrgus antipodarum	Microphallus sp.	Shift in foraging time	Levri and Lively 1996
Helisoma trivolvis	Echinostoma revolutum	Reduced attraction to lettuce	Boland and Fried 1984
Australorbis glabratus	Schistosoma mansoni	Reduced sensitivity to wheat germ	Etges 1963
Biomphalaria glabrata		Changed time and increased frequency of feeding	Williams and Gilbertson 1983
Fly (Simulium exiguum)	Lecithodendriid	Reduced feeding?	Lewis and Wright 1962
Chaetognath (Sagitta spp.)	Hemiurid	Increased foraging on surface prey	Pearre 1976, 1979
Dace (Leuciscus leuciscus)	Diplostomum spathaceum	Increased time and difficulty foraging	Crowden and Broom 1980

continued

Table 3.9. *(continued)*

HOST	PARASITE	BEHAVIOR	SOURCE
Three-spined stickleback (*Gasterosteus aculeatus*)	*Diplostomum* spp.	Reduced reactive distance to prey; bias right/left vision	Owen et al. 1993
Ethiopian baboons	*Schistosoma* sp.	Consumption of *Balanites aegyptiaca* leaves/berries (medicinal?) correlated with exposure to parasite	Phillips-Conroy 1986; but see Newton and Nishida 1990; see also Janzen 1978 for use by elephants
CESTODA			
Copepod (*Cyclops strenuus abyssorum*)	*Diphyllobothrium* spp.	Feeding rate decreased	Pasternak et al. 1995
Beetles (*Tribolium confusum, Tenebrio molitor*)	*Hymenolepis diminuta*	Preferential feeding on infected feces?	Evans et al. 1992; Pappas et al. 1995; but see Shostak and Smyth 1998
Ant (*Leptothorax nylanderi*)	*Anomotaenia brevis*	Inhibited foraging; solicitation of food	Plateaux 1971; Gabrion et al. 1976
Three-spined stickleback (*Gasterosteus aculeatus*)	*Schistocephalus solidus*	Altered choice of profitable prey; increased likelihood of foraging in presence of predator; foraging rate constrained	Giles 1983; Milinski 1984, 1985; Godin and Sproul 1988; Jakobsen et al. 1988; Ranta 1995
NEMATODA/ACANTHOCEPHALA			
Mite (*Ornithonyssus bacoti*)	Nematode *Litomosoides carinii*	Increased probing time; difficulty in engorging	Molyneux and Jefferies 1986, citing Jefferies 1984
Bumblebee (*Bombus* spp.)	Nematode *Sphaerularia bombi*	Inefficient foraging	Lundberg and Svensson 1975

Host	Parasite	Effect	Reference
Ant (*Pheidole dentata*)	*Mermis* sp.	Foraging and brood-feeding inhibited; solicitation of food	Wheeler 1907
Fly (*Chrysops silacea*)	Nematode (*Loa loa*)	Inhibited feeding	Lavoipierre 1958
Simuliid (*Prosimulium mixtum*)	Mermithid nematode	Inhibited bite	Colbo and Porter 1980
Sandfly (*Phlebotomus papatasi*, *P. sergenti*)	Tetradonematid nematode	Female feeding inhibited	Killick-Kendrick et al. 1989
Mosquito (*Aedes communis*) larva	Nematode (*Hydromermis churchillensis*)	Reduced feeding	Welch 1960
Ae. aegypti	Nematode (*Brugia pahangi*, *Dirofilaria immitis*)	Feeding inhibition Reduced meal size	Lindsay and Denham 1985 Courtney et al. 1985
Anopheles togoi	Nematode (*Brugia malayi*)	Reduced feeding	Husain and Kershaw 1971
An. gambiae, *Culex fatigans*	Nematode (*Wuchereria bancrofti*)	Feeding inhibition	Zielke 1976
Red grouse	Nematode (*Trichostrongylus tenuis*)	Decreased food intake	Shaw 1990
Rat (*Rattus norvegicus*)	Nematode (*Syphacia muris*)	Increased time feeding	Webster 1994
Rat	Nematode (*Nippostrongylus brasiliensis*)	Food intake decreased	Crompton et al. 1981; Ovington 1985
Reindeer (*Rangifer tarandus*)	Gastrointestinal nematodes	Reduced food intake	Arneberg et al. 1996
Humans	*Ascaris lumbricoides*	Food intake increased after treatment	Hadju et al. 1996
Amphipod (*Gammarus pulex*)	Acanthocephala (*Pomphorhynchus laevis*)	Reduced food intake	McCahon et al. 1988, 1989, 1991b; A. F. Brown and Pascoe 1989

continued

Table 3.9. (continued)

HOST	PARASITE	BEHAVIOR	SOURCE
		ECTOPARASITES/PARASITOIDS	
Atlantic cod (*Gadus morhua*)	Copepod (*Lernaeocera branchialis*)	Reduced food intake	Khan 1988
Caterpillar (*Pieris brassicae*)	Braconid (*Apanteles glomeratus*)	Decreased consumption	Führer 1981
P. rapae		Increased food intake and feeding period	Rahman 1970; Parker and Pinnell 1973
P. rapae	*Apanteles rubercula*	Reduced food intake and feeding period	Rahman 1970; Parker and Pinnell 1973
Spodoptera littoralis	*Chelonus* sp., *Microplitis* sp.	Reduced feeding	Führer 1981
Manduca sexta	*Apanteles congregatus*	Reduced feeding	Beckage and Riddiford 1978; Adamo et al. 1997
Trichoplusia ni	*Hypsoter exiguae*	Depressed feeding	Thompson 1982a,b, 1983, 1985
Heliothis virescens	*Copidosoma truncatellum*	Increased food intake	Hunter and Stoner 1975
	Cardiochiles nigriceps	Reduced food consumption	Guillot and Vinson 1973
Arctiid (*Platyprepia virginalis*)	Tachinid (*Thelaira americana*)	Increased likelihood of selecting hemlock as food	Karban and English-Loeb 1997

Host	Parasite	Effect	Reference
Diatraea saccharalis	*Lixophaga diatraenae*	Reduced food consumption	Brewer and King 1978
Bumblebee (*Bombus pascuorum*)	Conopid fly (*Physocephala rufipes, Sicus femigineus*)	Shift in flower species; reduced pollen collection	Schmid-Hempel and Schmid-Hempel 1990, 1991
B. humilis		Shift in pollen source, reduce flower constancy	Schmid-Hempel and Stauffer 1998
Ant (*Pheidole titanis*)	Phorid fly	Shift in foraging time	Feener 1988
Great tit (*Parus major*)	Hen flea (*Ceratophyllus gallinae*)	More intense begging in young; male parent increases foraging	Christe et al. 1996a
Mouse (*Peromyscus maniculatus*)	Botfly (*Cuterebra* sp.)	Increased food consumption	D. H. Smith 1978b

Table 3.10. Some Ways That Parasite Propagules Are Transmitted

PARASITE	"STRATEGY"	SOURCE
Nuclear polyhedrosis virus in forest tent caterpillar (*Malacosoma disstria*)	Attractive to flies; mechanically dispersed	Stairs 1965
NPV in cabbage loopers (*Trichoplusia ni*)	Caterpillars conspicuous to avian predators (dispersal agents)	Hostetter and Biever 1970
Protist (*Nosema locustae*)	Weakens grasshopper (*Melanoplus* spp.), which is cannibalized	Henry 1972
Nosema sp. in Portugese gypsy moth (*Lymantria dispar*)	Encased in edible silk	Jeffords et al. 1987
Some monogeneans (*Hexabothrium appendiculatum*, *Leptocotyle minor*)	Long "appendages" suspend egg in water column, increase host contact	Kearn 1986
Trematode (*Aporchis massiliensis*) cercariae	Encyst on the alga *Cystoseira amentacea*, food for gulls	Bartoli et al. 1997
Rattenkönigcercariae, Proterometra	Resemble prey	Combes et al. 1994
Leucochloridiids	Resemble caterpillars?	Wesenberg-Lund 1931
Acanthocephalan (*Leptorhynchoides thecatus*)	Filaments around egg entangle with algae	Barger and Nickol 1998

Table 3.11. Some Effects of Parasites on Host Reproduction

HOST/PARASITE	REPRODUCTIVE EFFECT	SOURCE
	PROTISTA/FUNGI	
Bush cricket (*Requena verticalis*)/gregarine	Males more choosey than infected females	Simmons 1994
Copepod (*Pseudocalanus elongatus*)/dinoflagellate (*Blastodinium contortum hyalinum*)	Feminization	Cattley 1948
Mosquito (*Anopheles stephensi*)/*Plasmodium yoelii nigeriensis*	Vitellogenin uptake inhibited	Jahan and Hurd 1998
Pied flycatcher (*Ficedula hypoleuca*)/trypanosomes	Parasitized males have shorter feathers/wings; arrive late at breeding ground	Rätti et al. 1993
Tengmalm's owl (*Aegolius funereus*)/blood parasites	Reduced nest defense	Hakkarainen et al. 1998
Diamondback moth (*Plutella xylostella*)/*Zoophthora radicans*	Reduced response to pheromone; reduced oviposition	Reddy et al. 1998
Carrot flies/*Entomophthora muscae*	Oviposit in unfavorable places	Eilenberg 1987
	MONOGENEA/TREMATODA	
Guppy (*Poecilia reticulata*)/*Gyrodactylus turnbulli*	Infected females less choosey	López 1999
Amphipod (*Corophium volutator*)/*Gynaecotyla aduncta*	Older females abort and eat young; younger females advance receptivity	McCurdy et al. 1999a
Snail (*Stagnicola elodes*)/*Plagiorchis elegans*	Permanent castration	Zakikhani and Rau 1999
Biomphalaria glabrata/*Schistosoma mansoni*	Mate less frequently; bias toward mating as male	Rupp 1996
Lymnaea stagnalis/*Trichobilharzia ocellata*	Castration	de Jong-Brink et al. 1990, 1999
Mouse/*Schistosoma mansoni*	Castration	Isserhoff et al. 1986

continued

Table 3.11. (*continued*)

HOST/PARASITE	REPRODUCTIVE EFFECT	SOURCE
	CESTODA	
Copepod (*Macrocyclops albidus*)/*Schistocephalus solidus*	Infected female more likely to have egg sac, but with fewer eggs	Wedekind 1997
Cyclops strenuus abyssurum/*Diphyllobothrium* spp.	Reduced egg production	Pasternak et al. 1995
Amphipod (*Gammarus zaddachi*)/*Diplocotyle* sp.	Ovaries absent	Stark 1965
Ant (*Leptothorax nylanderi*)/cestode	Decreased nest-mate egg production when parasitized ants present	Salzemann and Plateaux 1987
Beetle (*Tenebrio molitor*)/*Hymenolepis diminuta*	Reduced female fecundity, pheromone production; reduced male pheromone response.	Hurd and Parry 1991
	Reduced male odor attractiveness, parasitized male less fecund	Worden et al. 2000
Three-spined stickleback (*Gasterosteus aculeatus*)/*Schistocephalus solidus*	Distended abdomen may wreck nest during spawning; reduced egg production; most damaging effects may be postreproductive (life history adaptation?)	Arme and Owen 1967; McPhail and Peacock 1983; but see Heins et al. 1999
Several fish/*Ligula* sp.	Interference with gonadal development	Dence 1958; Orr 1966; Arme and Owen 1968; Sweeting 1976
Rats/*Taenia taeniaeformis*	Reduced fertility, both sexes	Lin et al. 1990
Rats/*Hymenelepis diminuta*	Reduced time until pup retrieval, reduced wariness	Willis and Poulin 1999
Mice/*Taenia crassiceps*	Reduced testosterone; do not mate readily	Morales et al. 1996

NEMATODA/ACANTHOCEPHALA

California five-spined engraver (*Ips confusus*)/ *Aphelenchulus elongatus*	Reduced egg production	Massey 1960
Fir engraver beetle (*Scolytus ventralis*)/*Parasitylenchus elongatus, P. scrutillis*	Short galleries, no eggs	Massey 1964
Shot hole borer (*Scolytus rugulosus*)/*Neoparasitylenchus rugulosi*	Horizontal gallery; no eggs	Nickle 1971
Southern pine beetle (*Dendroctonus frontalis*)/ *Contortylenchus brevicorni*	Fewer eggs; shorter galleries	MacGuidwin et al. 1980
Douglas fir beetle (*Dendroctonus pseudotsugae*)/ *Contortylenchus* sp.	Fewer eggs	Thong and Webster 1975
Prosimulium mixtum/Neomesomermis flumenalis	Reduced blood feeding, mock oviposition	Colbo and Porter 1980
Many insects/many nematodes	Castration, false oviposition	Chironomids: Rempel 1940; Wülker 1985; simuliids: Hunter and Moorhouse 1976; Colbo and Porter 1980; Molloy 1981; emphemopterans: Flecker and Allan 1988; Vance 1996; see also Poinar 1965; Nappi 1973
Blackflies/mermithid nematodes	Intersex formation	Rempel 1940; Rubtsov 1958; Hunter and Moorhouse 1976; Molloy 1981

continued

Table 3.11. (*continued*)

HOST/PARASITE	REPRODUCTIVE EFFECT	SOURCE
Chironomids, ceratopogonids/mermithid nematodes	Intersex formation	Hunter and Moorhouse 1976
Mayflies/mermithid nematodes	Intersex formation	Flecker and Allan 1988; Vance 1996
Drosophila testacea/*Howardula aoronymphium*	Reduced fertility	Jaenike 1988; James and Jaenike 1992
Mosquito (*Armigeres subalbatus*)/*Brugia malayi*	In resistant mosquitoes, egg development delayed	Ferdig et al. 1993
Red grouse (*Lagopus lagopus scoticus*)/ *Trichostrongylus tenuis*	Reduced, delayed egg laying	Shaw 1990
American kestrel/*Trichinella pseudospiralis*	Reduced fecundity; eggs placed in inappropriate locations	Saumier et al. 1986
Shrews (*Sorex* spp.)/*Mammanidula asperocutis*	Milk production inhibited	Okhotina and Nadtochy 1970
Mouse/*Trichinella spiralis*	Early offspring dispersal; reduced mating	Rau 1985; but see Edwards 1987; Edwards and Barnard 1987; Sorci et al. 1994
Isopod (*Asellus aquaticus*)/polymorphid acanthocephalan	Intersex formation	Munro 1953
	ARTHROPODA	
Porcellanid crab (*Petrolisthes cabrilloi*)/ barnacle *Lernaeodiscus porcellanae*	Mock parental care	Ritchie and Høeg 1981

Host (parasite)	Effect	Reference
Commercial sand crab (*Portunus pelagicus*)/barnacle (*Sacculina granifera*)	Mock parental care	Bishop and Cannon 1979
Atlantic cod (*Gadus morhua*)/*Lernaeocera branchialis*	Reduced fecundity	Khan 1988
Great tit (*Parus major*)/hen flea (*Ceratophyllus gallinae*)	Increased nest mass	Heeb et al. 1996
Damselfly (*Coenagrion puella*)/mites (*Arrenurus aspidator*)	Fewer, larger offspring (from heavily parasitized female)	Rolff 1999
Drosophila nigrospiracula/mites (*Macrocheles subbadius*)	Reduced number of eggs; nutritional drain	Polak 1996
Membracid (*Thelia bimaculata*)/parasitoid (*Aphelopus theliae*)	Feminization	Kornhauser 1919
Field cricket (*Gryllus integer*)/parasitoid (*Euphasiopteryx ochracea*)	Reduced calling; females have trouble locating parasitized males	Cade 1984, 1991; Cade and Cade 1992
Mice (*Peromyscus leucopus*)/botfly (*Cuterebra augustifrons*)	Testes do not descend	Wecker 196

Table 3.12. Some Effects of Parasites on Host Responses to Light

HOST	PARASITE	BEHAVIOR	SOURCE
VIRUSES/BACTERIA			
Noctuid (*Aletia oxygala luteopallens*)	*Borrelinavirus luteopallens*	Do not seek cover; do not return to diurnal refugia	Smirnoff 1965
Bollworm (*Heliothis armigera*)	Granulosis virus	Positively phototropic	Whitlock 1974
Black-legged tick (*Ixodes scapularis*)	Lyme disease bacterium (*Borrelia burgdorferi*)	Photophilia (nymph)	Lefcort and Durden 1996
TREMATODA			
Ant (*Camponotus* spp.)	*Brachylecithum mosquensis* metacercaria	Not photophobic; unresponsive to changes in light intensity	Carney 1969
Snails (*Oxyloma retusa, Quickella* sp.)	*Neoleucochloridium problematicum* sporocysts	Photophilic; pulsation in response to light; seek exposure	Kagan 1952; Lewis 1977
Succinea ovalis, Oxyloma retusa	*Leucochloridium variae*	Pulsation in response to light	Lewis 1977
Limicolaria	*Dicrocoelium hospes*	Not photophobic	Frank et al. 1984
Amphipods (*Gammarus insensibilis, G. aequicauda*)	*Microphallus papillorobustus*	Attracted to light; negative geotaxis also an influence	Helluy 1983a,b; F. Thomas et al. 1996a,b
CESTODA			
Beetle (*Tribolium confusum*)	*Raillietina cesticillus*	Photophobia lost	Graham 1966
Tenebrio molitor	*Hymenolepis diminuta*	Photophobia decreased	Hurd and Fogo 1991

NEMATODA AND ACANTHOCEPHALA

Isopod (*Armadillidium vulgare*)	Nematoda (*Dispharynx nasuta*)	Increased photophilia	Moore and Lasswell 1986
Mouse	*Toxocara canis*	Time in light increased	Cox and Holland 1998
Isopod (*Asellus aquaticus*)	*Acanthocephalus anguillae*	Increased photophilia?	Lyndon 1996
Ant (*Pheidole dentata*)	*Mermis* sp.	Increased negative phototropism	Wheeler 1907
Amphipods (*Gammarus lacustris*)	Acanthocephala (*Polymorphus minutus, P. marilis, P. paradoxus*)	Altered reaction to light (various)	Hindsbo 1972; Bethel and Holmes 1973, 1974
G. pulex	*Pomphorhynchus laevis*	Prefer lighted areas less photophobic	Kennedy et al. 1978; A. F. Brown and Thompson 1986; Bakker et al. 1997;
Echinogammarus stammeri		Prefer lighted areas	Bellettato 1997; Maynard et al. 1998
Hyalella azteca	*Corynosoma constrictum*	Increased phototaxis, photophilia	Bethel and Holmes 1973, 1974
Cockroaches (*Periplaneta americana*)	*Moniliformis moniliformis*	Increased phototaxis	Moore 1983b; but see Gotelli and Moore 1992
P. brunnea		Delayed response to sudden light	Carmichael and Moore 1991
Supella longipalpa		Less movement in response to light, tendency to avoid light	Moore and Gotelli 1992
Ostracods (*Cypridopsis vidua, Physocypria pustulosa*)	*Octospiniferoides chandleri, Neoechinorhynchus chandleri*	Positive response to light	DeMont and Corkum 1982

continued

Table 3.12. (continued)

HOST	PARASITE	BEHAVIOR	SOURCE
		PARASITOIDS	
Common armyworm (*Leucania separata*)	*Apanteles kariyai*	Shift from nocturnal to diurnal	Sato et al. 1983
Spruce budworm (*C. fumiferana*)	*A. fumiferanae, Glypta fumiferanae*	Negative phototaxis absence of photic response	Lewis 1960
Potato aphid (*Macrosiphum euphorbiae*)	*Aphidius nigripes*	Negative phototaxis, diapausing; positive phototaxis, non-diapausing	Brodeur and McNeil 1990, 1991

Table 3.13. Some Effects of Parasites on Mental Ability

PARASITE/HOST	EFFECT	SOURCE
Streptococcal infection/children	Sydenham's chorea; obsessive-compulsive behaviors	P. Brown 1997; Schwartz 1997
Protist (*Eimeria vermiformis*)/mouse	Poor performance in water maze	Kavaliers et al. 1995
Toxoplasma gondii/rodents	Reduced learning	Hay and Hutchison 1983
T. gondii/humans	Low IQ?	Alford et al. 1974; Hutchison et al. 1980a
	Personality traits altered (gender influenced)	Flegr and Hrdý 1994; Flegr et al. 1996
Trematode (*Schistosoma mansoni*)/rats	Slow to learn mazes	Stretch et al. 1960a
Schistosoma spp./humans	Cognitive impairment?	Kvalsig 1988; Kvalsig et al. 1991
Nematode (*Heligmosomoides polygyrus*)/mouse	Poor performance in water maze	Kavalier and Colwell 1995e; see also Braithwaite et al. 1998
Toxocara canis/rodents	Poor performance in water maze, reduced response to aversive conditioning	Olson and Rose 1966, Dolinsky et al. 1981
Ectoparasite (*Stomoxys calcitrans*)/mouse	Poor performance in water maze	Kavaliers and Colwell 1995d

Table 4.1. Some Ways That Animals Avoid Parasites

HOST/PARASITE	AVOIDANCE BEHAVIOR	SOURCE
Scarab beetle larvae/fungi	Avoid mycelia	Hajek and St. Leger 1994
Termite (*Reticulitermes* sp.)/ fungus (*Metarhyizium anisopliae*)	Avoid dead conspecifics; no cannibalism (do groom living infected termites)	Marikovsky 1962; Kramm et al. 1982
Some mushroom-breeding *Drosophila*/nematodes	Amatin tolerance allows escape from amatin-intolerant nematode parasites	Jaenike 1985
Leaf-cutter ant (*Atta cephalotes*)/parasitoid (*Neodohrniphora curvinervis*)	Night foraging avoids diurnal parasitoid	Orr 1992
Ants/several spp. fungi	Leave nest vicinity	Evans 1982
Mosquito (*Aedes aegypti*)/trematode (*Plagiorchis elegans*)	Avoid oviposition in water containing mosquito larvae parasitized by *P. elegans*	Lowenberger and Rau 1994b
Honey bee/American foulbrood	Some genetic strains will remove killed larvae/pupae	Woodrow and Holst 1942; Rothenbuhler 1964
Honey bee/mite (*Acarapis woodi*)	Swarming has secondary effect of reducing mite density	Royce et al. 1991
Bluegill sunfish (*Lepomis macrochirus*)/ fungus (*Saprolegnia*)	Colonial males spend more time fanning eggs	Côté and Gross 1993
Marine iguana/tick (*Ornithodoros*)	Group at night and avoid parasites	Wikelski 1999
Heron/mosquito	Stealthy hunting behavior (lower activity) increases susceptibility to mosquitoes	Webber and Edman 1972; Edman et al. 1984
Parasitic cliff swallow (*Hirundo pyrrhonota*)/ swallow bugs, bird fleas	Choose least infected nests	C. R. Brown and Brown 1991
Barn swallow (*Hirundo rustica*)/nest parasites	Avoid parasitized nests	Barclay 1988

Species/parasite	Behavior	Reference
Sea birds (*Phalacrocorax bougainvillii*, *Sula variegata*, *Pelecanus occidentalis thagus*)/ticks	Desert eggs, young at high parasite levels	Duffy 1983
Rats/nematode (*Nippostrongylus brasiliensis*)	Avoid food flavors administered with infective *N. brasiliensis* larvae	Keymer et al. 1983
European badgers/ectoparasites	Change nest site	Butler and Roper 1996
Porcupine/mosquito	Climb trees to avoid mosquitoes	Marshall et al. 1962
Domestic cattle/mosquito	Stampede in response to mosquito harassment	Ralley et al. 1993
Cattle/tick larvae	Avoid grass with tick larvae	Sutherst et al. 1986
Friesian calves/biting fly (*Stomoxys calcitrans*)	Foot stamps, head swings, tail flicks reduced fly attacks	Warnes and Finlayson 1987
Reindeer/biting flies	Seek high windy places, snow, inland areas	Espmark and Langvatn 1979, citing Darling 1937; Halvorsen 1986b, citing Ballari 1986 and White et al. 1983
Red deer/biting flies	Callow; seek high, windy places	Espmark and Langvatn 1979, citing Darling 1937
Feral island horses/biting flies	Move to areas with fewer flies without changing herd size/social structure; endoparasite avoidance may be more important than parasite avoidance	Rubenstein and Hohmann 1989
Camargue horses/horseflies	Rest on bare ground in summer and reduce attack	Duncan and Cowtan 1980
Feral horses/tabanids	Island horses retreat to beaches, bays; desert horses move to ridges, snow patches	Keiper and Berger 1982–1983; Rutberg 1987

continued

Table 4.1. (*continued*)

HOST/PARASITE	AVOIDANCE BEHAVIOR	SOURCE
Baboon (*Papio cynocephalus*)/fecally-transmitted pathogens	Do not return to sleeping grove for 9 days, possibly long enough to kill some accumulated pathogens	Hausfater and Meade 1982; Hausfater and Sutherland 1984
Baboon/ticks	Change sleeping trees in response to ticks	Hausfater and Sutherland 1984
Red howler monkey (*Alouatta seniculus*)/fecally-transmitted pathogens	Defecate from perches above forest gaps; do not contaminate food leaves	Gilbert 1997
Various primates/directly transmitted pathogens	Group handing of infants more frequent in small groups—less contagion; in all groups, may have immunizing effect	Freeland 1976; Loehle 1995
Variety of animals/infectious agents	Cannibalism taboo may reduce exposure to pathogens that killed conspecific; infanticide may limit contagion if it is practiced on ailing offspring	Hart 1990; Pfennig et al. 1998

Table 4.2. Some Effects of Parasites on Host Social Behavior

HOST	PARASITE	BEHAVIOR	SOURCE
VIRUSES/BACTERIA/FUNGI			
Poplar sawfly (*Trichiocampus viminalis*)	*Borrelinavirus viminalis*	Cease gregarious behavior	Smirnoff 1965
Willow sawfly (*T. irregularis*)	Virus adapted from *T. viminalis*		
Various hosts	Rabies	Increase in aggression	Steck and Wandeler 1980; Steck 1982; Yuill 1987
Bee (*Apis cerana*)	Apis iridescent virus	Form clusters	Bailey and Ball 1978
African buffalo	Rinderpest	Increase in aggression, herds fragment while seeking water	Plowright 1982, citing Branagan 1966 and Carmichael 1983
Termites (*Reticulitermes* sp.)	Fungus (*Metarhizum anisoplia*)	Attract grooming	Kramm et al. 1982
Badger (*Meles meles*)	*Mycobacterium bovis*	Leave social groups	Cheeseman and Mallinson 1981
PROTISTA			
Honeybee (*Apis mellifera*)	*Nosema apis*	Altered division of labor	Wang and Moeller 1970
Western fence lizard (*Sceloporus occidentalis*)	*Plasmodium mexicanum*	Less time in social interactions; subordinate; less aggressive	Schall and Dearing 1987; Schall and Sarni 1987; Schall and Houle 1992
Mouse	*Toxoplasma gondii*	Increased companion investigation; altered grooming; increased territorial aggression and dominance	Arnott et al. 1990
Humans		Personality changes	Flegr and Hrdý 1994; Flegr and Havlíček 1999

continued

Table 4.2. (continued)

HOST	PARASITE	BEHAVIOR	SOURCE
MONOGENEA/TREMATODA			
Guppie (*Poecilia reticulata*)	Monogenea (*Gyrodactylus bullatarudis*)	Heavily infected fish attract others	Scott 1985
Dogwhelk (*Nucella lapillus*)	Trematode (*Parorchis acanthus*)	Failure to aggregate	Feare 1971
Minnow (*Pimephales promelas*)	*Ornithodiplostomum ptychocheilus*	Schools less compact	Radabaugh 1980b
Banded killifish (*Fundulus diaphanus*)	*Crassiphiala bulboglossa*	Less time in shoals; fewer neighbors; more frequently on periphery; shoal choice influenced	Krause and Godin 1994, 1996
CESTODA			
Common shiner (*Notropis cornutus frontalis*)	*Ligula intestinalis* plerocercoid	Separate from schools	Dence 1958
Rudd (*Scardinius erythrophthalmus*)	*Ligula intestinalis* plerocercoid	Do not form shoals	Orr 1966
Beetle (*Tenebrio molitor*)	*Hymenolepis diminuta*	Loss of response to aggregation pheromone	Hurd and Fogo 1991
Ant (*Leptothorax nylanderi*)	*Anomotaenia brevis*	Some intolerance against infected ants	Trabalon et al. 1994
Three-spined stickleback (*Gasterosteus aculeatus*)	*Schistocephalus solidus*	Reduced competitive ability; did not join shoals	Milinski 1984; Barber et al. 1995

Red jungle fowl (*Gallus gallus*)	Nematode (*Ascaridia galli*)	Reduced social rank (females)	Zuk et al. 1998b
Mouse	*Trichinella spiralis*	Subordinate status; attenuated social attention	Rau 1983b, 1984b; Edwards 1988
	Heligmosomoides polygyrus	Subordinate status	Freeland 1981
	Toxocara canis	Reduced aggression; subordinate status	Cox 1996; Cox and Holland 1998
Cockroach (*Periplaneta americana*)	Acanthocephala (*Moniliformis moniliformis*)	Not attracted to aggregation pheromone	Wilson and Edwards 1986

ECTOPARASITES AND PARASITOIDS

Commercial sand crab (*Portunus pelagicus*)	*Sacculina granifera*	Reduced aggression, decreased dominance (males)	Bishop and Cannon 1979
Three-spined stickleback (*Gasterosteus aculeatus*)	*Argulus canadensis*	Avoid schools in which individuals are parasitized	Dugatkin et al. 1994
		Increase shoal size when parasite present	Poulin and Fitzgerald 1989b
Menhaden (*Brevoortia* sp.)	Isopod (*Olincira praegustator*; debilitating)	May school separately	Guthrie and Kroger 1974
	(*Lernaeenicus radiatus*; less pathogenic)	School with parasitized fish	
Field cricket (*Gryllus* spp.)	Tachinid (*Ormia ochracea*)	Agonistic encounters increase	Adamo et al. 1995
Reindeer	Mosquitoes	Move closer together	Kuhmunen unpublished data, cited in Espmark and Langvatn 1979;
	Tabanids, oestrids	Disperse	
	Insects	Insect density influences group size	Halvorsen 1986b

Table 5.1. Some Examples of Defenses Against Parasites

HOST/PARASITE	DEFENSE	SOURCE
Cricket (*Acheta domesticus*)/*Rickettsiella grylli*	Prefer increased temperature	Adamo 1998a
Japanese beetle (*Popillia japonica*)/nematode (*Heterorhabditis bacteriophora*)	Use of legs and abrasive raster to dislodge and kill nematodes	Gaugler et al. 1994
Ant (*Solenopsis invicta*)/fungus (*Beauveria bassiana*)	Cadavers removed and buried	Pereira and Stimac 1992; Siebeneicher et al. 1992
Ant (*Myrmecia nigriscapa*)/various fungi	Metapleural gland secretions inhibit hyphal growth	Beattie et al. 1985, 1986
Bee (*Apis mellifera*)/mites (*Tropilaelaps clareae*)	Parasitized larvae removed	Burgett et al. 1983
Bee (*Apis mellifera*)/mites (*Varroa jacobsoni*)	Infected brood removed	Boecking and Drescher 1992
Rats/microparasites	Bacteriacidal saliva	Petrides et al. 1984; Dagogo-Jack et al. 1985
Dog/*Escherichia coli, Streptococcus canis*	Bacteriacidal saliva	Hart and Powell 1990
Gastropods/schistosome miracidia	Commensal oligochaete *Chaetogaster limnaei* ingests approaching miracidia	Michelson 1964; Sankurathri and Holmes 1976
Birds/lice	Birds with rich louse communities groom more	Cotgreave and Clayton 1994
Swallows/lice	Nestling preening reflects experimental manipulation of louse abundance	Møller 1991b
Birds/ectoparasites	Sunning behavior may inhibit ectoparasites	Murray 1990
Eastern screech owl (*Otus asio*)/insect larvae	Living blind snakes (*Leptotyphlops dulcis*) delivered to nest, resulted in lower mortality, faster growth	Gehlbach and Baldridge 1987
Great tits (*Parus major*)/hen fleas (*Ceratophyllus gallinae*)	Sacrifice sleep for nest sanitation	Christe et al. 1996b
Impala/ectoparasites	Grooming solicitation directed at inaccessible areas; impala have dental element specialized for grooming	Hart 1990; McKenzie 1990; Hart and Hart 1992

Bibliography

Able, D. J. (1996). The contagion indicator hypothesis for parasite-mediated selection. *Proceedings of the National Academy of Sciences (U.S.A.)* 93, 2229–33.

Adamo, S. A. (1997). How parasites alter the behavior of their insect hosts. In *Parasites and pathogens: effects on host hormones and behavior* (ed. N. E. Beckage), pp. 231–45. Chapman and Hall, New York.

Adamo, S. A. (1998a). The specificity of behavioral fever in the cricket *Acheta domesticus*. *Journal of Parasitology* 84, 529–33.

Adamo, S. A. (1998b). Feeding suppression in the tobacco hornworm, *Manduca sexta*: costs and benefits to the parasitic wasp *Cotesia congregata*. *Canadian Journal of Zoology* 76, 1634–40.

Adamo, S. A. (1999). Evidence for adaptive changes in egg laying in crickets exposed to bacteria and parasites. *Animal Behaviour* 57, 117–24.

Adamo, S. A., Linn, C. E., and Beckage, N. E. (1997). Correlation between changes in host behaviour and octopamine levels in the tobacco hornworm *Manduca sexta* parasitized by the gregarious braconid parasitoid wasp *Cotesia congregata*. *Journal of Experimental Biology* 200, 117–27.

Adamo, S. A., Robert, D., and Hoy, R. R. (1995). Effects of a tachinid parasitoid, *Ormia ochracea*, on the behaviour and reproduction of its male and female field cricket hosts (*Gryllus* spp). *Journal of Insect Physiology* 41, 269–77.

Aeby, G. S. (1991). Behavioral and ecological relationships of a parasite and its hosts within a coral reef system. *Pacific Science* 45, 263–69.

Aeby, G. S. (1992). The potential effect the ability of a coral intermediate host to regenerate has had on the evolution of its association with a marine parasite. *Proceedings of the Seventh International Coral Reef Symposium, Guam* 2, 809–15.

Aeby, G. S. (1998). A digenean metacercaria from the reef coral, *Porites compressa*, experimentally identified as *Podocotyloides stenometra*. *Journal of Parasitology* 84, 1259–61.

Agnew, P. and Koella, J. C. (1999). Constraints on the reproductive value of vertical transmission for a microsporidian parasite and its female-killing behaviour. *Journal of Animal Ecology* 68, 1010–19.

Alados, C. L. and Huffman, M. A. (2000). Fractal long-range correlations in behavioural sequences of wild chimpanzees: a non-invasive analytical tool for the evaluation of health. *Ethology* 106, 105–16.

Alekseev, A. N. (1991). Group and individual behavior of infected and noninfected arthropods—vectors of diseases. Contributions from the Zoological Institute, St. Petersburg, No. 2, 16 pp.

Alexander, R. D. (1974). The evolution of social behavior. *Annual Review of Ecology and Systematics* 5, 325–83.

Alford, C. A., Jr., Stagno, S., and Reynolds, D. W. (1974). Congenital toxoplasmosis: clinical, laboratory, and therapeutic considerations, with special reference to subclinical disease. *Bulletin of the New York Academy of Medicine* 50, 160–81.

Allander, K. (1998). The effects of an ectoparasite on reproductive success in the great tit: a 3-year experimental study. *Canadian Journal of Zoology* 76, 19–25.

Allely, Z., Moore, J., and Gotelli, N. J. (1992). *Moniliformis moniliformis* infection has no effect on some behaviors of the cockroach *Diploptera punctata*. *Journal of Parasitology* 78, 524–26.

Allen, G. E. and Buren, W. F. (1974). Microsporidan and fungal diseases of *Solenopsis invicta* Buren in Brazil. *New York Entomological Society* 82, 125–30.

Allen, G. R. (1998). Diel calling activity and field survival of the bushcricket, *Sciarasaga quadrata* (Orthoptera: Tettigoniidae): a role for sound-locating parasitic flies? *Ethology* 104, 645–60.

Allen, H. W. (1921). Notes on a bombylid parasite and a polyhedral disease of the southern grass worm, *Laphygma frugiperda*. *Journal of Economic Entomology* 14, 510–11.

Alleyne, M. and Beckage, N. E. (1997). Parasitism-induced effects on host growth and metabolic efficiency in tobacco hornworm larvae parasitized by *Cotesia congregata*. *Journal of Insect Physiology* 43, 407–24.

Altizer, S. M. and Oberhauser, K. S. (1999). Effects of the protozoan parasite *Ophryocystis elektroscirrha* on the fitness of the Monarch butterflies (*Danaus plexippus*). *Journal of Invertebrate Pathology* 74, 76–88.

Anderson, J. R. (1975). The behavior of nose bot flies (*Cephenemyia apicata* and C. *jellisoni*) when attacking black-tailed deer (*Odocoileus hemionus columbianus*) and the resulting reactions of the deer. *Canadian Journal of Zoology* 53, 977–92.

Anderson, R. A., J. C. Koella and H. Hurd. (1999). The effect of *Plasmodium yoelii nigeriensis* infection on the feeding persistence of *Anopheles stephensi* Liston throughout the sporogonic cycle. *Proceedings of The Royal Society of London* B266, 1729–733.

Anderson, R. C. (1971). Neurologic disease in reindeer (*Rangifer tarandus tarandus*) introduced into Ontario. *Canadian Journal of Zoology* 49, 159–66.

Anderson, R. M. and May, R. M. (1978). Regulation and stability of host-parasite population interactions. I. Regulatory processes. *Journal of Animal Ecology* 47, 219–47.

Anderson, R. M. and May, R. M. (1979). Population biology of infectious diseases: Part I. *Nature* 280, 361–67.

Anderson, R. M. and May, R. M. (1982). Coevolution of hosts and parasites. *Parasitology* 85, 411–26.

Anderson, R. M. and May, R. M. (1986). The invasion, persistence and spread of infectious diseases within animal and plant communities. *Philosophical Transactions of the Royal Society* B314, 533–70.

Anderson, R. M. and May, R. M. (1992). *Infectious diseases of humans.* Oxford University Press, Oxford.

Anderson, R. M., Whitfield, P. J., and Dobson, A. P. (1978). Experimental studies of infection dynamics: infection of the definitive host by the cercariae of *Transversotrema patialense. Parasitology* 77, 189–200.

Andersson, M. B. (1994). *Sexual selection.* Princeton University Press, Princeton, New Jersey.

Anez, N. and East, J. S. (1984). Studies on *Trypanosoma rangeli* Tejera, 1920. II. Its effect on feeding behaviour of triatomine bugs. *Acta Tropica* 41, 3–5.

Anokhin, I. A. (1966). Daily rhythm in ants infected with metacercariae of *Dicrocoelium lanceatum. Doklady Akademii Nauk SSSR* 166, 757–59.

Apanius, V. and Schad, G. A. (1994). Host behavior and the flow of parasites through host populations. In *Parasitic and infectious diseases: epidemiology and ecology* (ed. M. E. Scott and G. Smith), pp. 115–128. Academic Press, San Diego, CA.

Arme, C. and Owen, R. W. (1967). Infections of the three-spined stickleback, *Gasterosteus aculeatus* L., with the plerocercoid larvae of *Schistocephalus solidus* (Muller, 1776), with special reference to pathological effects. *Parasitology* 57, 301–14.

Arme, C. and Owen, R. W. (1968). Occurrence and pathology of *Ligula intestinalis* infections in British fishes. *Journal of Parasitology* 54, 272–80.

Armstrong, E. (1979). *Nosema whitei*: relationship of body weight gains and food consumption of *Tribolium castaneum* larvae. *Zeitschrift fur Parasitenkunde* 59, 27–29.

Arnal, C., Côté, I. M., Sasal, P., and Morand, S. (2000). Cleaner-client interactions on a Caribbean reef: influence of correlates of parasitism. *Behavioral Ecology and Sociobiology* 47, 353–58.

Arneberg, P., Folstad, I., and Karter, A. J. (1996). Gastrointestinal nematodes depress food intake in naturally infected reindeer. *Parasitology* 112, 213–19.

Arnott, M. A., Cassella, J. P., Aitken, P. P., and Hay, J. (1990). Social interactions of mice with congenital *Toxoplasma* infection. *Annals of Tropical Medicine and Parasitology* 84, 149–56.

Arnqvist, G. and Mäki, M. (1990). Infection rates and pathogenicity of trypanosomatid gut parasites in the water strider *Gerris odonogaster* (Zett.)(Heteroptera:Gerridae). *Oecologia* 84, 194–98.

Ashburn, P. M. (1947). *The ranks of death: a medical history of the conquest of America.* Coward-McCann, New York.

Ashraf, M. and Berryman, A. A. (1970). Biology of *Sulphuretylenchus elongatus* (Nematoda: Sphaerulariidae), and its effect on its host, *Scolytus ventralis* (Coleoptera: Scolytidae). *Canadian Entomologist* 102, 197–213.

Askew, R. R. and Shaw, M. R. (1986). Parasitoid communities: their size, structure and development. In *Insect Parasitoids* (ed. J. Waage, and D. Greathead), pp. 225–64. Academic Press, London.

Atkins, M. D. (1961). A study of the flight of the Douglas-fir beetle *Dendroctonus pseudotsugae* Hopk. (Coleoptera: Scolytidae) III Flight capacity. *Canadian Entomologist* 93, 467–74.

Axén, A. H. and Pierce, N. E. (1998). Aggregation as a cost-reducing strategy of lycaenid larvae. *Behavioral Ecology* 9, 109–15.

Bacot, A. W. (1914). Further notes on the mechanism of the transmission of plague by fleas. *Journal of Hygiene, Plague Supplement* 4, 774–76.

Bacot, A. W. and Martin, C.J. (1914). Observations on the mechanism of the transmission of plague by fleas. *Journal of Hygiene, Plague Supplement* 3, 423–39.

Bailey, W. S., ed. (1982). *Cues that influence behavior of internal parasites.* Agricultural Research Service, U.S. Department of Agriculture, New Orleans, LA.

Bailey, L. and Ball, B. V. (1978). *Apis* iridescent virus and "clustering disease" of *Apis cerana. Journal of Invertebrate Pathology* 31, 368–371.

Baines, D., DeSantis, T., and Downer, R. G. H. (1992). Octopamine and 5–hydroxytryptamine enhance the phagocytic and nodule formation activities of cockroach (*Periplaneta americana*) haemocytes. *Journal of Insect Physiology* 38, 904–14.

Baker, M. (1996). Fur rubbing: use of medicinal plants by capuchin monkeys (*Cebus capucinus*). *American Journal of Primatology* 38, 263–70.

Bakker, T. C. M., Mazzi, D., and Zala, S. (1997). Parasite-induced changes in behavior and color make *Gammarus pulex* more prone to fish predation. *Ecology* 78, 1098–1104.

Ballabeni, P. (1995). Parasite-induced gigantism in a snail: a host adaptation? *Functional Ecology* 9, 887–93.

Ballabeni, P., Benway, H., and Jaenike, J. (1995). Lack of behavioral fever in nematode-parasitized *Drosophila. Journal of Parasitology* 81, 670–74.

Bansemir, A. D. and Sukhdeo, M. V. K. (1996). Villus length influences habitat selection by *Heligmosomoides polygyrus. Parasitology* 113, 311–16.

Barber, I., Huntingford, F. A., and Crompton, D. W. T. (1995). The effect of hunger and cestode parasitism on the shoaling decisions of small freshwater fish. *Journal of Fish Biology* 47, 524–36.

Barber, I. and Ruxton, G. D. (1998). Temporal prey distribution affects the competitive ability of parasitized sticklebacks. *Animal Behaviour* 56, 1477–83.

Barclay, R. M. R. (1988). Variation in the costs, benefits, and frequency of nest reuse by barn swallows (*Hirundo rustica*). *Auk* 105, 53–60.

Barger, M. A. and B. B. Nickol. (1998). Structure of *Leptorhynchoides thecatus* and *Pomphorhynchus bulbocolli* (Acanthocephala) eggs in habitat partitioning and transmission. *Journal of Parasitology* 84, 534–37.

Barnard, C. J. (1990). Parasitic relationships. In *Parasitism and host behaviour* (ed. C. J. Barnard and J. M. Behnke), pp. 1–33. Taylor & Francis, London.

Barnard, D. R. (1989). Habitat use by cattle affects host contact with lone star ticks (Acari: Ixodidae). *Journal of Economic Entomology* 82, 854–59.

Barrett, J. and Butterworth, P. E. (1968). The carotenoids of *Polymorphus minutus* (Acanthocephala) and its intermediate host, *Gammarus pulex. Comparative Biochemistry and Physiology* 27, 575–81.

Barrow, J. H., Jr. (1955). Social behavior in fresh-water fish and its effect on resistance to trypanosomes. *Proceedings of the National Academy of Sciences (U.S.A.)* 41, 676–79.

Bartoli, P. (1976–1978). Modification de la croissance et du comportement de *Venereupis aurea* parasite par *Gymnophallus fossarus* P. Bartoli, 1965 (Trematoda, Digenea). *Haliotis* 7, 23–28.

Bartoli, P., Bourgeay-Causse, M., and Combes, C. (1997). Parasite transmission via a vitamin supplement. *BioScience* 47, 251–53.

Bartoli, P. and Combes, C. (1986). Stratégies de dissémination des cercaires de trematodes dans un écosystème marin littoral. *Acta Oecologica Oecologia Generalis* 7, 101–14.

Barton, R. (1985). Grooming site preferences in primates and their functional implications. *International Journal of Primatology* 6, 519–32.

Baudoin, M. (1975). Host castration as a parasitic strategy. *Evolution* 29, 335–52.

Baylis, M. and Nambiro, C. O. (1993). The effect of cattle infection by *Trypanosoma congolense* on the attraction, and feeding success, of the tsetse fly *Glossina pallidipes*. *Parasitology* 106, 357–61.

Beach, R., Kiilu, G., and Leeuwenburg, J. (1985). Modification of sand fly biting behavior by *Leishmania* leads to increased parasite transmission. *American Journal of Tropical Medicine and Hygiene* 34, 278–82.

Beattie, A. J., Turnbull, C. I., Hough, T., Jobson, S., and Knox, R. B. (1985). The vulnerability of pollen and fungal spores to ant secretions: evidence and some evolutionary implications. *American Journal of Botany* 72, 606–14.

Beattie, A. J., Turnbull, C. I., Hough, T., and Knox, R. B. (1986). Antibiotic production: a possible function for the metapleural glands of ants (Hymenoptera: Formicidae). *Annals of the Entomological Society of America* 79, 448–50.

Beaver, P. C. (1939). The morphology and life history of *Psilostomum ondatrae* Price, 1931 (Trematoda: Psilostomidae). *Journal of Parasitology* 25, 383–93.

Beckage, N. E. (1985). Endocrine interactions between endoparasitic insects and their host. *Annual Review of Entomology* 30, 371–413.

Beckage, N. E. (1991). Host-parasite hormonal relationships: a common theme? *Experimental Parasitology* 72, 332–38.

Beckage, N. E. (1993a). Endocrine and neuroendocrine host-parasite relationships. *Receptor* 3, 233–45.

Beckage, N. E. (1993b). Games parasites play: the dynamic roles of proteins and peptides in the relationship between parasite and host. In *Parasites and pathogens of insects* (eds. N. E. Beckage, S. N. Thompson, and B. A. Federici), vol. 1, pp. 25–57. Academic Press, New York.

Beckage, N. E. (1997). New insights: how parasites and pathogens alter the endocrine physiology and development of insect hosts. In *Parasites and pathogens: effects on host hormones and behavior* (ed. N. E. Beckage), pp. 3–36. Chapman and Hall, New York.

Beckage, N. E. (1998). Parasitoids and polydnaviruses. *BioScience* 48, 305–11.

Beckage, N. E. and Riddiford, L. M. (1978). Developmental interactions between the tobacco hornworm *Manduca sexta* and its braconid parasite *Apanteles congregatus*. *Entomologia Experimentalis et Applicata* 23, 139–51.

Beckage, N. E., Zitnan, D. and Schnal, F. (1994). Endocrine and neuroendocrine mechanisms of arrested host development in parasitized insects: lessons from the tobacco hornworm? In *Insect Neurochemistry and Neurophysiology* (ed. A. B. Borkovec and M. J. Loeb), pp. 123–26. CRC Press, Boca Raton, Florida.

Becker, K. and Kemper, H. (1964). Der Rattenkönig. Eine monographische Studie. Beiheft zur Zeitschrift fur angevandte Zoologie, Heft 2, pp. 1–99. Duncker & Humblot, Berlin.

Bellettato, C. M. (1997). Dinamica ed effetto dell'infestazione del parassita *Pomphorhynchus laevis* (Acanthocephala) sul suo ospite intermedio *Echinogammarus stammeri* (Crustacea: Amphipoda) nel fiume Brenta. M.Sc. Thesis, University of Ferrara, Italy.

Benton, M. J. and Pritchard, G. (1990). Mayfly locomotory responses to endoparasitic infection and predator presence: the effects on predatory encounter rate. *Freshwater Biology* 23, 363–71.

Berdoy, M., Webster, J. P., and MacDonald, D. W. (1995). Parasite-altered behaviour: is the effect of *Toxoplasma gondii* on *Rattus norvegicus* specific? *Parasitology* 111, 403–9.

Berdoy, M., Webster, J. P., and MacDonald, D. W. (2000). Fatal attraction in rats infected with *Toxoplasma gondii*. *Proceedings of the Royal Society of London B*. 267, 1591–594.

Bernard, G. R. (1959). Experimental trichinosis in the golden hamster. I. Spontaneous muscular activity patterns. *American Midland Naturalist* 62, 396–401.

Berry, W. J., Rowley, W. A., Christensen, B. M. (1986). Influence of developing *Brugia pahangi* on spontaneous flight activity of *Aedes aegypti* (Diptera: Culicidae). *Journal of Medical Entomology* 23, 441–45.

Berry, W. J., Rowley, W. A., and Christensen, B. M. (1987a). Influence of developing *Dirofilaria immitis* on the spontaneous flight activity of *Aedes aegypti* (Diptera: Culicidae). *Journal of Medical Entomology* 24, 699–701.

Berry, W. J., Rowley, W. A., and Christensen, B. M. (1988). Spontaneous flight activity of *Aedes trivittatus* infected with *Dirofilaria immitis*. *Journal of Parasitology* 74, 970–74.

Berry, W. J., Rowley, W. A., Clarke, J. L., III, Swack, N. S., and Hausler, W. J., Jr. (1987b). Spontaneous flight activity of *Aedes trivittatus* (Diptera: Culicidae) infected with Trivittatus virus (Bunyaviridae: California serogroup). *Journal of Medical Entomology* 24, 286–89.

Bethel, W. M. and Holmes, J. C. (1973). Altered evasive behavior and responses to light in amphipods harboring acanthocephalan cystacanths. *Journal of Parasitology* 59, 945–56.

Bethel, W. M. and Holmes, J. C. (1974). Correlation of development of altered evasive behavior in *Gammarus lacustris* (Amphipoda) harboring cystacanths of *Polymorphus paradoxus* (Acanthocephala) with the infectivity to the definitive host. *Journal of Parasitology* 60, 272–74.

Bethel, W. M. and Holmes, J. C. (1977). Increased vulnerability of amphipods to predation owing to altered behavior induced by larval acanthocephalans. *Canadian Journal of Zoology* 55, 110–15.

Beuret, J. and Pearson, J. C. (1994). Description of a new zygocercous cercaria (Opisthorchioidea: Heterophyidae) from prosobranch gastropods collected at Heron Island (Great Barrier Reef, Australia) and a review of zygocercariae. *Systematic Parasitology* 27, 105–25.

Billing, J. and Sherman, P. W. (1998). Antimicrobial functions of spices: why some like it hot. *Quarterly Review of Biology* 73, 3–49.

Bishop, R. K. and Cannon, L. R. G. (1979). Morbid behaviour of the commercial sand crab, *Protunus pelagicus* (L.), parasitized by *Sacculina ganifera* Boschma, 1973 (Cirripedia: Rhizocephala). *Journal of Fish Diseases* 2, 131–144.

Bizzell, W. E. and Ciordia, H. (1965). Dissemination of infective larvae of trichostrongylid parasites of ruminants from feces to pasture by the fungus, *Pilobolus* spp. *Journal of Parasitology* 51, 184.

Blanford, S., Thomas, M. B., and Langewald, J. (1998). Behavioural fever in the Senegalese grasshoper, *Oedaleus senegalensis*, and its implications for biological control using pathogens. *Ecological Entomology* 23, 9–14.

Blankespoor, H. D., Babiker, S. M., and Blankespoor, C. L. (1989). Influence of temperature on the development of *Schistosoma haematobium* in *Bulinus truncatus*. *Journal of Medical & Applied Malacology* 1, 123–31.

Blankespoor, C. L., Pappas, P. W., and Eisner, T. (1997). Impairment of the chemical defence of the beetle, *Tenebrio molitor*, by metacestodes (cysticercoids) of the tapeworm, *Hymenolepis diminuta*. *Parasitology* 115, 105–10.

Bloom, B. R. (1979). Games parasites play: how parasites evade immune surveillance. *Nature* 279, 21–26.

Boecking, O. and Drescher, W. (1992). The removal response of *Apis mellifera* L. colonies to brood in wax and plastic cells after artificial and natural infestation with *Varroa jacobsoni* Oud. and to freeze-killed brood. *Experimental & Applied Acarology* 16, 321–29.

Boland, L. M. and Fried, B. (1984). Chemoattraction of normal and *Echinostoma revolutum*-infected *Helisoma trivolvis* to romaine lettuce (*Lactuca sativa longefolia*). *Journal of Parasitology* 70, 436–39.

Bolles, R. C. (1960). Grooming behavior in the rat. *Journal of Comparative and Physiological Psychology* 53, 306–10.

Bomar, C. R., Lockwood, J. A., Pomerinke, M. A. and French, J. D. (1993). Multiyear evaluation of the effects of *Nosema locustae* (Microsporidia: Nosematidae) on rangeland grasshopper (Orthoptera: Acrididae) population density and natural biological controls. *Environmental Entomology* 22, 489–97.

Boorstein, S. M. and Ewald, P. W. (1987). Costs and benefits of behavioral fever in *Melanoplus sanguinipes* infected by *Nosema acridophagus*. *Physiological Zoology* 60, 586–95.

Bordat, D., Coquard, J., and Renand, M. (1984). Quelques moyens de lutte pour enrayer les nosémoses de trois foreurs des graminées élevés en laboratoire sur milieu nutritif artificiel. *L'Agronomie Tropicale* 39, 275–85.

Bouchet, F. (1995). Recovery of helminth eggs from archeological excavations of the Grand Louvre (Paris, France). *Journal of Parasitology* 81, 785–87.

Bouix-Busson, D., Rondelaud, D. and Combes, C. (1985). L'infestation de *Lymnea glabra* Müller par *Fasciola hepatica* L. *Annales de Parasitologie Humaine et Comparee* 60, 11–21.

Boulanger, D., Schneider, D., Sidikou, F., Capron, A., Chippaux, J-P. and Sellin, B. (1999). The oral route as a potential way of transmission of *Schistosoma bovis* in goats. *Journal of Parasitology* 85, 461–67.

Bourne, G. R. (1993). Differential snail-size predation by snail kites and limpkins. *Oikos* 68, 217–23.

Bowen, M. F. (1991). The sensory physiology of host-seeking behavior in mosquitoes. *Annual Review of Entomology* 35, 139–58.

Boyce, N. P. (1979). Effects of *Eubothrium salvelini* (Cestoda: Pseudophyllidea) on the growth and vitality of sockeye salmon, *Oncorhynchus nerka*. *Canadian Journal of Zoology* 57, 597–602.

Brafield, A. E. and Newell, G. E. (1961). The behaviour of *Macoma balthica* (L.). *Journal of the Marine Biological Association of the United Kingdom* 41, 81–87.

Braithwaite, V. A., Salkeld, D. J., McAdam, H. M., Hockings, C. G., Ludlow, A. M. and Read, A. F. (1998). Spatial discrimination learning in rodents infected with the nematode *Strongyloides ratti*. *Parasitology* 117, 145–54.

Brassard, P., Rau, M. E. and Curtis, M. A. (1982). Parasite-induced susceptibility to predation in diplostomiasis. *Parasitology* 85, 495–501.

Brattey, J. (1983). The effects of larval *Acanthocephalus lucii* on the pigmentation, reproduction, and susceptibility to predation of the isopod *Asellus aquaticus*. *Journal of Parasitology* 69, 1172–73.

Braude, S., Tang-Martinez, Z., and Taylor, G. T. (1999). Stress, testosterone, and the immunoredistribution hypothesis. *Behavioral Ecology* 10, 345–50.

Brewer, F. D. and King, E. G. (1978). Effects of parasitism by a tachinid, *Lixophaga diatraenae*, on growth and food consumption of sugarcane borer larvae. *Annals of the Entomological Society of America* 71, 19–22.

Brodeur, J. and McNeil, J. N. (1989). Seasonal microhabitat selection by an endoparasitoid through adaptive modification of host behavior. *Science* 244, 226–28.

Brodeur, J. and McNeil, J. N. (1990). Overwintering microhabitat selection by an endoparasitoid (Hymenoptera: Aphidiidae): induced phototactic and thigmokinetic responses in dying hosts. *Journal of Insect Behavior* 3, 751–63.

Brodeur, J. and McNeil, J. N. (1991). The effect of host plant architecture on the distribution and survival and *Aphidius nigripes* (Hymenoptera: Aphidiidae). *Redia* 74, 251–58.

Brodeur, J. and McNeil, J. N. (1992). Host behaviour modification by the endoparasitoid *Aphidius nigripes*: a strategy to reduce hyperparasitism. *Ecological Entomology* 17, 97–104.

Brodeur, J. and Vet, L. E. M. (1994). Usurpation of host behaviour by a parasitic wasp. *Animal Behaviour* 48, 187–92.

Brønseth, T. and Folstad, I. (1997). The effect of parasites on courtship dance in threespine sticklebacks: more than meets the eye? *Canadian Journal of Zoology* 75, 589–94.

Bronstein, S. M. and Conner, W. E. (1984). Endotoxin-induced behavioural fever in the Madagascar cockroach, *Gromphadorhina portentosa*. *Journal of Insect Physiology* 30, 327–30.

Brooke, M. de L. (1985). The effect of allopreening on tick burdens of molting eudyptid penguins. *Auk* 102, 893–95.

Brooks, D. R. and McLennan, D. A. (1991). *Phylogeny, ecology, and behavior: a research program in comparative biology*. University of Chicago Press, Chicago.

Brooks, D. R. and McLennan, D. A. (1993). *Parascript: parasites and the language of evolution*. Smithsonian Institution Press, Washington, DC.

Brown, A. F. and Pascoe, D. (1989). Parasitism and host sensitivity to cadmium: an acanthocephalan infection of the freshwater amphipod *Gammarus pulex*. *Journal of Applied Ecology* 26, 473–87.

Brown, A. F. and Thompson, D. B. A. (1986). Parasite manipulation of host behaviour: acanthocephalans and shrimps in the laboratory. *Journal of Biological Education* 20, 121–27.

Brown, C. R. and Brown, M. B. (1986). Ectoparasitism as a cost of coloniality in cliff swallows (*Hirundo pyrrhonota*). *Ecology* 67, 1206–18.

Brown, C. R. and Brown, M. B. (1991). Selection of high-quality host nests by parasitic cliff swallows. *Animal Behaviour* 41, 457–65.

Brown, C. R. and Brown, M. B. (1992). Ectoparasitism as a cause of natal dispersal in cliff swallows. *Ecology* 73, 1718–23.

Brown, C. R. and Brown, M. B. (1996). *Coloniality in the cliff swallow*. The University of Chicago Press, Chicago.

Brown, L. (1965). Botfly parasitism in the brush mouse and white-footed mouse in the Ozarks. *The Journal of Parasitology* 51, 302–304.

Brown, P. (1997). Over and over and over . . . *New Scientist* (2 Aug), 27–31.

Brusca, R. C. and Gilligan, M. R. (1983). Tongue replacement in a marine fish (*Lutjanus guttatus*) by a parasitic isopod (Crustacea: Isopoda). *Copeia*, 813–16.

Bull, C. M. and Burzacott, D. (1993). The impact of tick load on the fitness of their lizard hosts. *Oecologia* 96, 415–419.

Bull, J. J., Molineux, I. J., and Rice, W. R. (1991). Selection of benevolence in a host-parasite system. *Evolution* 45, 875–882.

Bundy, D. A. P. (1988). Gender-dependent patterns of infection and disease. *Parasitology Today* 4, 186–189.

Bundy, D. A. P. and Blumenthal, U. J. (1990). Human behaviour and the epidemiology of helminth infections: the role of behaviour in exposure to infection. In *Parasitism and host behaviour* (ed. C. J. Barnard and J. M. Behnke), pp. 264–289. Taylor & Francis, London.

Bundy, D. A. P. and Medley, G. F. (1992). Immuno-epidemiology of human geohelminthiasis: ecological and immunological determinants of worm burden. *Parasitology* 104, S105–19.

Bünzli, G. H. and Büttiker, W. W. (1959). Fungous diseases of lamellicorn larvae in southern Rhodesia. *Bulletin of Entomological Research* 50, 89–96.

Burgett, D. M., P. A. Rossignol and C. Kitprasert. (1990). A model of dispersion and regulation of brood mite (*Tropilaelaps clareae*) parasitism on the giant honeybee (*Apis dorsata*). *Canadian Journal of Zoology* 68, 1423–1427.

Burgett, M., Akratanakul, P., and Morse, R. A. (1983). *Tropilaelaps clareae*: a parasite of honeybees in south-east Asia. *Bee World* 64, 25–28.

Burkot, T. R. (1988). Non-random host selection by anopheline mosquitoes. *Parasitology Today* 4, 156–62.

Burkot, T. R., Graves, P. M., Paru, R., and Lagog, M.(1988). Mixed blood feeding by the malaria vectors in the *Anopheles punctulatus* complex (Diptera: Culicidae). *Journal of Medical Entomology* 25, 205–13.

Burkot, T. R., Molineaux, L., Graves, P. M., Paru, R., Battistutta, D., Dagoro, H., Barnes, A., Wirtz, R. A., and Garner, P. (1990). The prevalence of naturally acquired multiple infections of *Wuchereria bancrofti* and human malarias in anophelines. *Parasitology* 100, 369–75.

Burkot, T. R., Narara, A., Paru, R., Graves, P. M., and Garner, P. (1989). Human host selection by anophelines: no evidence for preferential selection of malaria or microfilariae-infected individuals in a hyperendemic area. *Parasitology* 98, 337–42.

Burright, R. G., Donovick, P. J., Dolinsky, Z., Hurd, Y., and Cypess, R. (1982). Behavioral changes in mice infected with *Toxocara canis*. *Journal of Toxicology and Environmental Health* 10, 621–26.

Butler, J. A. and Millemann, R. E. (1971). Effect of the "salmon poisoning" trematode, *Nanophyetus salmincola*, on the swimming ability of juvenile salmonid fishes. *Journal of Parasitology* 57, 860–65.

Butler, J. M. and Roper, T. J. (1996). Ectoparasites and sett use in European badgers. *Animal Behaviour* 52, 621–29.

Cabanac, M. (1989). Fever in the leech, *Nephelopsis obscura* (Annelida). *Journal of Comparative Physiology B* 159, 281–85.

Cabanac, M. (1990). Phylogeny of fever. In *Thermoreception and temperature regulation* (ed. J. Bligh and K. Voigt), pp. 284–96. Springer-Verlag, Berlin.

Cabanac, M. and Rossetti, Y. (1987). Fever in snails, reflection on a negative result. *Comparative Biochemistry and Physiology* 87A, 1017–20.

Cabaret, J. (1984). Influence de l'infestation par les protostrongylides sur l'activité géotaxique des mollusques terrestres. *Annales de Parasitologie Humaine et Comparée* 59, 529–30.

Cade, W. H. (1975). Acoustically orienting parasitoids: fly phonotaxis to cricket song. *Science* 190, 1312–13.

Cade, W. H. (1984). Effects of fly parasitoids on nightly calling duration in field crickets. *Canadian Journal of Zoology* 62, 226–28.

Cade, W. H. (1991). Inter- and intraspecific variation in nightly calling duration in field crickets, *Gryllus integer* and *G. rubens* (Orthoptera: Gryllidae). *Journal of Insect Behavior* 4, 185–94.

Cade, W. H. and Cade, E. S. (1992). Male mating success, calling and searching behaviour at high and low densities in the field cricket, *Gryllus integer*. *Animal Behaviour* 43, 49–56.

Cade, W. H. and Wyatt, D. R. (1984). Factors affecting calling behaviour in field crickets, *Teleogryllus* and *Gryllus* (age, weight, density, and parasites). *Behaviour* 88, 61–75.

Camp, J. W. and Huizinga, H. W. (1979). Altered color, behavior and predation susceptibility of the isopod *Asellus intermedius* infected with *Acanthocephalus dirus*. *Journal of Parasitology* 65, 667–69.

Campbell, J. F. and Kaya, H. K. (1999) How and why a parasitic nematode jumps. *Nature* 397, 485–86.

Campbell, W. C. (1977). Can alcoholic beverages provide protection against trichinosis? *Proceedings of the Helminthological Society of Washington* 44, 120–25.

Carmichael, L. M. and Moore, J. (1991). A comparison of behavioral alterations in the brown cockroach, *Periplaneta americana*, infected with the acanthocephalan, *Moniliformis moniliformis*. *Journal of Parasitology* 77, 931–36.

Carmichael, L. M., Moore, J., and Bjostad, L. B. (1993). Parasitism and decreased response to sex pheromones in male *Periplaneta americana* (Dictyoptera: Blattidae). *Journal of Insect Behavior* 6, 25–32.

Carner, G. R. (1980). *Entomophthora lampyridarum*, a fungal pathogen of the soldier beetle, *Chauliognathus pennsylvanicus*. *Journal of Invertebrate Pathology* 36, 394–98.

Carney, W. P. (1969). Behavioral and morphological changes in carpenter ants harboring dicrocoeliid metacercariae. *American Midland Naturalist* 82, 605–11.

Carruthers, R. I., Larkin, T. S., Firstencel, H. (1992). Influence of thermal ecology on the mycosis of a rangeland grasshopper. *Ecology* 73, 190–204.

Cattley, J. G. (1948). Sex reversal in copepods. *Nature* 161, 937.

Cavanaugh, D. C. (1971). Specific effect of temperature upon transmission of the plague bacillus by the oriental rat flea, *Xenopsylla cheopis*. *American Journal of Tropical Medicine and Hygiene* 20, 264–73.

Chau, A. and Mackauer, M. (1997). Dropping of pea aphids from feeding site: a consequence of parasitism by the wasp, *Monoctonus paulensis*. *Entomologia Experimentalis et Applicata* 83, 247–52.

Cheeseman, C. L. and Mallinson, P. J. (1981). Behaviour of badgers (*Meles meles*) infected with bovine tuberculosis. *Journal of Zoology* 194, 284–87.

Cheng, T. C. (1973). *General Parasitology*. Academic Press, New York.

Cheng, T. C. (1986). *General parasitology*. (2nd ed.). Academic Press, New York.

Chernin, E. (1967). Behavior of *Biomphalaria glabrata* and of other snails in a thermal gradient. *Journal of Parasitology* 53, 1233–40.

Chigusa, Y. and Otieno, L. H. (1988). Longevity and feeding behaviour of *Glossina morsitans morsitans* infected with *Trypanosoma brucei brucei*. *Japanese Journal of Sanitary Zoology* 39, 71–75.

Christe, P., Oppliger, A., and Richner, H. (1994). Ectoparasite affects choice and use of roost sites in the great tit, *Parus major*. *Animal Behaviour* 47, 895–98.

Christe, P., Richner, H., and Oppliger, A. (1996a). Begging, food provisioning, and nestling competition in great tit broods infested with ectoparasites. *Behavioral Ecology* 7, 127–31.

Christe, P., Richner, H., and Oppliger, A. (1996b). Of great tits and fleas: sleep baby sleep. . . *Animal Behaviour* 52, 1087–92.

Christie, M. G. (1963). The disintegration of sheep dung and the pre-parasitic stages of trichostrongylids. *Journal of Comparative Pathology* 73, 416–23.

Clark, C. C., Clark, L., and Clark, L. (1990). "Anting" behavior by common grackles and European starlings. *Wilson Bulletin* 102, 167–69.

Clark, L. (1990). Starlings as herbalists: countering parasites and pathogens. *Parasitology Today* 6, 358–60.

Clark, L. (1991). The nest protection hypothesis: the adaptive use of plant secondary compounds by European starlings. In *Bird–parasite interactions: ecology, evolution and behaviour* J. E. Loye and M. Zuk (ed.), pp. 205–21. Oxford University Press, Oxford.

Clark, L. and Mason, J. R. (1985). Use of nest material as insecticidal and anti-pathogenic agents by the European starling. *Oecologia* 67, 169–76.

Clark, L. and Mason, J. R. (1987). Olfactory discrimination of plant volatiles by the European starling. *Animal Behaviour* 35, 227–35.

Clark, L. and Mason, J. R. (1988). Effect of biologically active plants used as nest material and the derived benefit to starling nestlings. *Oecologia* 77, 174–80.

Clay, K. (1991). Parasitic castration of plants by fungi. *Trends in Ecology and Evolution* 6, 162–66.

Clayton, D. H. (1990). Mate choice in experimentally parasitized rock doves: lousy males lose. *American Zoologist* 30, 251–62.

Clayton, D. H. (1991). The influence of parasites on host sexual selection. *Parasitology Today* 7, 329–34.

Clayton, D. H. and Tompkins, D. M. (1994). Ectoparasite virulence is linked to mode of transmission. *Proceedings of the Royal Society of London* B 256, 211–17.

Clayton, D. H. and Tompkins, D. M. (1995). Comparative effects of mites and lice on the reproductive success of rock doves (*Columba livia*). *Parasitology* 110, 195–206.

Clayton, D. H. and Vernon, J. G. (1993). Common grackle anting with lime fruit and its effect on ectoparasites. *Auk* 110, 95–102.

Clayton, D. H., Pruett-Jones, S. G., and Lande, R. (1992). Reappraisal of the inter-specific prediction of parasite-mediated sexual selection: opportunity knocks. *Journal of Theoretical Biology* 157, 95–108.

Clayton, D.H. and Wolfe, N. D. (1993). The adaptive significance of self-medication. *Trends in Evolution and Ecology* 6, 60–63.

Colbo, M. H. and Porter, G. N. (1980). Distribution and specificity of Mermithidae (Nematoda) infecting Simuliidae (Diptera) in Newfoundland. *Canadian Journal of Zoology* 58, 1483–90.

Coleman, F. C. (1993). Morphological and physiological consequences of parasites encysted in the bulbus arteriosus of an estuarine fish, the sheepshead minnow, *Cyprinodon variegatus*. *Journal of Parasitology* 79, 247–54.

Coleman, R. E. and Edman, J. D. (1987). The influence of host behavior on sandfly (*Lutzomyia longipalpis*) feeding success on laboraory mice. *Bulletin of the Society for Vector Ecology* 12, 539–40.

Coleman, R. E. and Edman, J. D. (1988). Feeding-site selection of *Lutzomyia longipalpis* (Diptera: Psychodidae) on mice infected with *Leishmania mexicana amazonensis*. *Journal of Medical Entomology* 25, 229–33.

Coleman, R. E., Edman, J. D., and Semprevivo, L. H. (1988). Interactions between malaria (*Plasmodium yoelii*) and Leishmaniasis (*Leishmania mexicana amazonensis*): effect of concomitant infection on host activity, host body temperature, and vector engorgement success. *Journal of Medical Entomology* 25, 467–71.

Coltman, D. W., Pilkington, J. G., Smith, J. A. and Pemberton, J. M. (1999). Parasite-mediated selection against inbred Soay sheep in a free-living, island population. *Evolution* 53, 1259–67.

Collins, W. B. and Urness, P. J. (1982). Mule deer and elk responses to horsefly attacks. *Northwest Science* 56, 299–302.

Colwell, D. D. and Kavaliers, M. (1990). Exposure to mosquitoes, *Aedes togoi* (Theo.), induces and augments opioid-mediated analgesia in mice. *Physiology and Behavior* 48, 397–401.

Colwell, D. D. and Kavaliers, M. (1992). Evidence for activation of endogenous opioid systems in mice following short exposure to stable flies. *Medical and Veterinary Entomology* 6, 159–64.

Colwell, D. D. and Kavaliers, M. (1993). Evidence for involvement of endogenous opioid peptides in altered nociceptive responses of mice infected with *Eimeria vermiformis*. *Journal of Parasitology* 79, 751–56.

Combes, C. (1990). Where do human schistosomes come from? An evolutionary approach. *Trends in Ecology and Evolution* 5, 334–37.

Combes, C. (1991). Ethological aspects of parasite transmission. *American Naturalist* 138, 866–80.

Combes, C. (1997). Fitness of parasites: pathology and selection. *International Journal for Parasitology* 27, 1–10.

Combes, C. (1998). *Interactions Durables*, University of Chicago Press, Chicago.

Combes, C., Fournier, A., Moné, H., and Théron, A. (1994). Behaviours in trematode cercariae that enhance parasite transmission: patterns and processes. *Parasitology* 109, S3–13.

Conn, D. B. and Etges, F. J. (1983). Maternal transmission of asexual proliferative *Mesocestoides corti* tetrathyridia (Cestoda) in mice. *Journal of Parasitology* 69, 922–25.

Connors, V. A. and Nickol, B. B. (1991). Effects of *Plagiorhynchus cylindraceus* (Acanthocephala) on the energy metabolism of adult starling *Sturnus vulgaris*. *Parasitology* 103, 395–402.

Cooper, L. A., Larson, S. E., and Lewis, F. A. (1996). Male reproductive success of *Schistosoma mansoni*-infected *Biomphalaria glabrata* snails. *Journal of Parasitology* 82, 428–31.

Cooper, L. A., Richards, C. S., Lewis, F. A., and Minchella, D. J. (1994). *Schistosoma mansoni*: relationship between low fecundity and reduced susceptibility to parasite infection in the snail *Biomphalaria glabrata*. *Experimental Parasitology* 79, 21–28.

Corson, J. F. (1932). The results of successive bites of an infected tsetse fly. *Journal of Tropical Medicine and Hygiene* 35, 136–37.

Côté, I. M. and Gross, M. R. (1993). Reduced disease in offspring: a benefit of coloniality in sunfish. *Behavioral Ecology and Sociobiology* 33, 269–74.

Côté, I. M. and Poulin, R. (1995). Parasitism and group size in social animals: a meta-analysis. *Behavioral Ecology* 6, 159–65.

Cotgreave, P. and Clayton, D. H. (1994). Comparative analysis of time spent grooming by birds, in relation to parasite load and other factors. *Behaviour* 131, 171–87.

Cowen, R. (1990). Medicine on the wild side. *Science News* 138, 280–82.

Cox, D. (1996). The effect of *Toxocara canis* infection on the paratenic host behaviour of two strains of mice. Ph.D. diss., Trinity College, Dublin.

Cox, D. M. and Holland, C. V. (1998). The relationship between numbers of larvae re-
covered from the brain of *Toxocara canis*-infected mice and social behaviour and
anxiety in the host. *Parasitology* 116, 579–94.

Cram, E. R. (1931). Developmental stages of some nematodes of the Spiruroidea parasitic in poultry and game birds. USDA Technical Bulletin no. 227, U.S. Department of Agriculture, Beltsville, MD.

Crisler, L. (1956). Observations of wolves hunting caribou. *Journal of Mammalogy* 37,
337–46.

Crofton, H. D. (1958). Nematode parasite populations in sheep on lowland farms. VI.
Sheep behaviour and nematode infections. *Parasitology* 48, 251–60.

Crompton, D. W. T. (1984). Influence of parasitic infection on food intake. *Federation
Proceedings* 43, 239–45.

Crompton, D. W. T. (1999). How much human helminthiasis is there in the world?
Journal of Parasitology 85, 397–403.

Crompton, D. W. T., Walter, D. E., and Arnold, S. (1981). Changes in the food intake
and body weight of protein-malnourished rats infected with *Nippostrongylus
brasiliensis* (Nematoda). *Parasitology* 82, 23–38.

Crosland, M. W. J. (1988). Effect of a gregarine parasite on the color of *Myrmecia pi-
losula* (Hymenoptera: Formicidae). *Annals of the Entomological Society of America* 81, 481–84.

Courtney, C. C., Christensen, B. M., and Goodman, W. G. (1985). Effects of *Dirofilaria immitis* on blood meal size and fecundity in *Aedes aegypti* (Diptera: Culicidae). *Journal of Medical Entomology* 22, 398–400.

Crowden, A. E. and Broom, D. M. (1980). Effects of the eyefluke, *Diplostomum
spathaceum*, on the behaviour of dace (*Leuciscus leuciscus*). *Animal Behaviour* 28,
287–94.

Cully, J. F., Jr., Grieco, J. P., and Kissel, D. (1991). Defensive behavior of eastern chipmunks against *Aedes triseriatus* (Diptera: Culicidae). *Journal of Medical Entomology* 28, 410–16.

Curtis, L. A. (1985). The influence of sex and trematode parasites on carrion response
of the estuarine snail *Ilyanassa obsoleta*. *Biological Bulletin* 169, 377–90.

Curtis, L. A. (1987). Vertical distribution of an estuarine snail altered by a parasite.
Science 235, 1509–11.

Curtis, L. A. (1990). Parasitism and the movements of intertidal gastropod individuals. *Biological Bulletin* 179, 105–12.

Curtis, L. A. (1993). Parasite transmission in the intertidal zone: vertical migrations,
infective stages, and snail trails. *Journal of Experimental Marine Biology and Ecology* 173, 197–209.

Curtis, L. A. and Hurd, L. E. (1981). Nutrient procurement strategy of a deposit-feed-
ing estuarine neogastropod, *Ilyanassa obsoleta*. *Estuarine, Coastal and Shelf Science* 13, 277–85.

Curtis, L. A. and Hurd, L. E. (1983). Age, sex, and parasites: spatial heterogeneity in
a sandflat population of *Ilyanassa obsoleta*. *Ecology* 64, 819–28.

Dagogo-Jack, S., Atkinson, S., and Kendall-Taylor, P. (1985). Homologous radioim-
munoassay for epidermal growth factor in human saliva. *Journal of Immunoassay*
6, 125–36.

Dailey, M. D. and Walker, W. A. (1978). Parasitism as a factor (?) in single strandings
of southern California cetaceans. *Journal of Parasitology* 64, 593–96.

Damian, R. T. (1987). The exploitation of host immune responses by parasites. *Journal of Parasitology* 73, 3–13.

Daniel, T. L. and Kingsolver, J. G. (1983). Feeding strategy and the mechanics of blood
sucking in insects. *Journal of Theoretical Biology* 105, 661–77.

Daniels, C. B. (1985). The effect of infection by a parasitic worm on swimming and
diving in the water skink, *Sphenomorphus quoyii*. *Journal of Herpetology* 19,
160–62.

Davies, C. R. (1990). Reply [to Killick-Kendrick and Molyneux, 1990]. *Parasitology
Today* 6, 189.

Davies, C. R., Ayres, J. M., Dye, C., and Deane, L. M. (1991). Malaria infection rate
of Amazonian primates increases with body weight and group size. *Functional Ecology* 5, 655–62.

Dawkins, R. (1982). *The extended phenotype*. W. H. Freeman, Oxford.

Dawkins, R. (1990). Parasites, desiderata lists and the paradox of the organism. *Parasitology* 100, S63–S73.

Day, J. F., Ebert, K. M., and Edman, J. D. (1983). Feeding patterns of mosquitoes
(Diptera: Culicidae) simultaneously exposed to malarious and healthy mice, in-
cluding a method for separating blood meals from conspecific hosts. *Journal of
Medical Entomology* 20, 120–27.

Day, J. F. and Edman, J. D. (1983). Malaria renders mice susceptible to mosquito feeding when gametocytes are most infective. *Journal of Medical Entomology* 69,
163–70.

Day, J. F. and Edman, J. D. (1984a). Mosquito engorgement on normally defensive
hosts depends on host activity patterns. *Journal of Medical Entomology* 21, 732–40.

Day, J. F. and Edman, J. D. (1984b). The importance of disease induced changes in
mammalian body temperature to mosquito blood feeding. *Comparative Biochemistry and Physiology* 77A, 447–52.

de Garine-Wichatitsky, M., De Meeûs, T., Guégan, J.-F. and Renaud, F. (1999). Spatial and temporal distributions of parasites: can wild and domestic ungulates avoid
African tick larvae? *Parasitology* 119, 455–66.

DeGiusti, D. L. (1949). The life cycle of *Leptorhynchoides thecatus* (Linton), an acanthocephalan of fish. *Journal of Parasitology* 35, 437–60.

de Jong-Brink, M. (1990). How trematode parasites interfere with reproduction of their
intermediate hosts, freshwater snails. *Journal of Medical and Applied Malacology*
2, 101–33.

de Jong-Brink, M. (1995). How schistosomes profit from the stress responses they
elicit in their hosts. *Advances in Parasitology* 35, 178–256.

de Jong-Brink, M., Hoek, R. M., Lageweg, W., and Smit, A. B. (1997). Schistosome
parasites induce physiological changes in their snail host by interfering with two
regulatory systems, the internal defense system and the neuroendocrine system. In
Parasites and pathogens: effects on host hormones and behavior (ed. N. E. Beckage), pp. 57–75. Chapman and Hall, New York.

de Jong-Brink, M., Hordijk, P. L., Schallig, H. D. F. H., Bergamin-sassen, M. J. M.,
and Oosthoek, P. (1990). Possible mechanisms underlying parasitic castration in
trematode infected snails. *Advances in Invertebrate Reproduction* 5, 141–48.

De Jong-Brink, M., Reid, C. N., Tensen, C. P and Ter Maat, A. (1999). Parasites flicking the NPY gene on the host's switchboard: why NPY? *The FASEB Journal* 13, 1972–84.

DeMont, D. J. and Corkum, K. C. (1982). The life cycle of *Octospiniferoides chandleri* Bullock, 1957 (Acanthocephala: Neoechinorhynchidae) with some observations on parasite-induced, photophilic behavior in ostracods. *Journal of Parasitology* 68, 125–30.

Dence, W. A. (1958). Studies on *Ligula*-infected common shiners (*Notropis cornutus frontalis* Agassiz) in the Adirondacks. *Journal of Parasitology* 44, 334–58.

Desowitz, R. S. (1981). *New Guinea tapeworms and Jewish grandmothers, tales of parasites and people*. Avon Books, New York.

Desowitz, R. S. (1997). *Who gave Pinta to the Santa Maria?* W. W. Norton, New York.

DeVries, P. J. (1984). Butterflies and Tachinidae: does the parasite always kill its host? *Journal of Natural History* 18, 323–26.

Dezfuli, B. S., Rossetti, E., and Fano, E. A. (1994). Occurrence of larval *Acanthocephalus anguillae* (Acanthocephala) in *Asellus aquaticus* (Crustacea, Isopoda) from the River Brenta. *Bollettino di Zoologica* 61, 77–81.

Diamond, J. M. (1997). *Guns, germs, and steel: the fates of human societies*. W. W. Norton, New York.

Dirie, M. F., Ashford, R. W., Mungomba, L. M., Molyneux, D. H., and Green, E. E. (1990). Avian trypanosomes in *Simulium* and sparrow hawks (*Accipiter nisus*). *Parasitology* 101, 243–47.

Dixon, B. (1976). *Magnificent microbes*. Atheneum, New York.

Dobson, A. (1992). People and disease. In *The Cambridge encyclopedia of human evolution* (eds. S. Jones, D. Pilbeam and R. D. Martin), pp. 411–20. Cambridge University Press, Cambridge.

Dobson, A. P. (1988). The population biology of parasite-induced changes in host behavior. *Quarterly Review of Biology* 63, 139–65.

Dobson, A. P. and Carper, E. R. (1996). Infectious diseases and human population history. *BioScience* 46, 115–25.

Dobson, A. P. and Hudson, P. J. (1992). Regulation and stability of a free-living host-parasite system: *Trichostrongylus tenuis* in red grouse. II. Population models. *Journal of Animal Ecology* 61, 487–98.

Dobson, A. P. and McCallum, H. (1997). The role of parasites in bird conservation. In *Host-parasite evolution: general principles & avian models* (eds. D. H. Clayton and J. Moore), pp. 155–73. Oxford University Press, Oxford.

Dolinsky, Z. S., Burright, R. G., Donovick, P. J., Glickman, L. T., Babish, J., Summers, B., and Cypess, R. H. (1981). Behavioral effects of lead and *Toxocara canis* in mice. *Science* 213, 1142–44.

Dolinsky, Z. S., Hardy, C. A., Burright, R. G., and Donovick, P. J. (1985). The progression of behavioral and pathological effects of the parasite *Toxocara canis* in the mouse. *Physiology and Behavior* 35, 33–42.

Donovick, P. J., Dolinsky, Z. S., Perdue, V. P., Burright, R. G., Summers, B., and Cypess, R. H. (1981). *Toxocara canis* and lead alter consummatory behavior in mice. *Brain Research Bulletin* 7, 317–23.

Downes, C. M., Theberge, J. B., and Smith, S. M. (1986). The influence of insects on the distribution, microhabitat choice, and behavior of the Burwash caribou herd. *Canadian Journal of Zoology* 64, 622–29.

Draski, L. J., Summers, B., Cypess, R. H., Burright, R. G., and Donovick, P. J. (1987). The impact of single versus repeated exposure of mice to *Toxocara canis*. *Physiology and Behavior* 40, 301–6.

Dronen, N. O., Jr. (1973). Studies on the macrocercous cercariae of the Douglas Lake, Michigan area. *Transactions of the American Microscopical Society* 92, 641–48.

Dudley, R. and Milton, K. (1990). Parasite deterrence and the energetic costs of slapping in howler monkeys, *Alouatta palliata*. *Journal of Mammalogy* 71, 463–65.

Duffy, D. C. (1983). The ecology of tick parasitism on densely nesting Peruvian seabirds. *Ecology* 64, 110–19.

Dugatkin, L. A., FitzGerald, G. J., and Lavoie, J. (1994). Juvenile three-spined sticklebacks avoid parasitized conspecifics. *Environmental Biology of Fishes* 39, 215–18.

Duncan, P. and Cowtan, P. (1980). An unusual choice of habitat helps Camargue horses to avoid blood-sucking horse-flies. *Biology of Behaviour* 5, 55–60.

Duncan, P. and Vigne, N. (1979). The effect of group size in horses on the rate of attacks by blood-sucking flies. *Animal Behaviour* 27, 623–25.

Dunn, A. M., Adams, J., and Smith, J. E. (1993). Transovarial transmission and sex ratio distortion by a microsporidian parasite in a shrimp. *Journal of Invertebrate Pathology* 61, 248–52.

Dunn, A. M., Hatcher, M. J., Terry, R. S., and Tofts, C. (1995). Evolutionary ecology of vertically transmitted parasites: transovarial transmission of a microsporidian sex ratio distorter in *Gammarus duebeni*. *Parasitology* 111, S91–109.

Durden, L. A. (1987). Predator-prey interactions between ectoparasites. *Parasitology Today* 3, 306–9.

Durette-Desset, M.-C., Chabaud, A. G., and Moore, J. (1993). *Trichostrongylus cramae* n. sp. (Nematoda), a parasite of bob-white quail (*Colinus virginianus*). *Annales de Parasitologie Humaine et Comparee* 68, 43–48.

Durie, P. H. (1961). Parasitic gastro-enteritis of cattle: the distribution and survival of infective strongyle larvae on pasture. *Australian Journal of Agricultural Research* 12, 1200–11.

Durrer, S. and Schmid-Hempel, P. (1994). Shared use of flowers leads to horizontal pathogen transmission. *Proceedings of the Royal Society of London* B 258, 299–302.

Dwyer, G. (1991). The roles of density, stage, and patchiness in the transmission of an insect virus. *Ecology* 72, 559–74.

Dybdahl, M. F. and Lively, C. M. (1996). The geography of coevolution: comparative population structures for a snail and its trematode parasite. *Evolution* 50, 2264–75.

Dye, C. and Hasibeder, G. (1986). Population dynamics of mosquito-borne disease: effects of flies which bite some people more frequently than others. *Transactions of the Royal Society of Tropical Medicine and Hygiene* 80, 69–77.

Dye, C. and Williams, B. G. (1997). Multigenic drug resistance among inbred malaria parasites. *Proceedings of the Royal Society of London* B 264, 61–67.

Ebert, D. (1995). The ecological interactions between a microsporidian parasite and it host *Daphnia magna*. *Journal of Animal Ecology* 64, 361–69.

Ebert, D. and Hamilton, W. D. (1996). Sex against virulence: the coevolution of parasitic diseases. *Trends in Ecology and Evolution* 11, 79–82.

Edland, T. (1965). A granulosis of *Eupsilia transversa* Hufn. (Lepidoptera, Noctuidae) in west Norway. *Entomophaga* 10, 331–33.

Edman, J. D. (1988). Disease control through manipulation of vector-host interaction: some historical and evolutionary perspectives. In *The role of vector-host interactions in disease transmission* (eds. T. W. Scott and J. Grumstrup-Scott), pp. 43–50. Miscellaneous Publication no. 68, Entomological Society of America, College Park, Maryland.

Edman, J. D., Day, J. F., and Walker, E. D. (1984). Field confirmation of laboratory observations on the differential antimosquito behavior of herons. *Condor* 86, 91–92.

Edman, J. D., Day, J. F., and Walker, E. D. (1985). Vector-host interplay—Factors affecting disease transmission. In *Ecology of mosquitoes* (eds. L. P. Lounibos, J. R. Rey, and J. H. Frank), pp. 273–85. Florida Medical Entomology Laboratory, Vero Beach, FL.

Edman, J. D., and Kale, H. W., II. (1971). Host behavior: its influence on the feeding success of mosquitoes. *Annals of the Entomological Society of America* 64, 513–16.

Edman, J. D. and Scott, T. W. (1987). Host defensive behaviour and the feeding success of mosquitoes. *Insect Science and Its Application* 8, 617–22.

Edman, J. D. and Spielman, A. (1988). Blood-feeding by vectors: physiology, ecology, behavior and vertebrate defense. In *The arboviruses: epidemiology and ecology* (ed. T. P. Monath), vol. 1, pp. 153–89. CRC Press, Boca Raton, FL.

Edman, J. D., Webber, L. A., and Kale, H. W., II. (1972). Effect of mosquito density on the interrelationship of host behavior and mosquito feeding success. *American Journal of Tropical Medicine and Hygiene* 21, 487–91.

Edman, J. D., Webber, L. A., and Schmid, A. A. (1974). Effect of host defenses on the feeding pattern of *Culex nigripalpus* when offered a choice of blood sources. *Journal of Parasitology* 60, 874–83.

Edwards, J. C. (1987). Parasite-induced changes in host behaviour. PhD thesis, University of Nottingham.

Edwards, J. C. (1988). The effects of *Trichinella spiralis* infection on social interactions in mixed groups of infected and uninfected male mice. *Animal Behaviour* 36, 529–40.

Edwards, J. C. and Barnard, C. J. (1987). The effects of *Trichinella* infection on intersexual interactions between mice. *Animal Behaviour* 35, 533–40.

Egerter, D. E. and Anderson, J. R. (1989). Blood-feeding inhibition of *Aedes sierrensis* (Diptera: Culicidae) induced by the parasite *Lambornella clarki* (Ciliophora: Tetrahymenidae). *Journal of Medical Entomology* 26, 46–54.

Egerter, D. E., Anderson, J. R., and Washburn, J. O. (1986). Dispersal of the parasitic ciliate *Lambornella clarki*: Implications for ciliates in the biological control of mosquitoes. *Proceedings of the National Academy of Sciences (U.S.A.)* 83, 7335–39.

Eilenberg, J. (1987). Abnormal egg-laying behaviour of female carrot flies (*Psila rosae*) induced by the fungus *Entomophthora muscae*. *Entomologia Experimentalis et Applicata* 43, 61–65.

El-Aal, M. A. and Holdich, D. M. (1987). The occurrence of a rickettsial disease in British woodlice (Crustacea, Isopoda, Oniscidea) populations. *Journal of Invertebrate Pathology* 49, 252–58.

Emlen, J. T. (1986). Responses of breeding cliff swallows to nidicolous parasite infestations. *Condor* 88, 110–11.

Endler, J. A. (1986). *Natural selection in the wild*. Princeton University Press, Princeton, NJ.

English-Loeb, G. M., Brody, A., and Karban, R. (1993). Host-plant-mediated interactions between a generalist folivore and its tachinid parasitoid. *Journal of Animal Ecology* 62, 465–71.

Espmark, Y. and Langvatn, R. (1979). Lying down as a means of reducing fly harassment in red deer (*Cervus elaphus*). *Behavioral Ecology and Sociobiology* 5, 51–54.

Etges, F. J. (1963). Effects of *Schistosoma mansoni* infection on chemosensitivity and orientation of *Australorbis glabratus*. *American Journal of Tropical Medicine and Hygiene* 12, 696–700.

Evans, H. C. (1982). Entomogenous fungi in tropical forest ecosystems: an appraisal. *Ecological Entomology* 7, 47–60.

Evans, H. C. (1989). Mycopathogens of insects of epigeal and aerial habitats. In *Insect-Fungus Interactions* (eds. N. Wilding, N. M. Collins, P. M. Hammond and J. F. Webber), pp. 205–38. Academic Press, New York.

Evans, H. C. and Samson, R. A. (1982). *Cordyceps* species and their anamorphs pathogenic on ants (Formicidae) in tropical forest ecosystems I. The *Cephalotes* (Myrmicinae) complex. *Transactions of the British Mycological Society* 79, 431–53.

Evans, H. F. and Allaway, G. P. (1983). Dynamics of baculovirus growth and dispersal in *Mamestra brassicae* L. (Lepidoptera: Noctuidae) larval populations introduced into small cabbage plots. *Applied and Environmental Microbiology* 45, 493–501.

Evans, W. S., Hardy, M. C., Singh, R., Moodie, G. E., and Cote, J. J. (1992). Effect of the rat tapeworm, *Hymenolepis diminuta*, on the coprophagic activity of its intermediate host, *Tribolium confusum*. *Canadian Journal of Zoology* 70, 2311–14.

Evans, W. S., Wong, A., M. Hardy, Currie, R. W. and Vanderwel, D. (1998). Evidence that the factor used by the tapeworm, *Hymenolepis diminuta*, to direct the foraging of its intermediate host, *Tribolium confusum*, is a volatile attractant. *Journal of Parasitology* 84, 1098–1101.

Ewald, P. W. (1987). Transmission modes and evolution of the parasitism-mutualism continuum. *Annals of the New York Academy of Sciences* 503, 295–306.

Ewald, P. W. (1994). *Evolution of infectious disease.* Oxford University Press, Oxford.

Fauth, P. T., Krementz, D. G., and Hines, J. E. (1991). Ectoparasitism and the role of green nesting material in the European starling. *Oecologia* 88, 22–29.

Feare, C. J. (1971). Predation of limpets and dogwhelks by oystercatchers. *Bird Study* 18, 121–29.

Feener, D. H., Jr. (1981). Competition between ant species: outcome controlled by parasitic flies. *Science* 214, 815–17.

Feener, D. H., Jr. (1988). Effects of parasites on foraging and defense behavior of a termitophagous ant, *Pheidole titanis* Wheeler (Hymenoptera: Formicidae). *Behavioral Ecology and Sociobiology* 22, 421–27.

Feener, D. H., Jr., and Brown, B. V. (1993). Oviposition behavior of an ant-parasitizing fly, *Neodohrniphora curvinervis* (Diptera: Phoridae), and defense behavior by its leaf-cutting ant host *Atta cephalotes* (Hymenoptera: Formicidae). *Journal of Insect Behavior* 6, 675–88.

Feener, D. H., Jr., Jacobs, L. F. and Schmidt, J. O. (1996). Specialized parasitoid attracted to a pheromone of ants. *Animal Behaviour* 51, 61–66.

Feener, D. H., Jr., and Moss, K. A. G. (1990). Defense against parasites by hitchhikers in leaf-cutting ants: a quantitative assessment. *Behavioral Ecology and Sociobiology* 26, 17–29.

Ferdig, M. T., Beerntsen, B. T., Spray, F. J., Li, J., and Christensen, B. M. (1993). Reproductive costs associated with resistance in a mosquito-filarial worm system. *American Journal of Tropical Medicine and Hygiene* 49, 756–62.

Ferkin, M. H., Sorokin, E. S., and Johnston, R. E. (1996). Self-grooming as a sexually dimorphic communicative behaviour in meadow voles, *Microtus pennsylvanicus*. *Animal Behaviour* 51, 801–10.

Fernandez, J. and Esch, G. W. (1991). Effect of parasitism on the growth rate of the pulmonate snail, *Helisoma anceps*. *Journal of Parasitology* 77, 937–44.

Fialho, R. F. and Schall, J. J. (1995). Thermal ecology of a malarial parasite and its insect vector: consequences for the parasite's transmission success. *Journal of Animal Ecology* 64, 553–62.

Flecker, A. S. and Allan, J. D. (1988). Flight direction in some Rocky Mountain mayflies (Ephemeroptera), with observations of parasitism. *Aquatic Insects* 10, 33–42.

Flegr, J. and Havlíček, J. (1999). Changes in the personality profile of young women with latent toxoplasmosis. *Folia Parasitologica* 46, 22–28.

Flegr, J. and Hrdý, I. (1994). Influence of chronic toxoplasmosis on some human personality factors. *Folia parasitologica* 41, 122–26.

Flegr, J., Zitková, S., Kodym, P., and Frynta, D. (1996). Induction of changes in human behaviour by the parasitic protozoan *Toxoplasma gondii*. *Parasitology* 113, 49–54.

Folgarait, P. J. and Gilbert, L. E. (1999). Phorid parasitoids affect foraging activity of *Solenopsis richteri* under different availability of food in Argentina. *Ecological Entomology* 24, 163–73.

Folstad, I. and Karter, A. J. (1992). Parasites, bright males, and the immunocompetence handicap. *American Naturalist* 139, 603–22.

Folstad, I., Nilssen, A. C., Halvorsen, O., and Andersen, J. (1989). Why do male reindeer (*Rangifer t. tarandus*) have higher abundance of second and third instar larvae of *Hypoderma tarandi* than females? *Oikos* 55, 87–92.

Folstad, I., Nilssen, A. C., Halvorsen, O., and Andersen, J. (1991). Parasite avoidance: the cause of post-calving migrations in *Rangifer*? *Canadian Journal of Zoology* 69, 2423–29.

Forbes, M. R. L. (1993). Parasitism and host reproductive effort. *Oikos* 67, 444–50.

Forbes, M. R. L. and Baker, R. L. (1991). Condition and fecundity of the damselfly, *Enallagma ebrium* (Hagen): the importance of ectoparasites. *Oecologia* 86, 334–41.

Forsse, E. (1987). Flight duration in *Ips typographus* L.: Insensitivity to nematode infection. *Journal of Applied Entomology* 104, 326–28.

Foster, W. A. and Treherne, J. E. (1981). Evidence for the dilution effect in the selfish herd from fish predation on a marine insect. *Nature* 293, 466–67.

Fouad, K., Libersat, F., and Rathmayer, W. (1992). The venom of the sphecid wasp *Ampulex compressa* modulates the behaviour of its prey, the cockroach. In *Rhythmogenesis in neurons and networks* (eds. N. Elsner and D. Richter), p. 581. Proceedings of the 20th Gottingen Neurobiology Conference. Thieme, Stuttgart-New York.

Fouad, K., Libersat, F., and Rathmayer, W. (1996). Neuromodulation of the escape behavior of the cockroach *Periplaneta americana* by the venom of the parasitic wasp *Ampulex compressa*. *Journal of Comparative Physiology* 178, 91–100.

Fowler, M. and Robertson, J. S. (1972). Iridescent virus infection in field populations of *Wiseana cervinata* (Lepidoptera: Hepialidae) and *Witlesia* sp. (Lepidoptera: Pyralidae) in New Zealand. *Journal of Invertebrate Pathology* 19, 154–55.

Frank, W., Lucius, R., and Romig, T. (1984). Studies on the biology, pathology, ecology and epidemiology of *Dicrocoelium hospes* Looss, 1907 in West Africa (Ivory Coast). In *Recent German research on problems of parasitology, animal health and animal breeding in the tropics and subtropics* (eds. H. Markl and A. Bittner), pp. 38–58. Georg Hauser, Metzingen, Germany.

Freehling, M. and Moore, J. (1993). Susceptibility of 13 cockroach species to *Moniliformis moniliformis*. *Journal of Parasitology* 79, 442–44.

Freehling, M. and Moore, J. (1993). Host specificity of *Trichostrongylus tenuis* from red grouse and northern bobwhites in experimental infections of northern bobwhites. *Journal of Parasitology* 79, 538–41.

Freeland, W. J. (1976). Pathogens and the evolution of primate sociality. *Biotropica* 8, 12–24.

Freeland, W. J. (1977). Blood-sucking flies and primate polyspecific associations. *Nature* 269, 801–2.

Freeland, W. J. (1979). Primate social groups as biological islands. *Ecology* 60, 719–28.

Freeland, W. J. (1980). Mangabey (*Cercocebus albigena*) movement patterns in relation to food availability and fecal contamination. *Ecology* 61, 1297–1303.

Freeland, W. J. (1981). Parasitism and behavioral dominance among male mice. *Science* 213, 461–62.

Freeland, W. J. (1983). Parasites and the coexistence of animal host species. *American Naturalist* 121, 223–36.

Freier, J. E. and S. Friedman. (1976). Effect of host infection with *Plasmodium gallinaceum* on the reproductive capacity of *Aedes aegypti*. *Journal of Invertebrate Pathology* 28, 161–6.

Fritz, R. S. (1982). Selection for host modification by insect parasitoids. *Evolution* 36, 283–88.

Fromont, E., Courchamp, F., Artois, M. and Pontier, D. (1997). Infection strategies of retroviruses and social grouping of domestic cats. *Canadian Journal of Zoology* 75, 1994–2002.

Führer, E. (1981). Influence of braconid parasitism on host nutrition. *Proceedings of Symposium: IX International Congress of Plant Protection* 1, 96–99.

Fukase, T., Itagaki, H., Wakuui, H., Kano, S., Goris, R. C., and Kishida, R. (1986). Histopathological findings in snakes, *Elaphe quadrivirgata* (Reptilia: Colubridae), infected with plerocercoids of *Spirometra erinacei* (Cestoda: Diphyllobothriidae). *Japanese Journal of Herpetology* 11, 86–95.

Fukase, T., Itagaki, H., Wakui, S., Kano, Y., Goris, R. C., and Kishida, R. (1987). Parasitism of *Pharyngostomum cordatum* metacercariae (Trematoda: Diplostomatidae) in snakes, *Elaphe quadrivirgata* and *Rhabdophis tigrinus* (Reptilia: Colubridae). *Japanese Journal of Herpetology* 12, 39–44.

Gabrion, C., Plateaux, L., and Quentin, C. (1976). *Anomotaenia brevis* (Clerc, 1902) Fuhrmann, 1908 Cestode Cyclophyllide, parasite de *Leptothorax nylanderi* (Forster) Hymenoptere, Formicide. *Annales de Parasitologie* 51, 407–20.

Gannicott, A. M. and Tinsley, R. C. (1997). Egg hatching in the monogenean gill parasite *Discocotyle sagittata* from the rainbow trout (*Oncorhynchus mykiss*). *Parasitology* 114, 569–79.

Gargan, T. P., II, Bailey, C. L., Higbee, G. A., Gad, A., and El Said, S. (1983). The effect of laboratory colonization on the vector-pathogen interactions of Egyptian *Culex pipiens* and Rift Valley fever virus. *American Journal of Tropical Medicine and Hygiene* 32, 1154–63.

Garnick, E. and Margolis, L. (1990). Influence of four species of helminth parasites on orientation of seaward migrating sockeye salmon (*Oncorhynchus nerka*) smolts. *Canadian Journal of Fisheries and Aquatic Science* 47, 2380–89.

Garrett, L. (1994). *The coming plague: newly emerging diseases in a world out of balance*. Farrar, Straus and Giroux, New York.

Gaufler, R., Wang, Y. and Campbell, J. (1994). Aggressive and evasive behaviors in *Popillia japonica* (Coleoptera: Scarabaeidae) larvae: defenses against entomopathogenic nematode attack. *Journal of Invertebrate Pathology* 64, 193–99.

Gehlbach, F. R. and Baldridge, R. S. (1987). Live blind snakes (*Leptotyphlops dulcis*) in eastern screech owl (*Otus asio*) nests: a novel commensalism. *Oecologia* 71, 560–63.

Gemmill, A. W., Viney, M. E. and Read, A. F. (1997). Host immune status determines sexuality in a parasitic nematode. *Evolution* 51, 393–401.

Geraci, J. R. and St. Aubin, D. J. (1987). Effects of parasites on marine mammals. *International Journal for Parasitology* 17, 407–14.

Gilbert, K. A. (1997). Red howling monkey use of specific defecation sites as a parasite avoidance strategy. *Animal Behaviour* 54, 451–55.

Giles, N. (1983). Behavioural effects of the parasite *Schistocephalus solidus* (Cestoda) on an intermediate host, the three-spined stickleback, *Gasterosteus aculeatus* L. *Animal Behavior* 31, 1192–94.

Giles, N. (1987a). Predation risk and reduced foraging activity in fish: experiments with parasitized and non-parasitized three-spined sticklebacks, *Gasterosteus aculeatus* L. *Journal of Fish Biology* 31, 37–44.

Giles, N. (1987b). A comparison of the behavioural responses of parasitized and non-parasitized three-spined sticklebacks, *Gasterosteus aculeatus* L., to progressive hypoxia. *Journal of Fish Biology* 30, 631–38.

Ginsberg, H. S. and Ewing, C. P. (1989). Habitat distribution of *Ixodes dammini* (Acari: Ixodidae) and Lyme disease spirochetes on Fire Island, New York. *Journal of Medical Entomology* 26, 183–89.

Godfray, H. C. J. (1994). *Parasitoids: behavioural and evolutionary ecology*. Princeton University Press, Princeton, NJ.

Godin, J-G. J. and Sproul, C. D. (1988). Risk taking in parasitized sticklebacks under threat of predation: effects of energetic need and food availabiity. *Canadian Journal of Zoology* 66, 2360–67.

Goertz, J. W. (1966). Incidence of warbles in some Oklahoma rodents. *American Midland Naturalist* 75, 242–45.

Goff, L. J. (1982). Symbiosis and parasitism: another viewpoint. *BioScience* 32, 255–56.

Golderm T. K., Patel, N. Y., and Darji, N. (1987). The effect of *Trypanosoma brucei* infection on the localization of salivary gland cholinesterase in *Glossina morsitans morsitans*. *Acta Tropica* 44, 325–31.

Goldstein, B. (1929). A cytological study of the fungus *Massospora cicadina*, parasitic on the 17–year cicada, *Magicicada septendecim*. *American Journal of Botany* 16, 394–401.

Gompper, M. E. and Hoylman, A. M. (1993). Grooming with *Trattinnickia* resin: possible pharmaceutical plant use by coatis in Panama. *Journal of Tropical Ecology* 9, 533–40.

Goodchild, C. G. and Frankenberg, D. (1962). Voluntary running in the golden hamster, *Mesocricetus auratus* (Waterhouse, 1839), infected with *Trichinella spiralis* (Owen, 1835). *Transactions of the American Microscopical Society* 81, 292–98.

Goss-Custard, J. D. (1984). Intake rates and food supply in migrating and wintering shorebirds. In *Shorebirds: migration and foraging behavior* (eds. J. Burger and B. L. Olla), pp. 233–70. Plenum Press, New York.

Gotelli, N. J. and Moore, J. (1992). Altered host behaviour in a cockroach-acanthocephalan association. *Animal Behaviour* 43, 949–59.

Goulson, D. (1997). *Wipfelkrankheit*: modification of host behaviour during baculovirus infection. *Oecologia* 109, 219–28.

Graham, G. L. (1966). The behavior of beetles, *Tribolium confusum*, parasitized by the larval stage of a chicken tapeworm, *Raillietina cesticillus*. *Transactions of the American Microscopical Society* 85, 163.

Gravenor, M. B. and Kwiatkowski, D. (1998). An analysis of the temperature effects of fever on the intra-host population dynamics of *Plasmodium falciparum*. *Parasitology* 117, 97–105.

Grewal, M. S. (1957). Pathogenicity of *Trypanosoma rangeli* Tejera, 1920 in the invertebrate host. *Experimental Parasitology* 6, 123–30.

Grewal, M. S. (1969). Studies on *Trypanosoma rangeli*, a South American human trypanosome. *Research Bulletin (N. S.) of the Panjab University* 20, 449–86.

Grimstad, P. R., Ross, Q. E., and Craig, Jr., G. B. (1980). *Aedes triseriatus* (Diptera: Culicidae) and La Crosse virus. II. Modification of mosquito feeding behavior by virus infection. *Journal of Medical Entomology* 17, 1–7.

Gross, P. (1993). Insect behavioral and morphological defenses against parasitoids. *Annual Review of Entomology* 38, 251–73.

Grossman, C. J. (1985). Interactions between the gonadal steroids and the immune system. *Science* 227, 257–61.

Gruner, L. and Sauve, C. (1982). The distribution of trichostrongyle infective larvae on pasture and grazing behaviour in calves. *Veterinary Parasitology* 11, 203–13.

Guillot, F. S. and Vinson, S. B.(1973). Effect of parasitism by *Cardiochiles nigriceps* on food consumption and utilization by *Heliothis virescens*. *Journal of Insect Physiology* 19, 2073–82.

Gunter, G. and Ward, J. W. (1961). Some fishes that survive extreme injuries, and some aspects of tenacity of life. *Copeia*, 456–62.

Guthrie, J. F. and Kroger, R. L.(1974). Schooling habits of injured and parasitized menhaden. *Ecology* 55, 208–10.

Haas, W. (1994). Physiological analyses of host-finding behaviour in trematode cercariae: adaptations for transmission success. *Parasitology* 109, S15–29.

Hadju, V., Stephenson, L. S., Abadi, K., Mohammed, H. O., Bowman, D. D., and Parker, R. S. (1996). Improvements in appetite and growth in helminth-infected schoolboys three and seven weeks after a single dose of pyrantel pamoate. *Parasitology* 113, 497–504.

Hajek, A. E. and St. Leger, R. J. (1994). Interactions between fungal pathogens and insect hosts. *Annual Review of Entomology* 39, 293–322.

Hakkarainen, H., Ilmonen, P., Koivumen, V., and Korpimäki, E. (1998). Blood parasites and nest defense behaviour of Tengmalm's owls. *Oecologia* 114, 574–77.

Haldane, J. B. S. (1949). Disease and evolution. *La Ricerca Scientifica Supplementa* 19, 68–76.

Halvorsen, O. (1986a). On the relationship between social status of host and risk of parasitic infection. *Oikos* 47, 71–74.

Halvorsen, O. (1986b). Epidemiology of reindeer parasites. *Parasitology Today* 2, 334–39.

Halvorsen, O. and Skorping, A. (1982). The influence of temperature on growth and development of the nematode *Elaphostrongylus rangiferi* in the gastropods *Arianta arbustorum* and *Euconulus fulvus*. *Oikos* 38, 285–90.

Hamilton, W. D. (1971). Geometry for the selfish herd. *Journal of Theoretical Biology* 31, 295–311.

Hamilton, W. D. (1980). Sex versus non-sex versus parasite. *Oikos* 35, 282–90.

Hamilton, W. D. (1990). Mate choice near or far. *American Zoologist* 30, 341–52.

Hamilton, W. D., Axelrod, R., and Tanese, R. (1990). Sexual reproduction as an adaptation to resist parasites (a review). *Proceedings of the National Academy of Sciences (U.S.A.)*, 87, 3566–73.

Hamilton, W. D. and Zuk, M. (1982). Heritable true fitness and bright birds: a role for parasites? *Science* 218, 384–86.

Hamilton, W. J. III, Buskirk, R. E., and Buskirk, W. H. (1978). Omnivory and utilization of food resources by chacma baboons, *Papio ursinus*. *American Naturalist* 112, 911–24.

Hamilton, W. J. and Poulin, R. (1997). The Hamilton and Zuk hypothesis revisited: a meta-analytical approach. *Behaviour* 134, 299–320.

Harley, J. M. B. (1967). The influence of sampling method on the trypanosome infection rates of catches of *Glossina pallidipes* and *G. fuscipes*. *Entomologia Experimentalis et Applicata* 10, 240–52.

Harper, A. M. (1958). Notes on behaviour of *Pemphigus betae* Doane (Homoptera: Aphididae) infected with *Entomophthora aphidis* Hoffm. *Canadian Entomologist* 40, 439–40.

Hart, B. L. (1988). Biological basis of the behavior of sick animals. *Neuroscience and Biobehavioral Reviews* 12, 123–37.

Hart, B. L. (1990). Behavioral adaptations to pathogens and parasites: five strategies. *Neuroscience and Biobehavioral Reviews* 14, 273–94.

Hart, B. L. (1992). Behavioral adaptations to parasites: an ethological approach. *Journal of Parasitology* 78, 256–65.

Hart, B. L. (1994). Behavioral defense against parasites: interactions with parasite invasiveness. *Parasitology* 109 (suppl.), 139–51.

Hart, B. L. (1997a). Behavioural defence. In *Host-parasite evolution: general principles & avian models* (eds. D. H. Clayton and J. Moore), pp. 59–77. Oxford University Press, Oxford.

Hart, B. L. (1997b). Effects of hormones on behavioral defenses against parasites. In *Parasites and pathogens: effects on host hormones and behavior* (ed. N. E. Beckage), pp. 210–30. Chapman and Hall, New York.

Hart, B. L. and Hart, L. A. (1992). Reciprocal allogrooming in impala, *Aepyceros melampus*. *Animal Behaviour* 44, 1073–83.

Hart, B. L. and Hart, L. A. (1994). Fly switching by Asian elephants: tool use to control parasites. *Animal Behaviour* 48, 35–45.

Hart, B. L., Hart, L. A., and Mooring, M. S. (1990). Differential foraging of oxpeckers on impala in comparison with sympatric antelope species. *African Journal of Ecology* 28, 240–49.

Hart, B. L., Hart, L. A., Mooring, M. S., and Olubayo, R. (1992). Biological basis of grooming behaviour in antelope: the body-size, vigilance and habitat principles. *Animal Behaviour* 44, 615–31.

Hart, B. L., Korinek, E. and Brennan, P. (1987). Postcopulatory genital grooming in male rats: prevention of sexually transmitted infections. *Physiology and Behavior* 41, 321–25.

Hart, B. L. and Powell, K. L. (1990). Antibacterial properties of saliva: role in maternal periparturient grooming and in licking wounds. *Physiology and Behavior* 48, 383–86.

Hart, M. (1990). *Drumming at the edge of magic*. Harper, San Francisco.

Harwood, C. L. Young, I. S., Lee, D. L. and Tringham, J. D. (1996). The effect of *Trichinella spiralis* infection on the mechanical properties of the mammalian diaphragm. *Parasitology* 113, 535–43.

Haseeb, M. A. and Fried, B. (1988). Chemical communication in helminths. *Advances in Parasitology* 27, 170–207.

Hatalski, C. G. and Lipkin, W. I. (1997). Behavioral abnormalities and disease caused by viral infections of the central nervous system. In *Parasites and pathogens: effects on host hormones and behavior* (ed. N. E. Beckage), pp. 201–9. Chapman and Hall, New York.

Hausfater, G. and Meade, B. J. (1982). Alternation of sleeping groves by yellow baboons (*Papio cynocephalus*) as a strategy for parasite avoidance. *Primate* 23, 287–97.

Hausfater, G. and Sutherland, R. (1984). Little things that tick off baboons. *Natural History* 93, 55–60.

Hausfater, G. and Watson, D. F. (1976). Social and reproductive correlates of parasite ova emissions by baboons. *Nature* 262, 688–89.

Hawking, F. (1970). The clock of the malaria parasite. *Scientific American* 236, 123–31.

Hawking, F. and Worms, M. (1961). Transmission of filarioid nematodes. *Annual Review of Entomology* 6, 413–32.

Hawking, F., Worms, M. J., and Gammage, K. (1968). 24- and 48-hour cycles of malaria parasites in the blood: their purpose, production and control. *Transactions of the Royal Society of Tropical Medicine and Hygiene* 62, 731–60.

Hay, J. and Aitken, P. P. (1983). Parasitic infections and mental subnormality with special reference to congenital toxoplasmosis. *Ecology of Disease* 2, 203–9.

Hay, J. and Aitken, P. P. (1984). Experimental toxocariasis in mice and its effect on their behaviour. *Annals of Tropical Medicine and Parasitology* 78, 145–55.

Hay, J. and Hutchison, W. M.(1983). *Toxoplasma gondii*—an environmental contaminant. *Ecology of Disease* 2, 33–43.

Hay, J., Hutchison, W. M., and Aitken, P. P.(1983). The effect of *Toxocara canis* infection on the behaviour of mice. *Annals of Tropical Medicine and Parasitology* 77, 543–44.

Haye, P. A. and Ojeda, F. P. (1998). Metabolic and behavioral alterations in the crab *Hemigrapsus crenulatus* (Milne-Edwards 1837) induced by its acanthocephalan parasite *Profilicollis antarcticus* (Zdzitowiecki 1985). *Journal of Experimental Marine Biology and Ecology* 228, 73–82.

Hayunga, E. G. (1979). Observations on the intestinal pathology caused by three caryophyllid tapeworms of the white sucker *Catostomus commersoni* Laecpede. *Journal of Fish Diseases* 2, 239–48.

Hechtel, L. J., Johnson, C. L., and Juliano, S. A. (1993). Modification of antipredator behavior of *Caecidotea intermedius* by its parasite *Acanthocephalus dirus*. *Ecology* 74, 710–13.

Heeb, P., Werner, I., Richner, H., and Kölliker, M. (1996). Horizontal transmission and reproductive rates of hen fleas in great tit nests. *Journal of Animal Ecology* 65, 474–84.

Heinrich, B. (1993). *The hot-blooded insects: strategies and mechanisms of thermoregulation*. Harvard University Press, Cambridge, MA.

Heins, D. C., Singer, S. S., and Baker, J. A. (1999). Virulence of the cestode *Schistocephalus solidus* and reproduction in infected threspine stickleback, *Gasterosteus aculeatus*. *Canadian Journal of Zoology* 77, 1967–74.

Helluy, S. (1982). Relations hôtes-parasite du trematode *Microphallus papillorobustus* (Rankin, 1940). I. Pénétration des cercaires et rapports des metacercaires avec le tissu nerveux des *Gammarus*, hotes intermédiaires. *Annales de Parasitologie* 57, 263–70.

Helluy, S. (1983a). Un mode de favorisation de la transmission parasitaire: la manipulation du comportement de l'hôte intermédiaire. *Revue d'Écologie (Terre et Vie)* 38, 211–23.

Helluy, S. (1983b). Relations hôtes-parasites du trematode *Microphallus papilloro-bustus* (Rankin, 1940). II. Modifications du comportement des *Gammarus* hôtes intermédiaires et localisation des métacercaires. *Annales de Parasitologie Humaine et Comparée* 58, 1–17.

Helluy, S. (1984). Relations hôtes-parasites du trematode *Microphallus papillorobus-tus* (Rankin, 1940). III. Facteurs impliqués dans les modifications du comportement des *Gammarus* hôtes intermédiaires et tests de prédation. *Annales de Parasitologie Humaine et Comparée* 59, 41–56.

Helluy, S. and Holmes, J. C.(1990). Serotonin, octopamine, and the clinging behavior induced by the parasite *Polymorphus paradoxus* (Acanthocephala) in *Gammarus lacustris* (Crustacea). *Canadian Journal of Zoology* 68, 1214–20.

Hendrickson, G. L. and Kingston, N. (1974). *Cercaria laramiensis* sp. n., a freshwater zygocercous cercaria from *Physa gyrina* Say, with a discussion of cercarial aggregation. *Journal of Parasitology* 60, 777–81.

Henry, J. E. (1971). *Nosema cuneatum* sp. n. (Microsporida: Nosematidae) in grasshoppers (Orthoptera: Acrididae). *Journal of Invertebrate Pathology* 20, 66–69.

Henry, J. E. (1972). Epizootiology of infections by *Nosema locustae* Canning (Microsporida: Nosematidae) in grasshoppers. *Acrida* 1, 111–20.

Henry, J. E. and Oma, E. A. (1981). Pest control by *Nosema locustae*, a pathogen of grasshoppers and crickets. In *Microbial control of pests and plant diseases 1970–1980* (ed. H. D. Burges), pp. 573–86. Academic Press, London.

Hernandez, A. D. and Sukhdeo, M. V. K. (1995). Host grooming and the transmission strategy of *Heligmosomoides polygyrus*. *Journal of Parasitology*, 81, 865–69.

Herre, E. A. (1993). Population structure and the evolution of virulence in nematode parasites of fig wasps. *Science* 259, 1442–45.

Herrmann, B. (1988). Parasite remains from mediaeval latrine deposits: an epidemiologic and ecologic approach. *Actes des 3-emes Journées Anthropologiques, Notes et Monographies Techniques No. 24*, 135–42.

Herting, G. E. and Witt, A., Jr. (1967). The role of physical fitness of forage fishes in relation to their vulnerability to predation by bowfin (*Amia calva*). *Transactions of the American Fisheries Society* 96, 427–30.

Hews, D. K. and Moore, M. C. (1997). Hormones and sex-specific traits: critical questions. In *Parasites and pathogens: effects on host hormones and behavior* (ed. N. E. Beckage), pp. 277–92. Chapman and Hall, New York.

Hillgarth, N. and Wingfield, J. C. (1997a). Parasite-mediated sexual selection: endocrine aspects. In *Host-parasite evolution: General principles & avian models* (eds. D. H. Clayton and J. Moore), pp. 78–104. Oxford University Press, Oxford.

Hillgarth, N. and Wingfield, J. C. (1997b). Testosterone and immunosuppression in vertebrates: implications for parasite-mediated sexual selection. In *Parasites and pathogens: effects on host hormones and behavior* (ed. N. E. Beckage), pp. 143–55. Chapman and Hall, New York.

Hindsbo, O. (1972). Effects of *Polymorphus* (Acanthocephala) on colour and behaviour of *Gammarus lacustris*. *Nature* 238, 333.

Hinnebusch, B. J., Fischer, E.R., and Schwan, T. G. (1998). Evaluation of the role of the *Yersinia pestis* plasminogen activator and other plasmid-encoded factors in temperature-dependent blockage of the flea. *Journal of Infectious Diseases* 178, 1406–15.

Hinnebusch, B. J., Perry, R. D., and Schwan, T. C. (1996). Role of the *Yersinia pestis* hemin storage (hms) locus in the transmission of plague by fleas. *Science*, 273, 367–70.

Hirth, H. F. (1959). Small mammals in old field succession. *Ecology* 40, 417–25.

Hochberg, M. E. (1991). Viruses as costs to gregarious feeding behaviour in the Lepidoptera. *Oikos,* 61, 291–96.

Hockmeyer, W. T., Schiefer, B. A., Remington, B. C., and Eldridge, B. F. (1975). *Brugia pahangi*: effects upon the flight capability of *Aedes aegypti. Experimental Parasitology* 38, 1–5.

Hoek, R. M., van Kesteren, R. E., Smit, A. B., de Jong-Brink, M., and Geraerts, W. P. M. (1997). Altered gene expression in the host brain caused by a trematode parasite: neuropeptide genes are preferentially affected during parasitosis. *Proceedings of the National Academy of Sciences (U.S.A.)* 94, 14072–76.

Holdenried, R. (1952). Sylvatic plague studies. VIII. Notes on the alimentary and reproductive tracts of fleas, made during experimental studies of plague. *Journal of Parasitology* 38, 289–92.

Höller, C. (1991). Movement away from the feeding site in parasitized aphids: host suicide or an attempt by the parasitoid to escape hyperparasitism? In *Behaviour and impact of Aphidophaga* (eds. L. Polgar, R. J. Chambers, A. F. G. Dixon, and I. Hodek), pp. 45–49. SPB Academic Publishing, The Hague.

Holmes, J. C. and Bethel, W. M. (1972). Modification of intermediate host behavior by parasites. In *Behavioral aspects of parasite transmission* (eds. E. U. Canning and C. A. Wright), pp. 123–49. Academic Press, New York.

Holmes, J. C. and Zohar, S. (1990). Pathology and host behaviour. In *Parasitism and host behaviour* (eds. C. J. Barnard and J. M. Behnke), pp. 34–63. Taylor and Francis, London.

Hoogenboom, I. and Dijkstra, C. (1987). *Sarcocystis cernae*: a parasite increasing the risk of predation of its intermediate host, *Microtus arvalis. Oecologia* 74, 86–92.

Hoogland, J. L. (1979). Aggression, ectoparasitism, and other possible costs of prairie dog (Sciuridae, *Cynomys* spp.) coloniality. *Behaviour* 69, 1–35.

Hoogland, J. L. and Sherman, P. W. (1976). Advantages and disadvantages of bank swallow (*Riparia riparia*) coloniality. *Ecological Monographs* 46, 33–58.

Horne, P. D. (1985). A review of the evidence of human endoparasitism in the pre-Columbian New World through the study of coprolites. *Journal of Archaeological Science* 12, 299–310.

Horton, D. R. and Moore, J. (1993). Behavioral effects of parasites and pathogens in insect hosts. In *Parasites and pathogens of insects,* vol. 1 (eds. N. E. Beckage, S. N. Thompson, and B. A. Federici), pp. 107–24. Academic Press, New York.

Hostetter, D. L. and Biever, K. D. (1970). The recovery of virulent nuclear-polyhedrosis virus of the cabbage looper, *Trichoplusia ni*, from the feces of birds. *Journal of Invertebrate Pathology* 15, 173–76.

Houde, A. E. and Torio, A. J. (1992). Effect of parasitic infection on male color pattern and female choice in guppies. *Behavioral Ecology* 3, 346–51.

Howard, R. D. and Minchella, D.J. (1990). Parasitism and mate competition. *Oikos* 58, 120–22.

Howard, R. S. and Lively, C. M. (1994). Parasitism, mutation accumulation and the maintenance of sex. *Nature* 367, 554–57.

Hudson, P. J. and Dobson, A. P. (1991). The direct and indirect effects of the caecal nematode *Trichostrongylus tenuis* on red grouse. In *Bird–parasite interactions:* ecology, *evolution and behaviour* (eds. J. E. Loye and M. Zuk), pp. 49–68. Oxford University Press, Oxford.

Hudson, P. J. and Dobson, A. P. (1997). Host-parasite processes and demographic consequences. In *Host–parasite evolution: general principles and avian models* (eds. D. H. Clayton and J. Moore), pp. 128–54. Oxford University Press, Oxford.

Hudson, P. J., Dobson, A. P., and Newborn, D. (1992). Do parasites make prey vulnerable to predation? Red grouse and parasites. *Journal of Animal Ecology* 61, 681–92.

Huffman, M. A. (1997). Current evidence for self-medication in primates: a multidisciplinary perspective. *Yearbook of Physical Anthropology* 40, 171–200.

Huffman, M. A., Gotoh, S., Izutsu, D., Koshimizu, K., and Kalunde, M. S. (1993). Further observations on the use of the medicinal plant, *Vernonia amygdalina* (Del.), by a wild chimpanzee, its possible effect on parasite load, and its phytochemistry. *African Study Monographs* 14, 227–40.

Huffman, M. A., Koshimizu, K., and Ohigashi, H. (1996*b*). Ethnobotany and zoopharmacognosy of *Vernonia amygdalina,* a medicinal plant used by humans and chimpanzees. In *Proceedings of the Compositae conference, Kew, 1994, vol. 2. Biology and utilization* (ed. D. J. N. Hind), pp. 351–60. Royal Botanic Gardens, Kew, UK.

Huffman, M. A., Page, J. E., Sukhdeo, M. V. K., Gotoh, S., Kalunde, M. S., Chandrasiri, T., and Towers, G. H. N. (1996*a*) Leaf-swallowing by chimpanzees: a behavioral adaptation for the control of strongyle nematode infections. *International Jornal of Primatology* 17, 475–503.

Huffman, M. A. and Seifu, M. (1989). Observations on the illness and consumption of a possibly medicinal plant *Vernonia amygdalina* (Del.) by a wild chimpanzee in the Mahale Mountains National Park, Tanzania. *Primates* 30, 51–63.

Huffman, M. A. and Wrangham, R. W. (1994). Diversity of medicinal plant use by chimpanzees in the wild. In *Chimpanzee cultures* (eds. R. C. Wrangham, F. B. M. de Waal, and P. G. Heltne), pp. 129–48. Harvard University Press, Cambridge, MA.

Hulscher, J. B. (1973). Burying-depth and trematode infection in *Macoma balthica. Netherlands Journal of Sea Research* 6, 141–56.

Hulscher, J. B. (1982). The oystercatcher *Haematopus ostralegus* as a predator of the bivalve *Macoma balthica* in the Dutch Wadden Sea. *Ardea* 70, 89–152.

Hunter, D. M. and Moorhouse, D. E. (1976). Sexual mosaics and mermithid parasitism in *Austrosimulium bancrofti* (Tayl.) (Diptera, Simuliidae) *Bulletin of Entomological Research* 65, 549–53.

Hunter, D. M., Sadleir, R. M. F. S., and Webster, J. M. (1972). Studies on the ecology of cuterebrid parasitism in deer mice. *Canadian Journal of Zoology* 50, 25–29.

Hunter, K. W., Jr., and Stoner, A. (1975). *Copidosoma truncatellum:* effect of parasitization on food consumption of larval *Trichoplusia ni. Environmental Entomology* 4, 381–82.

Hurd, H. (1990). Physiological and behavioural interactions between parasites and invertebrate hosts. *Advances in Parasitology* 29, 271–318.

Hurd, H. (1993). Reproductive disturbances induced by parasites and pathogens of insects. In *Parasites and pathogens of insects*, vol. 1 (ed. N. E. Beckage, S. N. Thompson, and B. A. Federici), pp. 87–105. Academic Press, San Diego, CA.

Hurd, H. (1998). Parasite manipulation of insect reproduction: who benefits? *Parasitology* 116, S13–S24.

Hurd, H. and Arme, C. (1986). *Hymenolepis diminuta:* effect of metacestodes on production and viability of eggs in the intermediate host, *Tenebrio molitor. Journal of Invertebrate Pathology* 47, 225–30.

Hurd H. and Arme, C. (1987). *Hymenolepis diminuta*: effect of infection upon the patency of the follicular epithelium in the intermediate host, *Tenebrio molitor. Journal of Invertebrate Pathology* 49, 227–34.

Hurd, H. and Fogo, S. (1991). Changes induced by *Hymenolepis diminuta* (Cestoda) in the behaviour of the intermediate host *Tenebrio molitor* (Coleoptera). *Canadian Journal of Zoology* 69, 2291–94.

Hurd, H. and Parry, G. (1991). Metacestode-induced depression of the production of, and response to, sex pheromone in the intermediate host *Tenebrio molitor. Journal of Invertebrate Pathology* 58, 82–87.

Hurd, H. and Webb, T. (1997). The role of endocrinological versus nutritional influences in mediating reproductive changes in insect hosts and insect vectors. In *Parasites and pathogens: effects on host hormones and behavior* (ed. N. E. Beckage), pp. 179–97. Chapman and Hall, New York.

Husain, A. and Kershaw, W. E. (1971). The effect of filariasis on the ability of a vector mosquito to fly and feed and to transmit the infection. *Transactions of the Royal Society of Tropical Medicine and Hygiene*. 65, 617–19.

Hutcheson, D. P., Savage, D. C., Parker, D. S., Miles, R. D., and Bootwalla, S. M. (1991). Direct-fed microbials in animal production: a review of literature. National Feed Ingredients Association. West Des Moines, Iowa.

Hutchins, M. and Barash, D. P. (1976). Grooming in primates: implications for its utilitarian function. *Primates* 17, 145–50.

Hutchison, W. M., Aitken, P. P., and Wells, B. W. P. (1980a). Chronic *Toxoplasma* infections and familiarity-novelty discrimination in the mouse. *Annals of Tropical Medicine and Parasitology* 74, 145–50.

Hutchison, W. M., Bradley, M., Cheyne, W. M., Wells, B. W. P., and Hay, J. (1980b). Behavioural abnormalities in *Toxoplasma*-infected mice. *Annals of Tropical Medicine and Parasitology* 74, 337–45.

Hutchison, W. M., Aitken, P. P., and Wells, B. W. P. (1980c). Chronic *Toxoplasma* infections and motor performance in the mouse. *Annals of Tropical Medicine and Parasitology* 74, 507–10.

Hynes, H. B. N. and Nicholas, W. L. (1963). The importance of the acanthocephalan *Polymorphus minutus* as a parasite of domestic ducks in the United Kingdom. *Journal of Helminthology* 37, 185–98.

Ignoffo, C. M. (1981). The fungus *Nomuraea rileyi* as a microbial insecticide. In *Microbial control of pests and plant diseases 1970–1980* (ed. H. D. Burges), pp. 513–38. Academic Press, New York.

Ilmonen, P., Hakkarainen, H., Koivunen, V., Korpimäki, E., Mullie, A. and Shutler, D. (1999). Parental effort and blood parasitism in Tengmalm's owl: effects of natural and experimental variation in food abundance. *Oikos* 86, 79–86.

Inglis, G. D., Johnson, D. L., and Goettel, M. S. (1996). Effects of temperature and thermoregulation on mycosis by *Beauveria bassiana* in grasshoppers. *Biological Control* 7, 131–39.

Isseroff, H., Sylvester, P. W. and Held, W. A. (1986). Effects of *Schistosoma mansoni* on androgen regulated gene expression in the mouse. *Molecular and Biochemical Parasitology* 18, 401–12.

Jaenike, J. (1985). Parasite pressure and the evolution of amanitin tolerance in *Drosophila. Evolution* 39, 1295–1301.

Jaenike, J. (1988). Parasitism and male mating success in *Drosophila testacea. American Naturalist* 131, 774–80.

Jaenike, J. (1995). Interactions between mycophagous *Drosophila* and their nematode parasites: from physiological to community ecology. *Oikos* 72, 235–44.

Jaenike, J. (1996). Suboptimal virulence of an insect-parasite nematode. *Evolution* 50, 2241–47.

Jakobsen, P. J., Johnsen, G. H., and Larsson, P. (1988). Effects of predation risk and parasitism on the feeding ecology, habitat use, and abundance of lacustrine three-spine stickleback (*Gasterosteus aculeatus*). *Canadian Journal of Fisheries and Aquatic Sciences* 45, 426–31.

Jakobsen, P. J. and Wedekind, C. (1998). Copepod reaction to odor stimuli influenced by cestode infection. *Behavioral Ecology* 9, 414–18.

Jahan, N. and Hurd, H. (1998). Effect of *Plasmodium yoelii nigeriensis* (Haemosporidia: Plasmodiidae) on *Anopheles stephensi* (Diptera: Culicidae) vitellogenesis. *Journal of Medical Entomology* 35, 956–61.

James, A. A. and Rossignol, P. A. (1991). Mosquito salivary glands: parasitological and molecular aspects. *Parasitology Today* 7, 267–71.

James, A. C. and J. Jaenike. (1992). Determinants of mating success in wild *Drosophila testacea*. *Animal Behaviour* 44, 168–170.

James, B. L. (1968). The occurrence of larval Digenea in ten species of intertidal prosobranch molluscs in Cardigan Bay. *Journal of Natural History* 2, 329–43.

Janzen, D. H. (1977). Why fruits rot, seeds mold, and meat spoils. *American Naturalist* 111, 691–713.

Janzen, D. H. (1978). Complications in interpreting the chemical defenses of trees against tropical arboreal plant-eating vertebrates. In *The ecology of arboreal folivores* (ed. G. G. Montgomery), pp. 73–84. Smithsonian Press, Washington, DC.

Jarecka, L. (1961). Morphological adaptations of tapeworm eggs and their importance in the life cycles. *Acta Parasitologica Polonica* 9, 409–26.

Jeffords, M. R., Maddox, J. V., and O'Hayer, K. W. (1987). Microsporidian spores in gypsy moth larval silk: a possible route of horizontal transmission. *Journal of Invertebrate Pathology* 49, 332–33.

Jenkins, D., Watson, A., and Miller, G. R. (1963). Population studies on red grouse, *Lagopus lagopus scoticus* (Lath.) in north-east Scotland. *Journal of Animal Ecology* 32, 317–76.

Jenni, L., Molyneux, D. H. , Livesey, J. L., and Galun, R. (1980). Feeding behaviour of tsetse flies infected with salivarian trypanosomes. *Nature* 283, 383–85.

John, J. L. (1997). The Hamilton-Zuk theory and initial test: an examination of some parasitological criticisms. *International Journal for Parasitology* 27, 1269–88.

Johnson, D. L. and Pavlikova, E. (1986). Reduction of consumption of grasshoppers (Orthoptera: Acrididae) infected with *Nosema locustae* Canning (Microsporida: Nosematidae). *Journal of Invertebrate Pathology* 48, 232–38.

Johnson, P. T. J., Lunde, K. B., Ritchie, E. G., and Launer, A. E. (1999). The effect of trematode infection on amphibian limb development and survivorship. *Science* 284, 802–4.

Johnson, S. G. (1992). Parasite-induced parthenogenesis in a freshwater snail: stable, persistent patterns of parasitism. *Oecologia* 89, 533–41.

Jokela, J. and Lively, C. M. (1995). Parasites, sex, and early reproduction in a mixed population of freshwater snails. *Evolution* 49, 1268–71.

Jones, D. (1985). Endocrine interaction between host (Lepidoptera) and parasite (Cheloninae: Hymenoptera): is the host or the parasite in control? *Annals of the Entomological Society of America* 78, 141–48.

Jones, D., Jones, G., Rudnicka, M., Click, A., Reck-Malleczewen, V.,and Iwaya, M. (1986). Pseudoparasitism of host *Trichoplusiani* by *Chelonus* spp. as a new model system for parasite regulation of host physiology. *Journal of Insect Physiology* 32, 315–28.

Jones, L. D., Hodgson, E., and Nuttall, P. A. (1989). Enhancement of virus transmission by tick salivary glands. *Journal of General Virology* 70, 1895–98.

Jordan, P. and Randal, K. (1962). Bilharziasis in Tanganyika: observations on its effects and the effects of treatment in schoolchildren. *Journal of Tropical Medicine and Hygiene* 65, 1–6.

Judson, O. P. and Bennett, A. T. D. (1992). 'Anting' as food preparation: formic acid is worse on an empty stomach. *Behavioral Ecology and Sociobiology* 31, 427–39.

Kagan, I. G. (1951). Aspects in the life history of *Neoleucochloridium problematicum* (Magath, 1920) new comb. and *Leucochloridium cyanocittae* McIntosh, 1932 (Trematoda: Brachylaemidae). *Transactions of the American Microscopical Society* 70, 281–318.

Kagan, I. G. (1952). Further contributions to the life history of *Neoleucochloridium problematicum* (Magath, 1920) new comb. (Trematoda: Brachylaemidae). *Transactions of the American Microscopical Society* 71, 20–44.

Kale, H. W., Edman, J. D., and Webber, L. A. (1972). Effect of behavior and age of individual ciconiiform birds on mosquito feeding success. *Mosquito News* 12, 343–50.

Kalmakoff, J. and Moore, S. G. (1975). The ecology of Nucleopolyhedrosis virus in *Porina (Wiseana* spp.) (Lepidoptera: Hepialidae). *New Zealand Entomologist* 6, 73–76.

Karban, R. (1998). Caterpillar basking behavior and nonlethal parasitism by tachinid flies. *Journal of Insect Behavior* 11, 713–724.

Karban, R. and English-Loeb, G. (1997). Tachinid parasitoids affect host plant choice by caterpillars to increase caterpillar survival. *Ecology* 78, 603–11.

Kavaliers, M. and Colwell, D. D. (1992a). Parasitism, opioid systems and host behaviour. *Advances in Neuroimmunology* 2, 287–95.

Kavaliers, M. and Colwell, D. D. (1992b). Exposure to the scent of male mice infected with the protozoan parasite, *Eimeria vermiformis,* induces opioid- and nonopioid-mediated analgesia in female mice. *Physiology and Behavior* 52, 373–77.

Kavaliers, M. and Colwell, D. D. (1993a). Aversive responses of female mice to the odors of parasitized males: neuromodulatory mechanisms and implications for mate choice. *Ethology* 95, 202–12.

Kavaliers, M. and Colwell, D. D. (1993b). Multiple opioid system involvement in the mediation of parasitic-infection induced analgesia. *Brain Research* 623, 316–20.

Kavaliers, M. and D. D. Colwell. (1994). Parasite infection attenuates non-opioid mediated predator-induced analgesia in mice. *Physiology and Behavior* 55, 505–10.

Kavaliers, M. and Colwell, D. D. (1995a). Decreased predator avoidance in parasitized mice: neuromodulatory correlates. *Parasitology* 111, 257–63.

Kavaliers, M. and Colwell, D. D. (1995b). Discrimination by female mice between the odours of parasitized and non-parasitized males. *Proceedings of the Royal Society of London* B 261, 31–35.

Kavaliers, M. and Colwell, D. D. (1995c). Odours of parasitized males induce aversive responses in female mice. *Animal Behaviour* 50, 1161–69.

Kavaliers, M. and Colwell, D. D. (1995d). Exposure to stable flies reduces spatial learning in mice: involvement of endogenous opioid systems. *Medical and Veterinary Entomology* 9, 300–6.

Kavaliers, M. and Colwell, D. D. (1995e). Reduced spatial learning in mice infected with the nematode, *Heligmosomoides polygyrus*. *Parasitology* 110, 591–97.

Kavaliers, M., Colwell, D. D. and Choleris, E. (1998a). Analgesic responses of male mice exposed to the odors of parasitized females: effects of male sexual experience and infection status. *Behavioral Neuroscience* 112, 1001–11.

Kavaliers, M., Colwell, D. D., and Choleris, E. (1998b) Sex differences in opioid and *N*-methyl-D-aspartate mediated non-opioid biting fly exposure induced analgesia in deer mice. *Pain* 77, 163–71.

Kavaliers, M., Colwell, D. D. and Choleris, E. (1998c). Parasites and behavior: an ethopharmacological analysis and biomedical implications. *Neuroscience and Biobehavioral Reviews* 23, 1037–45.

Kavaliers, M., Colwell, D. D., Choleris, E. and Ossenkopp, K. P. (1999). Learning to cope with biting flies: rapid NMDA-mediated acquisition of conditioned analgesia. *Behavioral Neuroscience* 113, 126–35.

Kavaliers, M., Colwell, D. D., and Galea, L. A. M. (1995) Parasitic infection impairs spatial learning in mice. *Animal Behaviour* 50, 223–29.

Kavaliers, M., Colwell, D. D., and Perrot-Sinal, T. S. (1997a). Opioid and non-opioid NMDA-mediated predator-induced analgesia in mice and the effects of parasitic infection. *Brain Research* 766, 11–18.

Kavaliers, M., Colwell, D. D., Ossenkopp, K.-P., and Perrot-Sinal, T. S. (1997b). Altered responses to female odors in parasitized male mice: neuromodulatory mechanisms and relations to female choice. *Behavioral Ecology and Sociobiology* 40, 373–84.

Kavaliers, M. and Podesta, R. (1988). Opioid involvement in parasite-induced behavioural modifications: evidence from hamsters infected with *Schistosoma mansoni*. *Canadian Journal of Zoology* 66, 2653–57.

Kearn, G. C. (1980). Light and gravity responses of the oncomiracidium of *Entobdella soleae* and their role in host location. *Parasitology* 81, 71–89.

Kearn, G. C. (1986). Role of chemical substances from fish hosts in hatching and host-finding in monogeneans. *Journal of Chemical Ecology* 12, 1651–58.

Keiper, R. R. and Berger, J. (1982/1983). Refuge-seeking and pest avoidance by feral horses in desert and island environments. *Applied Animal Ethology* 9, 111–20.

Kelly, R. and Edman, J. D. (1992). Multiple transmission of *Plasmodium gallinaceum* (Eucoccida: Plasmodiidae) during serial probing by *Aedes aegypti* (Diptera: Culicidae) on several hosts. *Journal of Medical Entomology* 29, 329–31.

Kendall, S. B. and McCullough, F. S. (1951). The emergence of the cercariae of *Fasciola hepatica* from the snail *Limnaea truncatula*. *Journal of Helminthology* 25, 77–92.

Kennedy, C. R., Broughton, P. F., and Hine, P. M. (1978). The status of brown and rainbow trout, *Salmo trutta* and *S. gairdneri* as hosts of the acanthocephalan, *Pomphorhynchus laevis*. *Journal of Fish Biology* 13, 265–75.

Kent, S., Bluthé, R-M., Kelley, K. W. and Dantzer, R. (1992). Sickness behavior as a new target for drug development. *Trends in Pharmacological Sciences* 13, 24–28.

Kershaw, W. E. and Storey, D. M. (1976). Host-parasite relations in cotton rat filariasis. I. The quantitative transmission and subsequent development of *Litomosoides carinii* infections in cotton rats and other laboratory animals. *Annals of Tropical Medicine and Parasitology* 70, 303–12.

Keymer, A., Crompton, D. W. T., and Sahakian, B. J. (1983). Parasite-induced learned taste aversion involving *Nippostrongylus* in rats. *Parasitology* 86, 455–60.

Khan, R. A. (1988). Experimental transmission, development, and effects of a parasitic copepod, *Lernaeocera branchialis*, on Atlantic cod, *Gadus morhua*. *Journal of Parasitology* 74, 586–99.

Kightlinger, L. K., Seed, J. R., and Kightlinger, M. B. (1998). *Ascaris lumbricoides* intensity in relation to environmental, socioeconomic, and behavioral determinants of exposure to infection in children from southeast Madagascar. *Journal of Parasitology* 84, 480–84.

Killick-Kendrick, R., Killick-Kendrick, M., Qala, N. A., Nawi, I., Ashford, R. W., and Tang, Y. (1989). Preliminary observations on a tetradonematid nematode of phlebotomine sandflies of Afghanistan. *Annales de Parasitologie Humaine et Compareé* 64, 332–39.

Killick-Kendrick, R. and Molyneux, D. H. (1990). Interrupted feeding of vectors. *Parasitology Today* 6, 188–89.

Killick-Kendrick, R., Leaney, A. J., Ready, P. D., and Molyneux, D. H. (1977). *Leishmania* in phlebotomid sandflies. IV. The transmission of *Leishmania mexicana amazonensis* to hamsters by the bite of experimentally infected *Lutzomyia longipalpis*. *Proceedings of the Royal Society of London B* 196, 105–13.

Killick-Kendrick, R., Wallbanks, K. R., Molyneux, D. H., and Lavin, D. R. (1988). The ultrastructure of *Leishmania major* in the foregut and proboscis of *Phlebotomus papatasi*. *Parasitology Research* 74, 586–90.

King, D. S. and Humber, R. A. (1981). Identification of the Entomophthorales. In *Microbial control of pests and plant diseases 1970–1980* (ed. H. D. Burges), pp. 107–27. Academic Press, New York.

King, K. M. and Atkinson, N. J. (1928). The biological control factors of the immature stages of *Euxoa ochrogaster* Gn. (Lepidoptera, Phalaenidae) in Saskatchewan. *Annals of the Entomological Society of America* 21, 167–88.

Kingsolver, J. G. (1987). Mosquito host choice and the epidemiology of malaria. *American Naturalist* 130, 811–27.

Klein, S. L., Gamble, H. R. and Nelson, R. J. (1999). *Trichinella spiralis* infection in voles alters female odor preference but not partner preference. *Behavioral Ecology and Sociobiology* 45, 323–29.

Kliks, M. M. (1990). Helminths as heirlooms and souvenirs: a review of New World paleoparasitology. *Parasitology Today* 6, 93–100.

Klowden, M. J. (1988). Factors influencing multiple host contacts by mosquitoes during a single gonotrophic cycle. *Miscellaneous Publications of the Entomological Society of America* 68, 29–36.

Klowden, M. (1990). The endogenous regulation of mosquito reproductive behavior. *Experientia* 46, 660–70.

Klowden, M. J. (1995). Blood, sex and the mosquito. *BioScience* 45, 326–31.

Klowden, M. J. (1996). Vector behavior. In *The biology of disease vectors* (eds. B. J. Beaty and W. C. Marquardt), pp. 34–50. University Press of Colorado, Niwot, Colorado.

Klowden, M. J. and Lea, A. O. (1978). Blood meal size as a factor affecting continued host-seeking by *Aedes aegypti* (L.). *American Journal of Tropical Medicine and Hygiene* 27, 827–31.

Klowden, M. J. and Lea, A. O. (1979). Effect of defensive host behavior on the blood meal size and feeding successs of natural populations of mosquitoes (Diptera: Culicidae). *Journal of Medical Entomology* 15, 514–17.

Klowden, M. J. and Lea, A. O. (1981). Laboratory transmission of *Brugia pahangi* by nulliparous *Aedes aegypti* (Diptera: Culicidae). *Journal of Medical Entomology* 18, 383–85.

Kluger, M. J. (1979). Phylogeny of fever. *Federation Proceedings* 38, 30–34.

Kluger, M. J. (1986). Is fever beneficial? *Yale Journal of Biology and Medicine* 59, 89–95.

Kluger, M. J. (1991). Fever: role of pyrogens and cryogens. *Physiological Reviews* 71, 93–127.

Knell, R. J. (1999). Sexually transmitted disease and parasite-mediated sexual selection. *Evolution* 53, 957–61.

Knezevich, M. (1998). Geophagy as a therapeutic mediator of endoparasitism in a free-ranging group of rhesus macaques (*Macaca mulatta*). *American Journal of Primatology* 44, 71–82.

Knols, B. G. J. and De Jong, R. (1996). Limburger cheese as an attractant for the malaria mosquito *Anopheles gambiae* s. s. *Parasitology Today* 12, 159–61.

Knols, B. G. J., van Loon, J. J. A., Cork, A., Robinson, R. D., Adam, W., Meijerink, J., De Jong, R. and Takken, W. (1997). Behavioural and electrophysiological responses of the female malaria mosquito *Anopheles gambiae* (Diptera: Culicidae) to Limburger cheese volatiles. *Bulletin of Entomological Research* 87, 151–59.

Koella, J. C. (1999). An evolutionary view of the interactions between anopheline mosquitoes and malaria parasites. *Microbes and Infection* 1, 303–8.

Koella, J. C. and Agnew, P. (1997). Blood-feeding success of the mosquito *Aedes aegypti* depends on the transmission route of its parasite *Edhazardia aedis*. *Oikos* 78, 311–16.

Koella, J. C., Agnew, P. and Michalakis, Y. (1998b). Coevolutionary interactions between host life histories and parasite life cycles. *Parasitology* 116, S47–S55.

Koella, J. C. and Packer, M. J. (1996). Malaria parasites enhance blood-feeding of their naturally infected vector *Anopheles punctulatus*. *Parasitology* 113, 105–9.

Koella, J. C., Sørensen, F. L., and Anderson, R. A. (1998a). The malaria parasite, *Plasmodium falciparum*, increases the frequency of multiple feeding of its mosquito vector, *Anopheles gambiae*. *Proceedings of the Royal Society of London* B265, 763–68.

Konishi, E. (1989). Size of blood meals of *Aedes albopictus* and *Culex tritaeniorhynchus* (Diptera: Culicidae) feeding on an unrestrained dog infected with *Dirofilaria immitis* (Spirurida: Filariidae). *Journal of Medical Entomology* 26, 535–38.

Kornhauser, S. I. (1919). Sexual characteristics of the membracid, *Thelia bimaculata* (Fabr.). I. External changes induced by *Aphelopus theliae* (Gahan). *Journal of Morphology* 32, 531–636.

Kramm, K. R., West, D. F., and Rockenbach, P. G. (1982). Termite pathogens: transfer of the entomopathogen *Metarhizium anisopliae* between *Reticulitermes* sp. termites. *Journal of Invertebrate Pathology* 40, 1–6.

Krasnoff, S. B., Watson, D. W., Gibson, D. M., and Kwan, E. C. (1995). Behavioral effects of the entomopathogenic fungus, *Entomophthora muscae* on its host *Musca domestica*: postural changes in dying hosts and gated pattern of mortality. *Journal of Insect Physiology* 41, 895–903.

Krause, J. (1994). Differential fitness returns in relation to spatial position in groups. *Biological Reviews* 69, 187–206.

Krause, J. and Godin, J-G. J. (1994). Influence of parasitism on the shoaling behaviour of banded killifish, *Fundulus diaphanus*. *Canadian Journal of Zoology* 72, 1775–79.

Krause, J. and Godin, J-G. J. (1996). Influence of parasitism on shoal choice in the banded killifish (*Fundulus diaphanus*, Teleostei, Cyprinodontidae). *Ethology* 102, 40–49.

Kuris, A. M. (1974). Trophic interactions: similarity of parasitic castrators to parasitoids. *Quarterly Review of Biology* 49, 129–48.

Kuris, A. M. (1997). Host behavior modification: An evolutionary perspective. In *Parasites and pathogens: effects on host hormones and behavior* (ed. N. E. Beckage), pp. 293–315. Chapman and Hall, New York.

Kvalsvig, J. D. (1986). The effects of *Schistosomiasis haematobium* on the activity of school children. *Journal of Tropical Medicine and Hygiene* 89, 85–90.

Kvalsvig, J. D. (1988). The effects of parasitic infection on cognitive performance. *Parasitology Today* 4, 206–8.

Kvalsvig, J. D. and Becker, P. J. (1988). Selective exposure of active and sociable children to schistosomiasis. *Annals of Tropical Medicine and Parasitology* 82, 471–74.

Kvalsvig, J. D., Cooppan, R. M., and Connolly, K. J. (1991). The effects of parasite infections on cognitive processes in children. *Annals of Tropical Medicine and Parasitology* 85, 551–68.

Kvalsvig, J. D. and Schutte, C. J. (1986). The role of human water contact patterns in the transmission of schistosomiasis in an informal settlement near a major industrial area. *Annals of Tropical Medicine and Parasitology* 80, 13–26.

Kyriazakis, I., Oldham, J. D., Coop, R. L., and Jackson, F. (1994). The effect of subclinical intestinal nematode infection on the diet selection of growing sheep. *British Journal of Nutrition* 72, 665–77.

Kyriazakis, I., Tolkamp, B. J., and Hutchings, M. R. (1998). Towards a functional explanation for the occurrence of anorexia during parasitic infections. *Animal Behaviour* 56, 265–74.

Lackie, J. M. (1972). The effect of temperature on the development of *Moniliformis dubius* (Acanthocephala) in the intermediate host, *Periplaneta americana*. *Parasitology* 65, 371–77.

Lafferty, K. D. (1992). Foraging on prey that are modified by parasites. *American Naturalist* 140, 854–67.

Lafferty, K. D. (1997). The ecology of parasites in a salt marsh ecosystem. In *Parasites and pathogens: effects on host hormones and behavior* (ed. N. E. Beckage), pp 316–32. Chapman and Hall, New York.

Lafferty, K. D. and Morris, A. K. (1996). Altered behavior of parasitized killifish increases susceptibility to predation by bird final hosts. *Ecology* 77, 1390–97.

Laird, M. (1959). Malayan Protozoa I. *Plistophora collessi* n. sp. (Sporozoa: Microsporidia), an ovarian parasite of Singapore mosquitoes. *Journal of Protozoology* 6, 37–45.

Lambert, T. C. and Farley, J. (1968). The effect of parasitism by the trematode *Cryptocotyle lingua* (Creplin) on zonation and winter migration of the common periwinkle, *Littorina littorea* (L.). *Canadian Journal of Zoology* 46, 1139–47.

La Rosa, G., Pozio, E., Rossi, P., and Murrell, K. D.(1992). Allozyme analysis of

Trichinella isolates from various host species and geographical regions. *Journal of Parasitology* 78, 641–46.

Lassiere, O. L. and Crompton, D. W. T. (1988). Evidence for post-cyclic transmission in the life-history of *Neoechinorhynchus rutili* (Acanthocephala). *Parasitology* 97, 339–43.

Latta, B. (1987). Letter to *Nature*. *Nature* 330, 701.

Lauckner, G. (1984). Impact of trematode parasitism on the fauna of a North Sea tidal flat. *Helgolander Meeresuntersuchungen* 37, 185–99.

Lauckner, G. (1987). Ecological effects of larval trematode infestation on littoral marine invertebrate populations. *International Journal for Parasitology* 17, 391–98.

Lavine, M. D. and Beckage, N. E. (1995). Polydnaviruses: potent mediators of host insect immune dysfunction. *Parasitology Today* 11, 368–78.

Lavoiperre, M. M. J.(1958). Studies on the host-parasite relationships of filarial nematodes and their arthropod hosts. II. The arthropod as a host to the nematode: a brief appraisal of our present knowledge, based on a study of the more important literature from 1878 to 1957. *Annals of Tropical Medicine and Hygiene* 52, 326–45.

Lawlor, B. J., Read, A. F., Keymer, A. E., Parveen, G., and Crompton, D. W. T. (1990). Non-random mating in a parasitic worm: mate choice by males? *Animal Behaviour* 40, 870–76.

Lefcort, H. and Bayne, C. J. (1991). Thermal preferences of resistant and susceptible strains of *Biomphalaria glabrata* (Gastropoda) exposed to *Schistosoma mansoni* (Trematoda). *Parasitology* 103, 357–62.

Lefcort, H. and Blaustein, A. R. (1995). Disease, predator avoidance, and vulnerability to predation in tadpoles. *Oikos* 74, 469–74.

Lefcort, H. and Durden, L. A. (1996). The effect of infection with Lyme disease spirochetes (*Borrelia burgdorferi*) on the phototaxis, activity, and questing height of the tick vector *Ixodes scapularis*. *Parasitology* 113, 97–103.

Lefcort, H. and Eiger, S. M. (1993). Antipredatory behaviour of feverish tadpoles: implications for pathogen transmission. *Behaviour* 126, 13–27.

Lehmann, T. (1993). Ectoparasites: direct impact on host fitness. *Parasitology Today* 9, 8–13.

Leiby, P. D. and Dyer, W. G. (1971). Cyclophyllidean tapeworms of wild carnivores. In *Parasitic diseases of wild mammals* (eds. J. W. Davis and R. C. Anderson), pp. 175–234. Iowa State University Press, Ames, Iowa.

Lester, R. J. G. (1971). The influence of *Schistocephalus* plerocercoids on the respiration of *Gasterosteus* and a possible resulting effect on the behavior of the fish. *Canadian Journal of Zoology* 49, 361–66.

Levri, E. P. (1998). The influence of non-host predators on parasite-induced behavioral changes in a freshwater snail. *Oikos* 81, 531–37.

Levri, E. P. (1999). Parasite-induced change in host behavior of a freshwater snail: parasitic manipulation or byproduct of infection? *Behavioral Ecology* 10, 234–41.

Levri, E. P. and Lively, C. M. (1996). The effects of size, reproductive condition, and parasitism on foraging behaviour in a freshwater snail, *Potamopyrgus antipodarum*. *Animal Behaviour* 51, 891–901.

Lewis, D. J. and Wright, C. A. (1962). A trematode parasite of *Simulium*. *Nature* 193, 1311–12.

Lewis, F. B. (1960). Factors affecting assessment of parasitization by *Apanteles fumiferanae* Vier. and *Glypta fumiferanae* (Vier.) on spruce budworm larvae. *Canadian Entomologist* 42, 881–91.

Lewis, M. C., Welsford, I. G., and Uglem, G. L. (1989). Cercarial emergence of *Proterometra macrostoma* and *P. edneyi* (Digenea: Azygiidae): contrasting responses to light:dark cycling. *Parasitology* 99, 215–23.

Lewis, P. D., Jr. (1974). Helminths of terrestrial molluscs in Nebraska. II. Life cycle of *Leucochloridium variae* McIntosh, 1932 (Digenea: Leucochloridiidae). *Journal of Parasitology* 60, 251–55.

Lewis, P. D., Jr. (1977). Adaptations for the transmission of species of *Leucochloridium* from molluscan to avian hosts. *Proceedings of the Montana Academy of Sciences* 37, 70–81.

Li, X., Sina, B., and Rossignol, P. A. (1992). Probing behaviour and sporozoite delivery by *Anopheles stephensi* infected with *Plasmodium berghei*. *Medical and Veterinary Entomology* 6, 57–61.

Libersat, F. and Moore, J. (2000). An acanthocephalan alters the escape response of its cockroach intermediate host. *Journal of Insect Behavior.* 13, 103–110.

Lie, K. J. (1966). Antagonistic interaction between *Schistosoma mansoni* sporocysts and echinostome rediae in the snail *Australorbis glabratus*. *Nature* 211, 1213–15.

Lie, K. J., Basch, P. F., and Umathevy, T. (1965). Antagonism between two species of larval trematodes in the same snail. *Nature* 206, 422–23.

Lim, S. S. L. and Green, R. H. (1991). The relationship between parasite load, crawling behaviour, and growth rate of *Macoma balthica* (L.) (Mollusca, Pelecypoda) from Hudson Bay, Canada. *Canadian Journal of Zoology* 69, 2202–8.

Lima, S. L. (1992). Life in a multi-predator environment: some considerations for anti-predatory vigilance. *Annales Zoologici Fennici* 29, 217–26.

Lima, S. L. (1995). Collective detection of predatory attack by social foragers: fraught with ambiguity? *Animal Behaviour* 50, 1097–1108.

Lima, S. L. and Dill, L. M. (1990). Behavioral decisions made under the risk of predation: a review and prospectus. *Canadian Journal of Zoology* 68, 619–40.

Limbaugh, C. (1961). Cleaning symbiosis. *Scientific American* 205, 42–49.

Lin, Y. C., Rikihisa, Y., Kono, H., and Gu, Y. (1990). Effects of larval tapeworm (*Taenia taeniaeformis*) infection on reproductive functions in male and female host rats. *Experimental Parasitology* 70, 344–52.

Lindsay, S. W. and Denham, D. A. (1985). The effect of different types of skin surfaces on the transmission of *Brugia pahangi* infective larvae by the mosquito *Aedes aegypti*. *Transaction of the Royal Society of Tropical Medicine and Hygiene* 79, 56–59.

Lipsitch, M., Herre, E. A., and Nowak, M. (1995). Host population structure and the evolution of virulence: a "law of diminishing returns." *Evolution* 49, 743–48.

Lipsitch, M., Siller, S., and Nowak, M. (1996). The evolution of virulence in pathogens with vertical and horizontal transmission. *Evolution* 50, 1729–41.

Lively, C. M. (1987). Evidence from a New Zealand snail for the maintenance of sex by parasitism. *Nature* 328, 519–21.

Lively, C. M. (1992). Parthenogenesis in a freshwater snail: reproductive assurance versus parasitic release. *Evolution* 46, 907–13.

Lively, C. M. (1996). Host-parasite coevolution and sex. *BioScience* 46, 107–14.

Lively, C. M., Craddock, C., and Vrijenhoek, R. C. (1990). Red queen hypothesis supported by parasitism in sexual and clonal fish. *Nature* 344, 864–66.

LoBue, C. P. and Bell, M. A. (1993). Phenotypic manipulation by the cestode parasite *Schistocephalus solidus* of its intermediate host, *Gasterosteus aculeatus*, the three-spine stickleback. *American Naturalist* 142, 725–35.

Loehle, C. (1995). Social barriers to pathogen transmission in wild animal populations. *Ecology* 76, 326–35.

Loker, E. S. (1994). On being a parasite in an invertebrate host: a short survival course. *Journal of Parasitology* 80, 728–47.

Lombardo, M. P., Thorpe, P. A., and Power, H. W. (1999). The beneficial sexually transmitted microbe hypothesis of avian copulation. *Behavioral Ecology* 10, 333–37.

López, S. (1999). Parasitized females guppies do not prefer showy males. *Animal Behaviour* 57, 1129–34.

Louis, C., Jourdan, M., and Cabanac, M. (1986). Behavioral fever and therapy in a rickettsia-infected Orthoptera. *American Journal of Physiology* 250, R991–95.

Low, B. S. (1990). Marriage systems and pathogen stress in human societies. *American Zoologist* 30, 325–39.

Lowe-Jinde, L. and Zimmerman, A. M. (1991). Influence of *Cryptobia salmositica* on feeding, body composition, and growth in rainbow trout. *Canadian Journal of Zoology* 69, 1397–1401.

Lowenberger, C. A. and Rau, M. E. (1994*a*). *Plagiorchis elegans*: emergence, longevity and infectivity of cercariae, and host behavioural modifications during cercarial emergence. *Parasitology* 109, 65–72.

Lowenberger, C. A. and Rau. M. E. (1994*b*). Selective oviposition by *Aedes aegypti* (Diptera: Culicidae) in response to a larval parasite, *Plagiorchis elegans* (Trematoda: Plagiorchiidae). *Environmental Entomology* 23, 1269–1276.

Lower, H. F. (1954). A granulosis virus attacking the larvae of *Persectania ewingii* Westw. (Lepidoptera: Agrotidae) in South Australia. *Australian Journal of Biological Science* 7, 161–67.

Loye, J. E. and Carroll, S. P. (1991). Nest ectoparasite abundance and cliff swallow colony site selection, nestling development, and departure time. In *Bird–parasite interactions: ecology, evolution and behaviour* (eds. J. E. Loye and M. Zuk), pp. 222–42. Oxford University Press, Oxford.

Loye, J. and Carroll, S. (1995). Birds, bugs and blood: avian parasitism and conservation. *Trends in Ecology and Evolution*, 10, 232–35.

Lozano, G. A. (1991). Optimal foraging theory: a possible role for parasites. *Oikos* 60, 391–95.

Lozano, G. A. (1994). Carotenoids, parasites and sexual selection. *Oikos* 70, 309–311.

Lucius, R. and Frank, W. (1978). Beitrag zur Biologie von *Dicrocoelium hospes* Looss, 1907 (Trematodes, Dicrocoeliidae). *Acta Tropica* 35, 161–81.

Lucius, R., Romig, T., and Frank, W. (1980). *Camponotus compressiscapus* Andre (Hymenoptera, Formicidae) an experimental second intermediate host of *Dicrocoelium hospes* Looss, 1907 (Trematodes, Dicrocoeliidae). *Zeitschrift fur Parasitenkunde* 63, 271–75.

Lundberg, H. and Svensson, B. G. (1975). Studies on the behaviour of *Bombus* Latr. species (Hym., Apidae) parasitized by *Sphaerularia bombi* Dufour (Nematoda) in an alpine area. *Norwegian Journal of Entomology* 22, 129–34.

Lykouressis, D. P. and van Emden, H. F. (1983). Movement away from feeding site of the aphid *Sitobion avenae* (F.) (Hemiptera: Aphididae) when parasitized by *Aphelinus abdominalis* (Dalman) (Hymenoptera: Aphalinidae). *Entomologia Hellenica* 1, 59–63.

Lyndon, A. R. (1996). The role of acanthocephalan parasites in the predation of freshwater isopods by fish. In *Aquatic predators and their prey* (eds. S. P. R. Greenstreet and M. L. Tasker), pp. 26–32. Fishing News Books, Oxford.

MacFarland, C. G. and Reeder, W. G. (1974). Cleaning symbiosis involving Galapagos tortoises and two species of Darwin's finches. *Zeitschrift fur Tierpsychologie* 34, 464–83.

MacGuidwin, A. E., Smart, G. C., Jr., Wilkinson, R. C., and Allen, G. E. (1980). Effect of the nematode *Contortylenchus brevicorni* on gallery construction and fecundity of the southern pine beetle. *Journal of Nematology* 12, 278–82.

Mackiewicz, J. S. (1988). Cestode transmission patterns. *Journal of Parasitology* 74, 60–71.

MacKinnon, B. M. (1987). Sex attractants in nematodes. *Parasitology Today* 3, 156–58.

MacLeod, D. M, Tyrrell, D., Soper, R. S., and De Lyzer, A.J. (1973). *Entomophthora bullata* as a pathogen of *Sarcophaga aldrichi*. *Journal of Invertebrate Pathology* 22, 75–79.

Maema, M. (1986). Experimental infection of *Tribolium confusum* (Coleoptera) by *Hymenolepis diminuta* (Cestoda): host fecundity during infection. *Parasitology* 92, 405–12.

Maeyama, T., Terayama, M., and Matsumoto, T. (1994). The abnormal behavior of *Colobopsis* sp. (Hymenoptera: Formicidae) parasitized by *Mermis* (Nematoda) in Papua New Guinea. *Sociobiology* 24, 115–19.

Mahaney, W. C., Hancock, R. G. V., Aufreiter, S., and Huffman, M. A. (1996). Geochemistry and clay mineralogy of termite mound soil and a possible role of geophagy in chimpanzees of the Mahale Mountains, Tanzania. *Primates* 37, 121–34.

Mahaney, W. C., Zippin, J., Milncr, M W., Sanmugadas, K., Hancock, R. G. V., Aufreiter, S., Campbell, S., Huffman, M. A., Wink, M., Mallochs, D., and Kalm, V. (1999). Chemistry, mineralogy and microbiology of termite mound soil eaten by the chimpanzees of the Mahale Mountains, Western Tanzania. *Journal of Tropical Ecology* 15, 565–88.

Mahon, R. and Gibbs, A. (1982). Arbovirus-infected hens attract more mosquitoes. In *Viral diseases in south-east Asia and the western Pacific* (ed. J. S. Mackenzie), pp. 502–5. Academic Press, Sydney.

Maier, S. F. and Watkins, L. R. (1998). Cytokines for psychologists: implications of bidirectional immune-to-brain communication for understanding behavior, mood, and cognition. *Psychological Review* 105, 83–107.

Maier, S. F. and Watkins, L. R. (1999). Bidirectional communication between the brain and the immune system: implications for behaviour. *Animal Behaviour* 57, 741–51.

Maitland, D. P. (1994). A parasitic fungus infecting yellow dungflies manipulates host perching behaviour. *Proceedings of the Royal Society of London B* 258, 187–93.

Makumi, J. N. and Moloo, S. K. (1991). *Trypanosoma vivax* in *Glossina palpalis gambiensis* do not appear to affect feeding behaviour, longevity or reproductive performance of the vector. *Medical and Veterinary Entomology* 8, 35–42.

Marikovsky, P. I. (1962). On some features of behavior of the ants *Formica rufa* L. infected with fungous disease. *Insectes Sociaux* 9, 173–79.

Marquardt, W. C., Demaree, R. S. and Grieve, R. B. (2000). *Parasitology and Vector Biology* (2nd ed.). Academic Press, San Diego, CA.

Marshall, W. H., Gullion, G. G., and Schwab, R. G. (1962). Early summer activities of porcupines as determined by radio-positioning techniques. *Journal of Wildlife Management* 26, 75–79.

Massey, C. L. (1960). Nematode parasites and associates of the California five-spined engraver, *Ips confusus* (Lec.). *Proceedings of the Helminthological Society of Washington* 27, 14–22.

Massey, C. L. (1964). The nematode parasites and associates of the fir engraver beetle, *Scolytus ventralis* LeConte, in New Mexico. *Journal of Insect Pathology* 6, 133–55.

Mather, T. N., Duffy, D. C., and Campbell, S. R. (1993). An unexpected result from burning vegetation to reduce Lyme Disease transmission risks. *Journal of Medical Entomology* 30, 642–45.

May, R. M. (1985). Ecological aspects of disease and human populations. *American Zoologist* 25, 441–50.

May, R. M. and Anderson, R. M. (1978). Regulation and stability of host-parasite population interactions. II. Destabilizing processes. *Journal of Animal Ecology* 47, 249–67.

May, R. M. and Anderson, R. M. (1979). Population biology of infectious diseases: Part II. *Nature* 280, 455–63.

May, R. M. and Anderson, R. M. (1983a). Parasite-host coevolution. In *Coevolution* (eds. D. Futuyma and M. Slatkin), pp. 186–206. Sinauer, Sunderland, MA.

May, R. M. and Anderson, R. M. (1983b). Epidemiology and genetics in the coevolution of parasites and hosts. *Proceedings of the Royal Society of London B* 219, 281–313.

May, R. M. and R. M. Anderson. (1990). Parasite-host coevolution. *Parasitology* 100, S89–101.

Maynard, B. J., DeMartini, L., and Wright, W. G. (1996). *Gammarus lacustris* harboring *Polymorphus paradoxus* show altered patterns of serotonin-like immunoreactivity. *Journal of Parasitology* 82, 663–66.

Maynard, B. J., Wellnitz, T. A., Zanini, N., Wright, W. G., and Dezfuli, B. S. (1998). Parasite-altered behavior in a crustacean intermediate host: field and laboratory studies. *Journal of Parasitology* 84, 1102–6.

McAllister, M. K. and Roitberg, B. D. (1987). Adaptive suicidal behaviour in pea aphids. *Nature* 328, 797–99.

McAllister, M. K. and Roitberg, B. D. (1988). Assumptions about suicidal behaviour of aphids. *Nature* 332, 494–95.

McAllister, M. K., Roitberg, B. D., and Weldon, K. L. (1990). Adaptive suicide in pea aphids: decisions are cost sensitive. *Animal Behaviour* 40, 167–75.

McCahon, C. P., Brown, A. F., and Pascoe, D. (1988). The effect of the acanthocephalan *Pomphorhynchus laevis* (Muller 1776) on the acute toxicity of cadmium to its intermediate host, the amphipod *Gammarus pulex* (L.). *Archives of Environmental Contamination and Toxicology* 17, 239–43.

McCahon, C. P., Brown, A. F., Poulton, M. J., and Pascoe, D. (1989). Effects of acid, aluminium and lime additions on fish and invertebrates in a chronically acidic Welsh stream. *Water, Air, and Soil Pollution* 45, 345–59.

McCahon, C. P., Maund, S. J., and Poulton, M. J. (1991*a*). The effect of the acanthocephalan parasite (*Pomphorhynchus laevis*) on the drift of its intermediate host (*Gammarus pulex*). *Freshwater Biology* 25, 507–13.

McCahon, C. P., Poulton, M. J., Thomas, P. C., Xu, Q., Pasco, D. and Turner, C. (1991b). Lethal and sub-lethal toxicity of field simulated farm waste episodes to several freshwater invertebrate species. *Water Research* 25, 661–71.

McCallum, H. and Dobson, A. (1995). Detecting disease and parasite threats to endangered species and ecosystems. *Trends in Ecology and Evolution* 10, 190–94.

McCauley, D. E. (1994). Intrademic group selection imposed by a parasitoid-host interaction. *American Naturalist* 144, 1–13.

McClain, E., Magnuson, P., and Warner, S. J. (1988). Behavioural fever in a Namib desert tenebrionid beetle, *Onymacris plana*. *Journal of Insect Physiology* 34, 279–84.

McCurdy, D. G., Forbes, M. R. and Boates, J. S. (1999a). Testing alternative hypotheses for variation in amphipod behaviour and life history in relation to parasitism. *International Journal for Parasitology* 29, 1001–9.

McCurdy, D. G., Forbes, M. R. and Boates, J. S. (1999b). Evidence that the parasitic nematode *Skrjabinoclava* manipulates host *Corophium* behavior to increase transmission to the sandpiper, *Calidris pusilla*. *Behavioral Ecology* 10, 351–57.

McDonough, K. A. and Falkow, S. (1989). A *Yersinia pestis*-specific DNA fragment encodes temperature-dependent coagulase and fibrinolysin-associated phenotypes. *Molecular Microbiology* 3, 767–75.

McKenzie, A. A. (1990). The ruminant dental grooming apparatus. *Zoological Journal of the Linnean Society* 99, 117–28.

McLachlan, A. (1999). Parasites promote mating success: the case of a midge and a mite. *Animal Behaviour* 57, 1199–1205.

McLennan, D. A. and Brooks, D. R. (1991). Parasites and sexual selection: a macroevolutionary perspective. *Quarterly Review of Biology* 66, 255–86.

McNair, D. M. and Timmons, E. H. (1977). Effects of *Aspiculuris tetraptera* and *Syphacia obvelata* on exploratory behavior of an inbred mouse strain. *Laboratory Animal Science* 27, 38–42.

McNeill, W. H. (1976). *Plagues and peoples*. Anchor Press, Garden City, NY.

McNeill, W. H. (1979). *The human condition*. Princeton University Press, Princeton, NJ.

McPhail, J. D. and Peacock, S. D. (1983). Some effects of the cestode (*Schistocephalus solidus*) on reproduction in the threespine stickleback (*Gasterosteus aculeatus*): evolutionary aspects of a host-parasite interaction. *Canadian Journal of Zoology* 61, 901–8.

Meakins, R. H. and Walkey, M. (1975). The effects of parasitism by the plerocercoid of *Schistocephalus solidus* Muller 1776 (Pseudophyllidea) on the respiration of the three-spined stickleback *Gasterosteus aculeatus* L. *Journal of Fish Biology* 7, 817–24.

Mech, L. D. (1966). The wolves of Isle Royale. Fauna of the National Parks of the United States Fauna Series 7. U. S. Government Printing Office, Washington, DC.

Mech, L. D. (1970). *The wolf: the ecology and behavior of an endangered species*. Natural History Press, Garden City, NY.

Messier, F., Rau, M. E., and McNeill, M. A. (1989). *Echinococcus granulosus* (Cestoda: Taeniidae) infections and moose-wolf population dynamics in southwestern Quebec. *Canadian Journal of Zoology* 67, 216–19.

Michalakis, Y. and Hochberg, M. E. (1994). Parasitic effects on host life-history traits: a review of recent studies. *Parasite* 1, 291–94.

Michel, J. F. (1955). Parasitological significance of bovine grazing behaviour. *Nature* 175, 1088–89.

Michelson, E. H. (1964). The protective action of *Chaetogaster limnaei* on snails exposed to *Schistosoma mansoni*. *Journal of Parasitology* 50, 441–44.

Mikheev, N., Valtonen, E. T. and Rintamäki-Kinnunen, P. (1998). Host searching in *Argulus foliaceus* L. (Crustacea: Branchiura): the role of vision and selectivity. *Parasitology* 116, 425–30.

Milinski, M. (1984). Parasites determine a predator's optimal feeding strategy. *Behavioral Ecology and Sociobiology* 15, 35–37.

Milinski, M. (1985). Risk of predation of parasitized sticklebacks (*Gasterosteus aculeatus* L.) under competition for food. *Behaviour* 93, 203–16.

Milinski, M. (1990). Parasites and host decision-making. In *Parasitism and host behaviour* (eds. C. J. Barnard and J. M. Behnke), pp. 95–116. Taylor & Francis, London.

Milinski, M. and Bakker, T. C. M. (1990). Female sticklebacks use male coloration in mate choice and hence avoid parasitized males. *Nature* 344, 330–33.

Milinski, M. and Löwenstein, C. (1980). On predator selection against abnormalities of movement: a test of an hypothesis. *Zeitschrift fur Tierpsychologie* 53, 325–40.

Miller, D. H. and Getz, L. L. (1969). Botfly infections in a population of *Peromyscus leucopus*. *Journal of Mammalogy* 50, 277–83.

Miller, G. C. (1981). Helminths and the transmammary route of infection. *Parasitology* 82, 335–42.

Miller, L. A. and McClanahan, R. J. (1959). Note on occurrence of the fungus *Empusa muscae* Cohn on adults of the onion maggot, *Hylemya antiqua* (Meig.) (Diptera: Anthomyiidae). *Canadian Entomologist* 41, 525–26.

Milner, R. J., Holdom, D. G., and Glare, T. R. (1984). Diurnal patterns of mortality in aphids infected by entomopthoran fungi. *Entomologia Experimentalis et Applicata* 36, 37–42.

Mims, C. A., Day, M. F., and Marshall, I. D. (1966) Cytopathic effect of Semliki Forest virus in the mosquito *Aedes aegypti*. *American Journal of Tropical Medicine and Hygiene* 15, 775–84.

Minchella, D. J. (1985). Host life-history variation in response to parasitism. *Parasitology* 90, 205–16.

Minchella, D., Eddings, A. R., and Neel, S. T. (1994). Genetic, phenotypic, and behavior variation in North American sylvatic isolates of *Trichinella*. *Journal of Parasitology* 80, 696–704.

Minchella, D. J. and Loverde, P. T. (1981). A cost of increased early reproductive effort in the snail *Biomphalaria glabrata*. *American Naturalist* 118, 876–81.

Minchella, D. J. and Loverde, P. T. (1983). Laboratory comparison of the relative success of *Biomphalaria glabrata* stocks which are susceptible and insusceptible to infection with *Schistosoma mansoni*. *Parasitology* 86, 335–44.

Minchella, D. J. and Scott, M. E. (1991). Parasitism: a cryptic determinant of animal community structure. *Trends in Ecology and Evolution* 6, 250–54.

Mitchell, C. J., Bowen, G. S., Monath, T. P., Cropp, C. B., and Kerschner, J. (1979). St. Louis encephalitis virus transmission following multiple feeding of *Culex pipiens pipiens* (Diptera: Culicidae) during a single gonotrophic cycle. *Journal of Medical Entomology* 16, 254–58.

Mitchell, D., Laburn, H. P., Matter, M., and McClain, E. (1990). Fever in Namib and other ectotherms. In *Namib ecology: 25 years of Namib research* (ed. M. K. Seely), pp. 179–92. Transvaal Museum Monograph No. 7. Transvaal Museum, Pretoria.

Moffett, J. O. and Lawson, F. A. (1975). Effect of *Nosema* infection on O_2 consumption by honey bees. *Journal of Economic Entomology* 68, 627–29.

Møller, A. P. (1989). Parasites, predators and nest boxes: facts and artefacts in nest box studies of birds? *Oikos* 56, 421–23.

Møller, A P. (1991a). Parasite load reduces song output in a passerine bird. *Animal Behaviour* 41, 723–30.

Møller, A. P. (1991b). The preening activity of swallows, *Hirundo rustica*, in relation to experimentally manipulated loads of haematophagous mites. *Animal Behaviour* 42, 251–60.

Møller, A. P. (1992). Parasites differentially increase the degree of fluctuating asymmetry in secondary sexual characters. *Journal of Evolutionary Biology* 5, 691–99.

Møller, A. P. (1993). Fungus-infected dead domestic flies manipulate sexual behaviour of conspecifics. *Behavioral Ecology and Sociobiology* 33, 403–7.

Møller, A. P. (1994). *Sexual selection and the barn swallow.* Oxford University Press, Oxford.

Møller, A. P. (1997). Parasitism and the evolution of host life history. In *Host–parasite evolution: general principles and avian models* (eds. D. H. Clayton and J. Moore), pp. 105–27. Oxford University Press, Oxford.

Møller, A. P., Dufva, R., and Allander, K. (1993). Parasites and the evolution of host social behavior. *Advances in the Study of Behavior* 22, 65–102.

Møller, A. P. and Saino, N. (1994). Parasites, immunology of hosts, and host sexual selection. *Journal of Parasitology* 80, 850–58.

Molloy, D. P. (1981). Mermithid parasitism of black flies (Diptera: Simuliidae). *Journal of Nematology* 13, 250–56.

Moloo, S. K. (1983). Feeding behaviour of *Glossina morsitans morsitans* infected with *Trypanosoma vivax*, *T. congolense* or *T. brucei*. *Parasitology* 86, 51–56.

Moloo, S. K. and Dar, F. (1985). Probing by *Glossina morsitans centralis* infected with pathogenic *Trypanosoma* species. *Transactions of the Royal Society of Tropical Medicine and Hygiene* 79, 119.

Molyneux, D. H. (1980). Host-trypanosome interactions in *Glossina*. *Insect Science and Its Application* 1, 39–46.

Molyneux, D. H. and Elce, B. (1979). A possible relationship between salivarian trypanosomes and *Glossina labrum* mechano-receptors. *Annals of Tropical Medicine and Parasitology* 73, 287–90.

Molyneux, D. H. and Jefferies, D. (1986). Feeding behaviour of pathogen-infected vectors. *Parasitology* 92, 721–36.

Molyneux, D. H. and Jenni, L. (1981). Mechanoreceptors, feeding behaviour and trypanosome transmission in *Glossina*. *Transactions of the Royal Society of Tropical Medicine and Hygiene* 75, 160–63.

Moore, D., Reed, M., Le Patourel, G., Abraham, Y. J., and Prior, C. (1992). Reduction of feeding by the desert locust, *Schistocerca gregaria*, after infection with *Metarhizium flavoviride*. *Journal of Invertebrate Pathology* 60, 304–7.

Moore, J. (1981). Asexual reproduction and environmental predictability in cestodes (Cyclophyllidea: Taeniidae). *Evolution* 35, 723–41.

Moore, J. (1983a). Responses of an avian predator and its isopod prey to an acanthocephalan parasite. *Ecology* 64, 1000–15.

Moore, J. (1983b). Altered behavior in cockroaches (*Periplaneta americana*) infected with an archiacanthocephalan, *Moniliformis moniliformis*. *Journal of Parasitology* 69, 1174–76.

Moore, J. (1984a). Parasites and altered host behavior. *Scientific American* 250, 108–115.

Moore, J. (1984b). Altered behavioral responses in intermediate hosts—an acanthocephalan parasite strategy. *American Naturalist* 123, 572–77.

Moore, J. (1987). Some roles of parasitic helminths in trophic interactions: a view from North America. *Revista Chilena de Historia Natural* 60, 159–79.

Moore, J. (1993). Parasites and the behavior of biting flies. *Journal of Parasitology* 79, 1–16.

Moore J. (1995). The behavior of parasitized animals—when an ant is not an ant. *BioScience* 45, 89–96.

Moore, J. and Bell, D. H. (1983). Pathology (?) of *Plagiorhynchus cylindraceus* in the starling, *Sturnus vulgaris*. *Journal of Parasitology* 69, 387–90.

Moore, J. and Crompton, D. W. T. (1993). A quantitative study of the susceptibility of cockroach species to *Moniliformis moniliformis* (Acanthocephala). *Parasitology* 107, 63–69.

Moore, J., Freehling, M. and Gotelli, N. J. (1994). Altered behavior in two blattid cockroaches infected with *Moniliformis moniliformis* (Acanthocephala). *Journal of Parasitology* 80, 220–23.

Moore, J. and Gotelli, N. J. (1990). A phylogenetic perspective on the evolution of altered host behaviours: a critical look at the manipulation hypothesis. In *Parasitism and host behaviour* (eds. C. J. Barnard and J. M. Behnke), pp. 193–233. Taylor & Francis, London.

Moore, J. and Gotelli, N. J. (1992). *Moniliformis moniliformis* increases cryptic behaviors in the cockroach *Supella longipalpa*. *Journal of Parasitology* 78, 49–53.

Moore, J. and Gotelli, N J. (1996). Evolutionary patterns of altered behavior and susceptibility in parasitized hosts. *Evolution* 50, 807–19.

Moore, J. and Lasswell, J. (1986). Altered behavior in isopods (*Armadillidium vulgare*) infected with the nematode *Dispharynx nasuta*. *Journal of Parasitology* 72, 186–89.

Moore, J. and Simberloff, D. (1990). Gastrointestinal helminth communities of bobwhite quail. *Ecology* 71, 344–59.

Moore, J., Simberloff, D., and Freehling, M. (1988). Relationships between bobwhite quail social-group size and intestinal helminth parasitism. *American Naturalist* 131, 22–32.

Mooring, M. S. (1995). The effect of tick challenge on grooming rate by impala. *Animal Behaviour* 50, 377–92.

Mooring, M. S. and Hart, B. L. (1992). Animal grouping for protection from parasites: selfish herd and encounter-dilution effects. *Behaviour* 123, 173–93.

Mooring, M. S. and Hart, B. L. (1997). Self grooming in impala mothers and lambs: testing the body size and tick challenge principles. *Animal Behaviour* 53, 925–34.

Mooring, M. S. and Samuel, W. M. (1998a). Tick defense strategies in bison: the role of grooming and hair coat. *Behaviour* 135, 693–718.

Mooring, M. S. and Samuel, W. M. (1998b). Tick-removal grooming by elk (*Cervus elaphus*): testing the principles of the programmed-grooming hypothesis. *Canadian Journal of Zoology* 76, 740–50.

Morales, J., Larraide, C., Arteaga, M., Govezensky, T., Romano, M. C. and Morali, G. (1996). Inhibition of sexual behavior in male mice infected with *Taenia crassiceps* cysticerci. *Journal of Parasitology* 82, 689–93.

Morris, G. K., Kerr, G. E., and Gwynne, D. T. (1975). Ontogeny of phonotaxis in *Orchelimum gladiator* (Orthoptera: Tettigoniidae: Conocephalinae). *Canadian Journal of Zoology* 53, 1127–30.

Moss, J. and Vaughan, M. (1979). Activation of adenylate cyclase by choleragen. *Annual Review of Biochemistry* 48, 581–600.

Mouritsen, K. N. (1997). Crawling behaviour in the bivalve *Macoma balthica*: the parasite-manipulation hypothesis revisited. *Oikos* 79, 513–20.

Mouritsen, K. N. and Jensen, K. T. (1997). Parasite transmission between soft-bottom invertebrates: temperature mediated infection rates and mortality in *Corophium volutator. Marine Ecology Progress Series* 151, 123–34.

Mueller, J. F. (1963). Parasite-induced weight gain in mice. *Annals of the New York Academy of Science* 113, 217–33.

Mueller, J. F. (1980). A growth factor produced by a larval tapeworm and its biological activity. In *Growth and growth factors* (eds. K. Shizume and K. Takano), pp. 193–201. University Park Press, Baltimore, MD.

Mullens, B. A. (1990). *Entomophthora muscae* (Entomophthorales: Entomophthoraceae) as a pathogen of filth flies. In *Biocontrol of arthropods affecting livestock and poultry* (eds. D. A. Rutz and R. S. Patterson), pp. 231–45. Westview Press, Boulder, CO.

Mullens, B. A., and Gerhardt, R. R. (1979). Feeding behavior of some Tennessee Tabanidae. *Environmental Entomology* 8, 1047–51.

Mullens, B. A., Rodriguez, J. L., and Meyer, J. A. (1987). An epizootiological study of *Entomophthora muscae* in muscoid fly populations on southern California poultry facilities, with emphasis on *Musca domestica. Hilgardia* 55, 1–41.

Müller, C. B. (1994). Parasitoid induced digging behaviour in bumblebee workers. *Animal Behaviour* 48, 961–66.

Müller, C. B. and Schmid-Hempel, R. (1992). To die for host or parasite? *Animal Behaviour* 44, 177–79.

Müller, C. B. and Schmid-Hempel, P. (1993). Exploitation of cold temperature as defence against parasitoids in bumblebees. *Nature* 363, 65–66.

Müller, C. B., Völkl, W., and Godfray, H. C. J. (1997). Are behavioural changes in parasitised aphids a protection against hyperparasitism? *European Journal of Entomology* 94, 221–34.

Müller-Graf, C. D. M., Collins, D. A., and Woolhouse, M E. J. (1996). Intestinal parasite burden in five troops of olive baboons (*Papio cynocephalus anubis*) in Gombe Stream National Park, Tanzania. *Parasitology* 112, 489–97.

Munro, W. R. (1953). Intersexuality in *Asellus aquaticus* L. parasitized by a larval acanthocephalan. *Nature* 172, 313.

Murray, M. D. (1961). The ecology of the louse *Polyplax serrata* (Burm.) on the mouse *Mus musculus* L. *Australian Journal of Zoology* 9, 1–13.

Murray, M. D. (1990). The influence of host behaviour on some ectoparasites of birds and mammals. In *Parasitism and host behaviour* (eds. C. J. Barnard and J. M. Behnke), pp. 290–315. Taylor & Francis, London.

Murray, P. R., Kobayashi, G. S., Pfaller, M. A., and Rosenthal, K. S. (1994). *Medical microbiology* (2nd ed.) Mosby, St. Louis, MO.

Muzzall, P. M. and Rabalais, F. C. (1975). Studies on *Acanthocephalus jacksoni* Bullock, 1962 (Acanthocephala: Echinorhynchidae). III. The altered behavior of *Lirceus lineatus* (Say) infected with cystacanths of *Acanthocephalus jacksoni. Proceedings of the Helminthological Society of Washington* 42, 116–18.

Nadler, S. A. (1995). Microevolution and the genetic structure of parasite populations. *Journal of Parasitology* 81, 395–403.

Nair, K. V., Rajendran, M., and Nadakal, A. M. (1982). Certain aspects of the ecology and host-parasite relations of the larval *Raillietina tetragona* (Cestoda) infection in ant vectors. In *Vectors and vector-borne diseases* (eds. K. M. Alexander and R. J. Prasad), pp. 159–64. Proceedings of the All India Symposium. Trivandrum, Kerala State, India.

Nappi, A. J. (1973). Effects of parasitization by the nematode, *Heterotylenchus autumnalis*, on mating and oviposition in the host, *Musca autumnalis*. *Journal of Parasitology* 59, 963–69.

Nelson, B. C. and Murray, M. D. (1971). The distribution of Mallophaga on the domestic pigeon (*Columba livia*). *Intenational Journal for Parasitology* 1, 21–29.

Nelson, G. S. (1990). Human behaviour and the epidemiology of helminth infections: cultural practices and microepidemiology. In *Parasitism and host behaviour* (eds. C. J. Barnard and J. M. Behnke), pp. 234–63. Taylor & Francis, London.

Nelson, S., Greene, T., and Ernhart, C. B. (1996). *Toxocara canis* infection in preschool age children: risk factors and the cognitive development of preschool children. *Neurotoxicology and Teratology* 18, 167–74.

Ness, J. H. and Foster, S. A. (1999). Parasite-associated phenotype modifications in threespine stickleback. *Oikos* 85, 127–34.

Nesse, R. M. and Williams, G. C. (1994). *Why we get sick*. Random House, New York.

Newman, G. G. and Carner, G. R. (1974). Diel periodicity of *Entomophthora gammae* in the soybean looper. *Environmental Entomology* 3, 888–90.

Newton, P. N. and Nishida, T. (1990). Possible buccal administration of herbal drugs by wild chimpanzees, *Pan troglodytes*. *Animal Behaviour* 39, 798–801.

Nickle, W. R. (1971). Behavior of the shothole borer, *Scolytus rugulosus,* altered by the nematode parasite *Neoparasitylenchus regulosi*. *Annals of the Entomological Society of America* 64, 751.

Nirula, K. K. (1957). Observations on the green muscardine fungus in populations of *Oryctes rhinoceros* L. *Journal of Economic Entomology* 50, 767–70.

Nokes, C. and Bundy, D. A. P. (1994). Does helminth infection affect mental processing and educational achievement? *Parasitology Today* 10, 14–18.

Nokes, C., Grantham-McGregor, S. M., Sawyer, A. W., Cooper, E. S., Robinson, B. A., and Bundy, D. A. P. (1992). Moderate to heavy infections of *Trichuris trichiura* affect cognitive function in Jamaican school children. *Parasitology* 104, 539–47.

Norris, K., Anwar, M., and Read, A. F. (1994). Reproductive effort influences the prevalence of haematozoan parasites in great tits. *Journal of Animal Ecology* 63, 601–10.

Norval, R. A. I. (1992). Host susceptibility to infestation with *Amblyomma hebraeum*. *Insect Science and Its Application* 13, 489–94.

Novak, M. (1979). Environmental temperature and the growth of *Mesocestoides corti* populations in mice. *International Journal for Parasitology* 9, 429–33.

Novak, M., Koschinsky, M., Smith, T., and Evans, W. S. (1986a). Growth and development of *Hymenolepis nana* in mice maintained at different environmental temperatures. *International Journal for Parasitology* 16, 13–17.

Novak, M., McMillan, E., and Evans, W. S. (1986b). The effects of environmental temperature and *Hymenolepis nana* on histopathology of the small intestine of mice. *Canadian Journal of Zoology* 64, 996–1000.

Nuttall, P. A. (1998). Displaced tick-parasite interactions at the host interface. *Parasitology* 116, S65–S72.

Obin, M. S. and Vander Meer, R. K. (1985). Gaster flagging by fire ants (*Solenopsis* spp.): functional significance of venom dispersal behavior. *Journal of Chemical Ecology* 11, 1757–88.

Obrebski, S. (1975). Parasite reproductive strategy and evolution of castration of hosts by parasites. *Science* 188, 1314–16.

Ödberg, F. O. and Francis-Smith, K. (1977). Studies on the formation of ungrazed eliminative areas in fields used by horses. *Applied Animal Ethology* 3, 27–34.

Oetinger, D. F. and Nickol. B. B. (1981). Effects of acanthocephalans on pigmentation of freshwater isopods. *Journal of Parasitology* 67, 672–84.

Oetinger, D. F. and Nickol, B. B. (1982a). Spectrophotometric characterization of integumental pigments from uninfected and *Acanthocephalus dirus*-infected *Asellus intermedius*. *Journal of Parasitology* 68, 270–75.

Oetinger, D. F. and Nickol, B. B. (1982b). Developmental relationships between acanthocephalans and altered pigmentation in freshwater isopods. *Journal of Parasitology* 68, 463–9.

Ohigashi, H., Huffman, M. A., Izutsu, D., Koshimizu, K., Kawanaka, M., Sugiyama, H., Kirby, G. C., Warhurst, D. C., Allen, D., Wright, C. W., Phillipson, J. D., Timon-David, P., Delmas, F., Elias, R., and Balansard, G. (1994). Toward the chemical ecology of medicinal plant use in chimpanzees: the case of *Vernonia amygdalina*, a plant used by wild chimpanzees possibly for parasite-related diseases. *Journal of Chemical Ecology* 20, 541–53.

Okhotina, M. V. and Nadtochy, E. V. (1970). Effect of *Mammanidula asperocutis* Sadovskaja in Skrjabin, Sihobalova et Sulc, 1954 (Nematoda), on the population size of shrews of the genus *Sorex*. *Acta Parasitologica Polonica* 18, 81–84.

Olsen, O. W. (1974). *Animal parasites: their life cycles and ecology* (3rd ed.). University Park Press, Baltimore, MD.

Olsen, O. W. and Lyons, E. T. (1965). Life cycle of *Uncinaria lucasi* Stiles, 1901 (Nematoda: Ancylostomatidae) of fur seals, *Callorhinus ursinus* Linn., on the Pribilof Islands, Alaska. *Journal of Parasitology* 51, 689–700.

Olson, L. J. and Rose, J. E. (1966). Effect of *Toxocara canis* infection on the ability of white rats to solve maze problems. *Experimental Parasitology* 19, 77–84.

Oma, E. A. and Hewitt, G. B. (1984). Effect of *Nosema locustae* (Microsporida: Nosematidae) on food consumption in the differential grasshopper (Orthoptera: Acrididae). *Journal of Economic Entomology* 77, 500–1.

Oi, D. H. and Pereira, R. M. (1993). Ant behavior and microbial pathogens (Hymenoptera: Formicidae). *Florida Entomologist* 76, 63–73.

Oppliger, A., Christe, P., and Richner, H. (1996a). Clutch size and malaria resistance. *Nature* 381, 565.

Oppliger, A., Célérier, M. L., and Clobert, J. (1996b). Physiological and behaviour changes in common lizards parasitized by haemogregarines. *Parasitology* 113, 433–38.

Oppliger, A., Christe, P., and Richner, H. (1997). Clutch size and malarial parasites in female great tits. *Behavioral Ecology* 8, 148–52.

Oppliger, A., Richner, H., and Christe, P. (1994). Effect of an ectoparasite on lay date, nest-site choice, desertion, and hatching success in the great tit (*Parus major*). *Behavioral Ecology* 5, 130–34.

O'Rourke, F. J. (1956). The medical and veterinary importance of the Formicidae. *Insectes Sociaux* 3, 107–18.

Orr, M. R. (1992). Parasitic flies (Diptera: Phoridae) influence foraging rhythms and caste division of labor in the leaf-cutter ant, *Atta cephalotes* (Hymenoptera: Formicidae). *Behavioral Ecology and Sociobiology* 30, 395–402.

Orr, T. S. C. (1966). Spawning behaviour of rudd, *Scardinius erythrophthalmus* infested with plerocercoids of *Ligula intestinalis*. *Nature* 212, 736.

Ovington, K. S. (1985). Dose-dependent relationships between *Nippostrongylus brasiliensis* populations and rat food intake. *Parasitology* 91, 157–67.

Owen, S. F., Barber, I., and Hart, P. J. B. (1993). Low level infection by eye fluke, Diplostomum spp., affects the vision of three-spined sticklebacks, Gasterosteus aculeatus. Journal of Fish Biology 42, 803–6.

Page, J. E., Huffman, M. A., Smith, V., and Towers, G. H. N. (1997). Chemical basis for medicinal consumption of Aspilia (Asteraceae) leaves by chimpanzees: a reanalysis. Journal of Chemical Ecology 23, 2211–25.

Pappas, P. W. and Wardrop, S. M. (1997). Preliminary biochemical characterization of faeces from uninfected rats and rats infected with the tapeworm, Hymenolepis diminuta. Journal of Helminthology 71, 57–59.

Pappas, P. W., Marschall, E. A., Morrison, S. E., Durka, G. M., and Daniel, C. S. (1995). Increased coprophagic activity of the beetle, Tenebrio molitor, on feces containing eggs of the tapeworm, Hymenolepis diminuta. International Journal for Parasitology 25, 1179–84.

Park, T. (1948). Experimental studies of interspecific competition. I. Competition between populations of the flour beetles, Tribolium confusum Duval and Tribolium castaneum Herbst. Ecological Monographs 18, 265–308.

Parker, C.D. and Schneider, D. R. (1981). Microorganism adaptation to host defenses. In Infection: the physiologic and metabolic responses of the host (eds. M. C. Powanda and P. G. Canonico), pp. 297–318. Elsevier/North-Holland Biomedical Press, Amsterdam.

Parker, F. D. and Pinnell, R. E. (1973). Effect on food consumption of the imported cabbageworm when parasitized by two species of Apanteles. Environmental Entomology 2, 216–19.

Paschke, J. D. and Hamm, J. J. (1961). A nuclear polyhedrosis of Rachiplusia ou (Guenee). Journal of Insect Pathology 3, 333–34.

Pasternak, A. F., Huntingford, F. A., and Crompton, D. W. T. (1995). Changes in metabolism and behaviour of the freshwater copepod Cyclops strenuus abyssorum infected with Diphyllobothrium spp. Parasitology 110, 395–99.

Patel, K. J., Rueda, L. M., Axtell, R. C., and Stinner, R. E. (1991). Temperature-dependent development of the fungal pathogen Lagenidium giganteum (Oomycetes: Lagenidiales) in larvae of Culex quinquefasciatus (Diptera: Culicidae). Journal of Medical Entomology 28, 95–100.

Patel, N. Y., Otieno, L. H., and Golder, T. K. (1982). Effect of Trypanosoma brucei infection on the salivary gland secretions of the tsetse Glossina morsitans morsitans (Westwood). Insect Science and Its Application 3, 35–38.

Patrican, L. A., DeFoliart, G. R., and Yuill, T. M. (1985). La Crosse viremias in juvenile, subadult and adult chipmunks (Tamias striatus) following feeding by transovarially-infected Aedes triseriatus. American Journal of Tropical Medicine and Hygiene 34, 596–602.

Pearre, S., Jr. (1976). Gigantism and partial parasitic castration of chaetognatha infected with larval trematodes. Journal of the Marine Biological Association of the United Kingdom 56, 503–13.

Pearre, S., Jr. (1979). Niche modification in chaetognatha infected with larval trematodes (Digenea). Internationale Revue der Gesamten Hydrobiologie 64, 193–206.

Penn, D. and Potts, W. (1998a). Chemical signals and parasite-mediated sexual selection. Trends in Evolution and Ecology 13, 391–396.

Penn, D. and Potts, W. (1998b). MHC-disassortative mating preferences reversed by cross-fostering. Proceedings of the Royal Society of London B 265, 1299–1306.

Penn, D. and Potts, W. (1999). The evolution of mating preferences and major histocompatibility complex genes. The American Naturalist 153, 145–64.

Penn, D., Schneider, G., White, K., Slev, P., and Potts, W. (1998). Influenza infection neutralizes the attractiveness of male odour to female mice (*Mus musculus*). *Ethology* 104, 685–94.

Pereira, R. M. and Stimac, J. L. (1992). Transmission of *Beauveria bassiana* within nests of *Solenopsis invicta* (Hymenolepis: Formicidae) in the laboratory. *Environmental Entomology* 21, 1427–1432.

Perrin, N., Christe, P., and Richner, H. (1996). On host life-history response to parasitism. *Oikos* 75, 317–20.

Petrides, P. E., Bohlen, P., and Shively, J. E. (1984). Chemical characterization of the two forms of epidermal growth factor in murine saliva. *Biochemical and Biophysical Research Communications* 125, 218–28.

Pfennig, D. W., Ho, S. G., and Hoffman, E. A. (1998). Pathogen transmission as a selective force against cannibalism. *Animal Behaviour* 55, 1255–61.

Phares, C. K. (1992). Biological characteristics of the growth hormone-like factor from plerocercoids of the tapeworm *Spirometra mansonoides*. *Advances in Neuroimmunology* 2, 235–47.

Phares, C. K. (1996). An unusual host-parasite relationship: the growth hormone-like factor from plerocercoids of spirometrid tapeworms. *International Journal for Parasitology* 26, 575–88.

Phares, C. K. (1997). The growth hormone-like factor from plerocercoids of the tapeworm *Spirometra mansonoides* is a multifunctional protein. In *Parasites and pathogens: effects on host hormones and behavior* (ed. N. E. Beckage), pp. 99–112. Chapman and Hall, New York.

Phares, C. K. and Corkum, K. C. (1974). Effects of spirometrid plerocercoids on several species of lower vertebrates. *Comparative Biochemistry and Physiology.* 49A, 525–31.

Phares, C. K., Shaffer, J. L., and Heidrick, M. L. (1990). Characteristics, distribution, and possible evolutionary importance of the human growth hormone-like factor from plerocercoids of the tapeworm genus *Spirometra*. In *Immune recognition and evasion: molecular aspects of host-parasite interaction* (eds. L. H. Vander Ploeg, C. R. Cantor and H. J. Vogel), pp. 149–61. Academic Press, New York.

Phillips-Conroy, J. E. (1986). Baboons, diet, and disease: food plant selection and schistosomiasis. In *Current perspectives in primate social dynamics* (eds. D. M. Taub and F. A. King), pp. 287–304. Van Nostrand Reinhold, New York.

Pichon, G. (1981). Migrations des microfilaires detdes peuples océaniens. *Annales de Parasitologie* 56, 107–20.

Piek, T., Visser, J. H., and Veenendaal, R. I. (1984). Change in behaviour of the cockroach, *Periplaneta americana*, after being stung by the specid wasp *Ampulex compressa*. *Entomologia Experimentalis et Applicata* 35, 195–203.

Piersma, T. (1997). Do global patterns of habitat use and migration strategies co-evolve with relative investments in immunocompetence due to spatial variation in parasite pressure? *Oikos* 80, 623–31.

Piersma, T. and van Eerden, M. R. (1989). Feather eating in Great Crested Grebes *Podiceps cristatus*: a unique solution to the problems of debris and gastric parasites in fish-eating birds. *Ibis* 131, 477–86.

Pike, A. W. (1990). Interpreting parasite host location behaviour. *Parasitology Today* 6, 343–44.

Pilecka-Rapacz, M. (1984). Cystacanths of *Polymorphus contortus* (Bremser, 1821) (Acanthocephala, Polymorphidae) in *Asellus aquaticus* L. *Acta Parasitologica Polonica* 29, 107–10.

Pilecka-Rapacz, M. (1986). On the development of acanthocephalans of the genus *Acanthocephalus* Koelreuther, 1771, with special attention to their influence on intermediate host, *Asellus aquaticus* L. *Acta Parasitologica Polonica* 30, 233–50.

Plateaux, L. (1972). Sur les modifications produites ches une fourmi par la présence d'un parasite cestode. *Annales des Sciences Naturelles, Zoologie* 14, 203–20.

Plateaux, L. and Péru, L. (1987). Film presentation: *Leptothorax* ants parasitized by a *Cestoda*. In *Chemistry and biology of social insects* (eds. J. Eder and H. Rembold), pp. 46–47. Verlag J. Peperny, Munchen.

Platt, K. B., Linthicum, K. J., Myint, K. S. A., Innis, B. L., Lerdthusnee, K., and Vaughn, D. W. (1997). Impact of dengue virus infection on feeding behavior of *Aedes aegypti*. *American Journal of Tropical Medicine and Hygiene* 57, 119–25.

Plowright, W. (1982). The effects of rinderpest and rinderpest control on wildlife in Africa. In *Animal Disease in Relation to Animal Conservation* (ed. M. A. Edwards and U. McDonnell), pp. 1–28. Academic Press, London.

Poiani, A. (1992). Ectoparasitism as a possible cost of social life: a comparative analysis using Australian passerines (Passeriformes). *Oecologia* 92, 429–41.

Poiani, A., Goldsmith, A. R. and Evans, M. R. (2000). Ectoparasites of house sparrows (*Passer domesticus*): an experimental test of the immunocompetence handicap hypothesis and a new model. *Behavioral Ecology and Sociobiology* 47, 230–242.

Poinar, G. O., Jr. (1965). The bionomics and parasitic development of *Tripius sciarae* (Bovien) (Sphaerulariidae: Aphelenchoidea), a nematode parasite of sciarid flies (Sciaridae: Diptera). *Parasitology* 55, 559–69.

Poinar, G. O., Jr. (1991). Hairworm (Nematomorpha: Gordioidea) parasites of New Zealand wetas (Orthoptera: Stenopelmatidae). *Canadian Journal of Zoology* 69, 1592–99.

Poinar, G. O., Jr. and Benton, C. I. B., Jr. (1986). *Aranimermis aptispicula* n. g., n. sp. (Mermithidae: Nematoda), a parasite of spiders (Arachnida: Araneida). *Systematic Parasitology* 8, 33–38.

Poinar, G. O., Chabaud, A. G. and Bain, O. (1989). *Rabbium paradoxus* sp. n. (Seuratidae: Skrjabinelaziinae) maturing in *Camponotus castaneus* (Hymenoptera: Formicidae). *Proceedings of the Helminthological Society of Washington* 56, 120–24.

Poinar, G. O., Jr. and van der Laan, P. A. (1972). Morphology and life history of *Sphaerularia bombi*. Nematologica 18, 239–52.

Poinar, G. O., Jr., Lane, R. S., and Thomas, G. M. (1976). Biology and redescription of *Pheromermis pachysoma* (v. Linstow) n. gen., n. comb. (Nematoda: Mermithidae), a parasite of yellowjackets (Hymenoptera: Vespidae). Nematologica 22, 360–70.

Poirier, S. R., Rau, M. E., and Wang, X. (1995). Diel locomotory activity of deer mice (*Peromyscus maniculatus*) infected with *Trichinella nativa* or *Trichinella pseudospiralis*. *Canadian Journal of Zoology* 73, 1323–34.

Polak, M. (1996). Ectoparasitic effects on host survival and reproduction: the *Drosophila-Macrocheles* association. *Ecology* 77, 1379–89.

Poulin, R. (1991a). Group-living and infestation by ectoparasites in passerines. *Condor* 93, 418–23.

Poulin, R. (1991b). Group-living and the richness of the parasite fauna in Canadian freshwater fishes. *Oecologia* 86, 390–94.

Poulin, R. (1992). Altered behaviour in parasitized bumblebees: parasite manipulation or adaptive suicide? *Animal Behaviour* 44, 174–76.

Poulin, R. (1993a). Age-dependent effects of parasites on antipredatory responses in 2 New Zealand freshwater fish. *Oecologia* 96, 431–38.

Poulin, R. (1993b). A cleaner perspective on cleaning symbiosis. *Reviews in Fish Biology and Fisheries* 3, 75–79.

Poulin, R. (1994a). The evolution of parasite manipulation of host behaviour: a theoretical analysis. *Parasitology* 109, S109–18.

Poulin, R. (1995). "Adaptive" changes in the behaviour of parasitized animals: a critical review. *International Journal for Parasitology* 25, 1371–83.

Poulin, R. (1998). Evolution and phylogeny of behavioural manipulation of insect hosts by parasites. *Parasitology* 116, S3–S11.

Poulin, R. (1999). Parasitism and schoal size in juvenile sticklebacks: conflicting selection pressures from different ectoparasites? *Ethology* 105, 959–68.

Poulin, R. (2000). Manipulation of host behaviour by parasites: a weakening paradigm? *Proceedings of the Royal Society of London B* 267, 1–6.

Poulin, R., Brodeur, J., and Moore, J. (1994). Parasite manipulation of host behaviour: should hosts always lose? *Oikos* 70, 479–84.

Poulin, R., Curtis, M. A., and Rau, M. E. (1991a). Size, behaviour, and acquisition of ectoparasitic copepods by brook trout, *Salvelinus fontinalis*. *Oikos* 61, 169–74.

Poulin, R., Curtis, M. A., and Rau, M. E. (1992). Effects of *Eubothrium salvelini* (Cestoda) on the behaviour of *Cyclops vernalis* (Copepods) and its susceptibility to fish predators. *Parasitology* 105, 265–71.

Poulin, R. and FitzGerald, G. J. (1989a). Risk of parasitism and microhabitat selection in juvenile sticklebacks. *Canadian Journal of Zoology* 67, 14–18.

Poulin, R. and FitzGerald, G. J. (1989b). Shoaling as an anti-ectoparasite mechanism in juvenile sticklebacks (*Gasterosteus* spp.). *Behavioral Ecology and Sociobiology* 24, 251–55.

Poulin, R. and Grutter, A. S. (1996). Cleaning symbioses: proximate and adaptive explanations. *BioScience* 46, 512–17.

Poulin, R., Rau, M. E., and Curtis, M. A. (1991b). Infection of brook trout fry, *Salvelinus fontinalis,* by ectoparasitic copepods: the role of host behaviour and initial parasite load. *Animal Behaviour* 41, 467–76.

Poulin, R. and Thomas, F. (1999). Phenotypic variability induced by parasites: extent and evolutionary implications. *Parasitology Today* 15, 28–32.

Prescott, L. M., Harley, J. P., and Klein, D. A. (1990). *Microbiology*. Wm. C. Brown, Dubuque, Iowa.

Prest, D. B., Gilliam, M., Taber III, S. and Mills, J. P. (1974). Fungi associated with discolored honey bee, *Apis mellifera*, larvae and pupae. *Journal of Invertebrate Pathology* 24, 253–55.

Price, P. W. (1980). *Evolutionary biology of parasites*. Princeton University Press, Princeton, NJ.

Price, P. W., Westoby, M., and Rice, B. (1988). Parasite-mediated competition: some predictions and tests. *American Naturalist* 131, 544–55.

Price, P. W., Westoby, M., Rice, B., Atsatt, P. R., Fritz, R. S., Thompson, J. N., and Mobley, K. (1986). Parasite mediation in ecological interactions. *Annual Review of Ecology and Systematics* 17, 487–505.

Pulgar, J., Aldana, M., Vergara, E., and George-Nascimento, M. (1995). La conducta de la jaiba estuaria *Hemigrapsus crenulatus* (Milne-Edwards 1837) en relacion al paratismo por el acantocefala *Profillicollis antarcticus* (Zdzitowiecki 1985) en el sur de Chile. *Revista Chilena de Historia Natural* 68, 439–50.

Putnam, J. L. and Scott, T. W. (1995). The effect of multiple host contacts on the infectivity of Dengue-2 virus-infected *Aedes aegypti*. *Journal of Parasitology* 81, 170–74.

Quinn, S. C., Brooks, R. J., and Cawthorn, R. J. (1987). Effects of the protozoan parasite *Sarcocystis rauschorum* on open-field behaviour of its intermediate vertebrate host, *Dicrostonyx richardsoni*. *Journal of Parasitology* 73, 265–71.

Radabaugh, D. C. (1980a). Encystment site selection in the brain-inhabiting metacercariae of *Ornithodiplostomum ptychocheilus* (Trematoda: Strigeoidea). *Journal of Parasitology* 66, 183–84.

Radabaugh, D. C. (1980b). Changes in minnow, *Pimephales promelas* Rafinesque, schooling behaviour associated with infections of brain-encysted larvae of the fluke, *Ornithodiplostomum ptychocheilus*. *Journal of Fish Biology* 16, 621–8.

Rahman, M. (1970). Effect of parasitism on food consumption of *Pieris rapae*. *Journal of Economic Entomology* 63, 820–21.

Ralley, W. E., Galloway, T. D., and Crow, G. H. (1993). Individual and group behaviour of pastured cattle in response to attack by biting flies. *Canadian Journal of Zoology* 71, 725–34.

Randolph, S. (1991). The effect of *Babesia microti* on feeding and survival in its tick vector, *Ixodes trianguliceps*. *Parasitology* 102, 9–16.

Randolph, S. E. (1998). Ticks are not insects: consequences of contrasting vector biology for transmission potential. *Parasitology Today* 14, 186–92.

Randolph, S. E. and Nuttall, P. A. (1994). Nearly right or precisely wrong? Natural versus laboratory studies of vector-borne diseases. *Parasitology Today* 10, 458–62.

Randolph, S. E., Williams, B. G., Rogers, D. J., and Connor, H. (1992). Modelling the effect of feeding-related mortality on the feeding strategy of tsetse (Diptera: Glossinidae). *Medical and Veterinary Entomology* 6, 231–40.

Ranta, E. (1995). *Schistocephalus* infestation improves prey-size selection by three-spined sticklebacks, *GasTersteus aculeatus*. *Journal of Fish Biology* 46, 156–58.

Rasa, O. A. E. (1983). A case of invalid care in wild dwarf mongooses. *Zeitschrift fur Tierpsychologie* 62, 235–40.

Rasmussen, E. (1959). Behaviour of sacculinized shore crabs (*Carcinus maenas* Pennant). *Nature* 183, 479–80.

Rätti, O., Dufva, R., and Alatalo, R. V. (1993). Blood parasites and male fitness in the pied flycatcher. *Oecologia* 96, 410–14.

Rau, M. E. (1982). Behavioural pathology and parasite transmission: a speculative synthesis. In *Aspects of parasitology* (ed. E. Meerovitch), pp. 335–60. The Institute of Parasitology, McGill University, Montreal.

Rau, M. E. (1983a). The open-field behaviour of mice infected with *Trichinella spiralis*. *Parasitology* 86, 311–18.

Rau, M. E. (1983b). Establishment and maintenance of behavioural dominance in male mice infected with *Trichinella spiralis*. *Parasitology* 86, 319–22.

Rau, M. E. (1984a). The open-field behaviour of mice infected with *Trichinella pseudospiralis*. *Parasitology* 88, 415–19.

Rau, M. E. (1984b). Loss of behavioural dominance in male mice infected with *Trichinella spiralis*. *Parasitology* 88, 371–73.

Rau, M. E. (1985). The effects of *Trichinella spiralis* infection of pregnant mice on the future behavior of their offspring. *Journal of Parasitology* 71, 774–78.

Rau, M. E. and Caron, F. R. (1979). Parasite-induced susceptibility of moose to hunting. *Canadian Journal of Zoology* 57, 2466–68.

Rau, M. E. and Putter, L. (1984). Running responses of *Trichinella spiralis*-infected CD-1 mice. *Parasitology* 89, 579–83.

Read, A. F. (1990). Parasites and the evolution of host sexual behavior. In *Parasitism and host behaviour* (eds. C. J. Barnard and J. M. Behnke), pp. 117–57. Taylor & Francis, London.

Read, A. F. (1991). Passerine polygyny: a role for parasites? *American Naturalist* 138, 434–59.

Read, A. F. (1994). The evolution of virulence. *Trends in Microbiology* 2, 73–76.

Read, A. F. and Skorping, A. (1995). The evolution of tissue migration by parasitic nematode larvae. *Parasitology* 111, 359–71.

Read, C. P. (1970). *Parasitism and symbiology*. Ronald Press, New York.

Reddy, G. V. P., Furlong, M. J., Pell, J. K., and Poppy, G. M. (1998). *Zoophthora radicans* infection inhibits the response to and production of sex pheromone in the diamondback moth. *Journal of Invertebrate Pathology* 72, 167–69.

Reed, D. A. and N. E. Beckage. (1997). Inhibition of testicular growth and development in *Manduca sexta* larva parasitized by the braconid wasp *Cotesia congregata*. *Journal of Insect Physiology* 43, 29–38.

Reinhard, K. J. (1990). Archaeoparasitology in North America. *American Journal of Physical Anthropology* 82, 145–63.

Reinhard, K. J., Ambler, J. R., and McGuffie, M. (1985). Diet and parasitism at Dust Devil Cave. *American Antiquity* 50, 819–24.

Rempel, J. G. 1940. Intersexuality in Chironomidae induced by nematode parasitism. *Journal of Experimental Zoology* 84, 261–89.

Rennie, J. (1992). Living Together. *Scientific American* January 104–13.

Ressel, S. and Schall, J. J. (1989). Parasites and showy males: malarial infection and color variation in fence lizards. *Oecologia* 78, 158–64.

Retnakaran, A., Lauzon, H., and Fast, P. (1983). *Bacillus thuringiensis* induced anorexia in the spruce budworm, *Choristoneura fumiferana*. *Entomologia Experimentalis et Applicata* 34, 233–39.

Ribeiro, J. M. C. (1984). Role of mosquito saliva in blood vessel location. *Journal of Experimental Biology* 108, 1–7.

Ribeiro, J. M. C. (1995). Blood-feeding arthropods: live syringes or invertebrate pharmacologists? *Infectious Agents and Disease* 4, 143–52.

Ribeiro, J. M. C. (1996). Common problems of arthropod vectors of disease. In *The biology of disease vectors* (eds. B. J. Beaty and W. C. Marquardt), pp. 25–33. University Press of Colorado, Niwot, Colorado.

Ribeiro, J. M. C., Rossignol, P. A., and Spielman, A. (1984). Role of mosquito saliva in blood vessel location. *Journal of Experimental Biology* 108, 1–7.

Ribeiro, J. M. C., Rossignol, P. A., and Spielman, A. (1985). *Aedes aegypti*: model for blood finding strategy and prediction of parasite manipulation. *Experimental Parasitology* 60, 118–32.

Richner, H., Christe, P., and Oppliger, A. (1995). Paternal investment affects prevalence of malaria. *Proceedings of the National Academy of Sciences (U.S.A.) 92,* 1192–94.

Richner, H. and Heeb, P. (1995). Are clutch and brood size patterns in birds shaped by ectoparasites? *Oikos* 73, 435–41.

Risser, A. C., Jr. (1975). Experimental modification of reproductive performance by density in captive starlings. *Condor* 77, 125–32.

Ritchie, L. E. and Høeg, J. T. (1981). The life history of *Lernaeodiscus porcellanae* (Cirripedia: Rhizocephala) and co-evolution with its porcellanid host. *Journal of Crustacean Biology* 1, 334–37.

Ritter, R. C. and Epstein, A. N. (1974). Saliva lost by grooming: a major item in the rat's water economy. *Behavioral Biology* 11, 581–85.

Robb, T. and Reid, M. L. (1996). Parasite-induced changes in the behaviour of cestode-infected beetles: adaptation or simple pathology? *Canadian Journal of Zoology 74*, 1268–74.

Robert, D., Amoroso, J., and Hoy, R. R. (1992). The evolutionary convergence of hearing in a parasitoid fly and its cricket host. *Science* 258, 1135–37.

Roberts, L. S. and Janovy, J., Jr. (2000). *Foundations of parasitology* (6th ed.). Wm. Brown Publ., Dubuque, Iowa.

Roberts, L. W. (1981). Probing by *Glossina morsitans morsitans* and transmission of *Trypanosoma (Nannomonas) congolense*. *American Journal of Tropical Medicine and Hygiene* 30, 948–51.

Robinson, J. (1962). *Pilobolus* spp. and the translation of the infective larvae of *Dictyocaulus viviparus* from faeces to pastures. *Nature* 193, 353–54.

Rockwood, L. P. (1950). Entomogenous fungi of the family Entomophthoraceae in the Pacific Northwest. *Journal of Economic Entomology* 43, 704–7.

Rodriguez, E., Aregullin, M., Nishida, T., Uehara, S., Wrangham, R., Abramowski, Z., Finlayson, A., and Towers, G. H. N. (1985). Thiarubrine A, a bioactive constituent of *Aspilia* (Asteraceae) consumed by wild chimpanzees. *Experientia* 41, 419–20.

Roffey, J. (1968). The occurrence of the fungus *Entomophthora grylli* Fresenius on locusts and grasshoppers in Thailand. *Journal of Invertebrate Pathology* 11, 237–41.

Rolff, J. (1999). Parasitism increases offspring size in a damselfly: experimental evidence for parasite-mediated maternal effects. *Animal Behaviour* 58, 1105–8.

Romig, T., Lucius, R., and Frank, W. (1980). Cerebral larvae in the second intermediate host of *Dicrocoelium dendriticum* (Rudolphi, 1819) and *Dicrocoelium hospes* Looss, 1907 (Trematoda, Dicrocoeliidae). *Zeitschrift fur Parasitenkunde* 63, 277–86.

Rossignol, P. A. (1988). Parasite modification of mosquito probing behavior. *Miscellaneous Publications of the Entomological Society of America* 68, 25–28.

Rossignol, P. A., Ribeiro, J. M. C., Jungery, M., Turell, M. J., Spielman, A., and Bailey, C. L. (1985). Enhanced mosquito blood-finding success on parasitemic hosts: evidence for vector-parasite mutualism. *Proceedings of the National Academy of Sciences (U. S. A.)* 82, 7725–27.

Rossignol, P. A., Ribeiro, J. M. C., and Spielman, A. (1984). Increased intradermal probing time in sporozoite-infected mosquitoes. *American Journal of Tropical Medicine and Hygiene* 33, 17–20.

Rossignol, P. A., Ribeiro, J. M. C., and Spielman, A. (1986). Increased biting rate and reduce fertility in sporozoite-infected mosquitoes. *American Journal of Tropical Medicine and Hygiene* 35, 277–79.

Rossignol, P. A. and Rossignol, A. M. (1988). Simulations of enhanced malaria transmission and host bias induced by modified vector blood location behaviour. *Parasitology* 97, 363–72.

Rothenbuhler, W. C. (1964). Behavior genetics of nest cleaning in honey bees. IV. Responses of F1 and backcross generations to disease-killed brood. *American Zoologist* 4, 111–23.

Rothschild, M. (1936). Gigantism and variation in *Peringia ulvae* Pennant 1777, caused by infection with larval trematodes. *Journal of the Marine Biological Association of the United Kingdom* 20, 537–46.

Rothschild, M. (1940). *Cercaria pricei*, a new trematode, with remarks on the specific characters of the "Prima" group of Xiphidiocercariae. *Journal of the Washington Academy of Sciences* 30, 437–48.

Rothschild, M. (1941a). Observations on the growth and trematode infections of *Peringia ulvae* (Pennant) 1777 in pool in the Tamar saltings, Plymouth. *Parasitology* 33, 406–15.

Rothschild, M. (1941b). The effect of trematode parasites on the growth of *Littorina neritoides* (L.) *Journal of the Marine Biological Association* 25, 69–80.

Rothschild, M. (1962). Changes in behaviour in the intermediate hosts of trematodes. *Nature* 193, 1312–13.

Rothschild, M. (1965a). Fleas. *Scientific American* 213, 44–53.

Rothschild, M. (1965b). The rabbit flea and hormones. *Endeavour* 24, 162–68.

Rothschild, M. (1969). Notes on fleas, with the first record of a mermithid nematode from the order. *Proceedings and Transactions of the British Entomological and Natural History Society* 1, 9–16.

Rothschild, M. and Ford, B. (1966). Hormones of the vertebrate host controlling ovarian regression and copulation of the rabbit flea. *Nature* 211, 261–66.

Rowland, M. and Boersma, E. (1988). Changes in the spontaneous flight activity of the mosquito *Anopheles stephensi* by parasitization with the rodent malaria *Plasmodium yoelii*. *Parasitology* 97, 221–27.

Rowland, M. W. and Lindsay, S. W. (1986). The circadian flight activity of *Aedes aegypti* parasitized with the filarial nematode *Brugia pahangi*. *Physiological Entomology* 11, 325–34.

Roy, H. E., Pell, J. K. and Alderson, P. G. (1999). Effects of fungal infection on the alarm response of pea aphids. *Journal of Invertebrate Pathology* 74, 69–75.

Royce, L. A., Rossignol, P. A., Burgett, D. M., and Stringer, B. A. (1991). Reduction of tracheal mite parasitism of honey bees by swarming. *Philosophical Transactions of the Royal Society of London B* 331, 123–29.

Rubenstein, D. I. and Hohmann, M. E. (1989). Parasites and social behavior of island feral horses. *Oikos* 55, 312–20.

Rubtsov, I. A. 1958. On the gynandromorphs and intersexes in black-flies (Simuliidae, Diptera). *Zoologicheskii Zhurnal* 37, 458–61.

Rumpus, A. E. and Kennedy, C. R. (1974). The effect of the acanthocephalan *Pomphorhynchus laevis* upon the respiration of its intermediate host, *Gammarus pulex*. *Parasitology* 68, 271–84.

Rupp, J. C. (1996). Parasite-altered behaviour: impact of infection and starvation on mating in *Biomphalaria glabrata*. *Parasitology* 113, 357–65.

Russell, L. R. (1980). Effects of *Truttaedacnitis truttae* (Nematoda: Culcullanidae) on growth and swimming of rainbow trout, *Salmo gairdneri*. *Canadian Journal of Zoology* 58, 1220–26.

Rutberg, A. T. (1987). Horse fly harassment and the social behavior of feral ponies. *Ethology* 75, 145–54.

Ryan, I. (1984). The effect of trypanosome infection on a natural population of *Glossina longipalpis* Wiedemann (Diptera: Glossinidae) in Ivory Coast. *Acta Tropica* 41, 355–59.

Salzemann, A. and Plateaux, L. (1987). Reduced egg laying by workers of the ant *Lep-*

tothorax nylanderi in presence of workers parasitized by a Cestoda. In *Chemistry and biology of social insects* (eds. J. Eder and H. Rembold), p. 45. Verlag J. Peperny, Munchen.

Samson, R. A., Evans, H. C., and van de Klashorst, G. (1981). Notes on entomogenous fungi from Ghana. V. The genera *Stilbella* and *Polycephalomyces*. *Proceedings of the Koninklijka Nederlandse Akademie van Wetenschappen ser C.* 84, 289–301.

Samuel, W. M. (1991). Grooming by moose (*Alces alces*) infested with the winter tick, *Dermacentor albipictus* (Acari): a mechanism for premature loss of winter hair. *Canadian Journal of Zoology* 69, 1255–60.

Sanchez-Peña, S. R., Buschinger, A. and Humber, R. A. (1993). *Myrmicinosporidium durum*, an enigmatic fungal parasite of ants. *Journal of Invertebrate Pathology* 61, 90–96.

Sankurathri, C. S. and Holmes, J. H. (1976). Effects of thermal effluents on parasites and commensals of *Physa gyrina* Say (Mollusca: Gastropoda) and their interactions of Lake Wabamun, Alberta. *Canadian Journal of Zoology* 54, 1742–53.

Sapp, J. (1994). *Evolution by association: a history of symbiosis.* Oxford University Press, Oxford.

Sato, Y., Tanaka, T., Imafuku, M., and Hidaka, T. (1983). How does diurnal *Apanteles kariyai* parasitize and egress from a nocturnal host larva? *Kontyu* 51, 128–39.

Saumier, M. D., Rau, M. E., and Bird, D. M. (1986). The effect of *Trichinella pseudospiralis* infection on the reproductive success of captive American kestrels (*Falco sparverius*). *Canadian Journal of Zoology* 64, 2123–25.

Saumier, M. D., M. E. Rau and D. M. Bird. (1988). The influence of *Trichinella pseudospiralis* infection on the behaviour of captive, nonbreeding American Kestrels (*Falco sparverius*). *Canadian Journal of Zoology* 66, 1685–1692.

Saumier, M. D., Rau, M. E., and Bird, D. M. (1991). Behavioural changes in breeding American kestrels infected with *Trichinella pseudospiralis*. In *Bird–parasite interactions: ecology, evolution and behaviour* (eds. J. E. Loye and M. Zuk), pp. 290–313. Oxford University Press, Oxford.

Schad, G. A. and Anderson, R. M. (1985). Predisposition to hookworm infection in humans. *Science* 228, 1537–40.

Schad, G. A., Nawalinski, T. A., and Kochar, V. (1983). Human ecology and the distribution and abundance of hookworm populations. In *Human ecology and infectious diseases* (eds. N. A. Croll and J. H. Cross), pp. 187–223. Academic Press, New York.

Schall, J. J. (1982). Lizards infected with malaria: physiological and behavioral consequences. *Science* 217, 1057–59.

Schall, J. J. (1990). Virulence of lizard malaria: the evolutionary ecology of an ancient parasite-host association. *Parasitology* 100, S35–52.

Schall, J. J. (1996). Malarial parasites of lizards: diversity and ecology. *Advances in Parasitology* 37, 256–333.

Schall, J. J., Bennett, A. F., and Putnam, R. W. (1982). Lizards infected with malaria: physiological and behavioral consequences. *Science* 217, 1057–59.

Schall, J. J. and Dearing, M. D. (1987). Malarial parasitism and male competition for mates in the western fence lizard, *Sceloporus occidentalis*. *Oecologia* 73, 389–92.

Schall, J. J. and Houle, P. R. (1992). Malarial parasitism and home range and social status of male western fence lizards, *Sceloporus occidentalis*. *Journal of Herpetology* 26, 74–76.

Schall, J. J. and Sarni, G. A. (1987). Malarial parasitism and the behavior of the lizard, *Sceloporus occidentalis. Copeia*, 84–93.

Schaub, G. A. (1992). The effects of trypanosomatids on insects. *Advances in Parasitology* 31, 255–319.

Schaub, G. A. and Schnitker, A. (1988). Influence of *Blastocrithidia triatomae* (Trypanosomatidae) on the reduviid bug *Triatoma infestans*: alterations in the Malpighian tubules. *Parasitology Research* 75, 88–97.

Schiefer, B. A., Ward, R. A., and Eldridge, B. F. (1977). *Plasmodium cynomolgi*: effects of malaria infection on laboratory flight performance of *Anopheles stephensi* mosquitoes. *Experimental Parasitology* 41, 397–404.

Schmid-Hempel, P. (1998). *Parasites in Social Insects*. Princeton University Press, Princeton, N. J.

Schmid-Hempel, P. and Durrer S. (1991). Parasites, floral resources and reproduction in natural populations of bumblebees. *Oikos* 62, 342–50.

Schmid-Hempel, P. and Koella, J. C. (1994). Variability and its implications for host-parasite interactions. *Parasitology Today* 10, 98–102.

Schmid-Hempel, R. and Müller, C. B. (1991). Do parasitized bumblebees forage for their colony? *Animal Behaviour* 41, 910–12.

Schmid-Hempel, P. and Schmid-Hempel, R. (1990). Endoparasitic larvae of conopid flies alter pollination behavior of bumblebees. *Naturwissenschaften* 77, 450–52.

Schmid-Hempel, R. and Schmid-Hempel, P. (1991). Endoparasitic flies, pollen-collection by bumblebees and a potential host-parasite conflict. *Oecologia* 87, 227–32.

Schmid-Hempel, P. and Stauffer, H.-P. (1998). Parasites and flower choice of bumblebees. *Animal Behaviour* 55, 819–25.

Schmid-Hempel, P. and Tanner, M. (1990). The evolution, ecological effects and health impact of parasites. *Parasitology Today* 6, 278–80.

Schmidt, G. D. (1964). Life cycle and development of *Prosthorhynchus formosus* (Van Cleave, 1918) Travassos, 1926, an acanthocephalan parasite of birds. Dissertation, Colorado State University, Fort Collins.

Schmidt, G. D. (1986). *Handbook of tapeworm identification*. CRC Press, Boca Raton, FL.

Schmidt, G. D. and L. S. Roberts. (1989). *Foundations of Parasitology* (4th ed.). Times Mirror/ Mosby College Publ., St. Louis, Missouri, 750 pp.

Schmidtmann, E. T. and Valla, M. E. (1982). Face-fly pest intensity, fly-avoidance behavior (bunching) and grazing time in Holstein heifers. *Applied Animal Ethology* 8, 429–38.

Schwartz J. M. (1997). Obsessive-compulsive disorder. *Science and Medicine* (March/ April), 14–23.

Scott, M. E. (1985). Experimental epidemiology of *Gyrodactylus bullatarudis* (Monogenea) on guppies (*Poecilia reticulata*): short- and long-term studies. In *Ecology and genetics of host-parasite interactions* (eds. D. Rollinson and R. M. Anderson), pp. 21–38. Academic Press, London.

Scott, M. E. (1988). The impact of infection and disease on animal populations: implications for conservation biology. *Conservation Biology* 2, 40–56.

Scott, T. W. and Edman, J. D. (1991). Effects of avian host age and arbovirus infection on mosquito attraction and blood-feeding success. In *Bird–parasite interactions: ecology, evolution and behaviour* (eds. J. E. Loye and M. Zuk), pp. 179–204. Oxford University Press, Oxford.

Scott, T. W., Edman, J. D., and Lorenz, L. H. (1988). Effects of disease on vertebrates' ability behaviorally to repel host-seeking mosquitoes. In *The role of vector-host interactions in disease transmission* (ed. T. W. Scott and J. Grumstrup-Scott), pp. 9–17. Miscellaneous Publication No. 68, Entomological Society of America College Park, Maryland.

Scott, T. W., Lorenz, L. H., and Edman, J. D. (1990). Effects of house sparrow age and arbovirus infection on attraction of mosquitoes. *Journal of Medical Entomology* 27, 856–63.

Seed, J. R. and Khalili, N. (1971). The changes in locomotor rhythms of *Microtus montanus* infected with *Trypanosoma gambiense. Journal of Interdisciplinary Cycle Research* 2, 91–99.

Seidenberg, A. J. (1973). Ecology of the acanthocephalan, *Acanthocephalus dirus* (Van Cleave, 1931), in its intermediate host, *Asellus intermedius* Forbes (Crustacea: Isopoda). *Journal of Parasitology* 59, 957–62.

Semel, M. 1956. Polyhedrosis wilt of cabbage looper on Long Island. *Journal of Economic Entomology* 49, 420–21.

Sengupta, S. (1981). Adaptive significance of the use of *Margosa* leaves in nests of house sparrows *Passer domesticus. Emu* 81, 114–15.

Serpell, J. (1986). *In the company of animals.* Cambidge University Press, Cambridge.

Shapiro, A. M. (1976). Beau geste! *American Naturalist* 110, 900–2.

Shaw, J. L. (1990). Effects of the caecal nematode *Trichostrongylus tenuis* on egg-laying by captive red grouse. *Research in Veterinary Science* 48, 59–63.

Sheldon, B. C. (1993). Sexually transmitted disease in birds: occurrence and evolutionary significance. *Philosophical Transactions of the Royal Society of London* B 339, 491–97.

Sherman, P. W. and Billing, J. (1999). Darwinian gastronomy: why we use spices. Bio-Science 49, 453–63.

Shoham, S., Davenne, D., Cady, A. B., Dinarello, C. A., and Krueger, J. M. (1987). Recombinant tumor necrosis factor and interleukin 1 enhance slow-wave sleep. *American Journal of Physiology* 253, R142–49.

Shoop, W. L. (1988). Trematode transmission patterns. *Journal of Parasitology* 74, 46–59.

Shoop, W. L. (1994). Vertical transmission in the Trematoda. *Journal of the Helminthological Society of Washington* 61, 153–61.

Shoop, W. L. and Corkum. K. C. (1984). Pathway of mesocercariae of *Alaria marcianae* (Trematoda) through the mammary glands of lactating mice. *Journal of Parasitology* 70, 333–36.

Shoop, W. L. and Corkum, K. C. (1987). Maternal transmission by *Alaria marcianae* (Trematoda) and the concept of amphiparatenesis. *Journal of Parasitology* 73, 110–15.

Shortt, H. E., Barraud, P. J., and Craighead, A. C. (1926). The life-history and morphology of *Herpetomonas donovani* in the sandfly *Phlebotomus argentipes. Indian Journal of Medical Research* 13, 947–59.

Shostak, A. W. and Esch, G. W. (1990). Photocycle-dependent emergence by cercariae of *Halipegus occidualis* from *Helisoma anceps*, with special reference to cercarial emergence patterns as adaptations for transmission. *Journal of Parasitology* 76, 790–95.

Shostak, A. W. and Smyth, K. A. (1998). Activity of flour beetles (*Tribolium confusum*) in the presence of feces from rats infected with rat tapeworm (*Hymenolepis diminuta*). *Canadian Journal of Zoology* 76, 1472–79.

Shykoff, J. A. and Widmer, A. (1996). Parasites and carotenoid-based signal intensity: how general should the relationship be? *Naturwissenschaften* 83, 113–21.

Siebeneicher, S. R., Vinson, S. B., and Kenerley, C. M. (1992). Infection of the red imported fire ant by *Beauveria bassiana* through various routes of exposure. *Journal of Invertebrate Pathology* 59, 280–85.

Sigerist, H. E. (1965). *Civilization and disease.* University of Chicago Press, Chicago.

Simberloff, D. and Moore, J. (1997). Communities of parasites and free-living animals. In *Host–parasite evolution: general principles and avian models* (eds. D. H. Clayton and J. Moore), pp. 174–97. Oxford University Press, Oxford.

Simmons, L. W. (1994). Courtship role reversal in bush crickets: another role for parasites? *Behavioral Ecology* 5, 259–66.

Sindermann, C. J. (1960). Ecological studies of marine dermatitis-producing schistosome larvae in northern new England. *Ecology* 41, 678–84.

Sindermann, C. J. and Farrin, A. E. (1962). Ecological studies of *Cryptocotyle lingua* (Trematoda: Heterophyidae) whose larvae cause "pigment spots" of marine fish. *Ecology* 43, 69–75.

Skaife, S. H. 1925. The locust fungus, *Empusa grylli*, and its effects on its host. *South African Journal of Science* 22, 298–308.

Smirnoff, W. A. (1965). Observations on the effect of virus infection on insect behavior. *Journal of Invertebrate Pathology* 7, 387–88.

Smith, D. H. (1978a). Effects of bot fly (*Cuterebra*) parasitism on activity patterns of *Peromyscus maniculatus* in the laboratory. *Journal of Wildlife Diseases* 14, 28–39.

Smith, D. H. (1978b). Vulnerability of bot fly (*Cuterebra*) infected *Peromyscus maniculatus* to shorttail weasel predation in the laboratory. *Journal of Wildlife Diseases* 14, 40–51.

Smith, E. N. (1984). Alteration of behavior by parasites: a problem for evolutionists. *Creation Research Society Quarterly* 21, 124.

Smith, J. E. and Dunn, A. M. (1991). Transovarial transmission. *Parasitology Today* 7, 146–48.

Smith, R. S. and Kramer, D. L. (1987). Effects of a cestode (*Schistocephalus* sp.) on the response of ninespine sticklebacks (*Pungitius pungitius*) to aquatic hypoxia. *Canadian Journal of Zoology* 65, 1862–65.

Smith Trail, D. R. (1980). Behavioral interactions between parasites and hosts: host suicide and the evolution of complex life cycles. *American Naturalist* 116, 77–91.

Sodeinde, O. A. and Goguen, J. D. (1988). Genetic analysis of the 9.5–kilobase virulence plasmid of *Yersinia pestis. Infection and Immunity* 56, 2743–48.

Soper, R. S. (1963). *Massospora levispora*, a new species of fungus pathogenic to the cicada, *Okanagana rimosa. Canadian Journal of Botany* 41, 875–78.

Soper, R. S., Delyzer, A. J., and Smith, L. F. R. (1976). The genus *Massospora* entomopathogenic for cicadas. Part II. Biology of *Massospora levispora* and its host *Okanagana rimosa*, with notes on *Massospora cicadina* on the periodical cicadas. *Annals of the Entomological Society of America* 69, 89–95.

Sorci, G., Massot, M., and Clobert, J. (1994). Maternal parasite load increases sprint speed and philopatry in female offspring of the common lizard. *The American Naturalist* 144, 153–64.

Sousa, W. P. (1983). Host life history and the effect of parasitic castration on growth: a field study of *Cerithidea californica* Haldeman (Gastropoda: Prosobranchia) and its trematode parasites. *Journal of Experimental Marine Biology and Ecology* 73, 273–96.

Sparks, A. K. and Chew, K. K. (1966). Gross infestation of the littleneck clam, *Venerupis staminea*, with a larval cestode (*Echeneibothrium* sp.). *Journal of Invertebrate Pathology* 8, 413–16.

Speare, A. T. (1921). *Massospora cicadina* Peck, a fungous parasite of the periodical cicada. *Mycologia* 13, 72–82.

Spindler, E.-M., Zahler, M., and Loos-Frank, B. (1986). Behavioural aspects of ants as second intermediate hosts of *Dicrocoelium dendriticum*. *Zeitschrift fur Parasitenkunde* 72, 689–92.

Sprengel, G. and Lüchtenberg, H. (1991). Infection by endoparasites reduces maximum swimming speed of European smelt *Osmerus eperlanus* and European eel *Anguilla anguilla*. *Diseases of Aquatic Organisms* 11, 31–35.

Spriggs, D. R., Sherman, M. L., Michie, H., Arthur, K. A., Imamura, K., Wilmore, D., Frei, E., and Kufe, D. W. (1988). Recombinant human tumor necrosis factor administered as a 24–hour intravenous infusion. A phase I and pharmacologic study. *Journal of the National Cancer Institute 80*, 1039–44.

Stadnyk, A. W. and Gauldie, J. (1991). The acute phase protein response during parasitic infection. *Immunology Today* 12, A7–12.

Stairs, G. R. (1965). Artificial initiation of virus epizootics in forest tent caterpillar populations. *Canadian Entomologist* 97, 1059–62.

Stambaugh, J. E. and McDermott, J. J. (1969). The effects of trematode larvae on the locomotion of naturally infected *Nassarius obsoletus* (Gastropoda). *Proceedings of the Pennsylvania Academy of Science* 43, 226–31.

Stamp, N. E. (1981). Behavior of parasitized aposematic caterpillars: advantageous to the parasitoid or the host? *American Naturalist* 118, 715–25.

Stamp, N. E. (1982). Behavioral interactions of parasitoids and Baltimore checkerspot caterpillars (*Euphydryas phaeton*). *Environmental Entomology* 11, 100–4.

Stamp, N. E. (1984). Interactions of parasitoids and checkerspot caterpillars *Euphydryas* spp. (Nymphalidae). *Journal of Research on the Lepidoptera* 23, 2–18.

Stark, G. T. C. (1965). *Diplocotyle* (Eucestoda), a parasite of *Gammarus zaddachi* in the estuary of the Yorkshire Esk, Britain. *Parasitology* 55, 415–20.

Stark, K. R. and A. A. James. (1996). Anticoagulants in rector arthropods. *Parasitology Today* 12, 430–437.

Steck, F. (1982). Rabies in wildlife. *Symposium of the Zoological Society of London* 50, 57–75.

Steck, F. and Wandeler, A. (1980). The epidemiology of fox rabies in Europe. *Epidemiologic Reviews* 2, 71–96.

Stibbs, H. H. (1984). Neurochemical and activity changes in rats infected with *Trypanosoma brucei gambiense*. *Journal of Parasitology* 70, 428–32.

Stibbs, H. H. and Curtis, D. A. (1987). Neurochemical changes in experimental African trypanosomiasis in voles and mice. *Annals of Tropical Medicine and Parasitology* 81, 673–79.

Stirewalt, M. A. 1954. Effect of snail maintenance temperatures on development of *Schistosoma mansoni*. *Experimental Parasitology* 3, 504–16.

Stone, W. and Smith, F. W. (1973). Infection of mammalian hosts by milk-borne nematode larvae: a review. *Experimental Parasitology* 34, 306–12.

Strangways-Dixon, J. and Lainson, R. (1966). The epidemiology of dermal leishmaniasis in British Honduras. Part III. The transmission of *Leishmania mexicana* to man by *Phlebotomus pessoanus*, with observations on the development of the parasite in different species of *Phlebotomus*. *Transactions of the Royal Society of Tropical Medicine and Hygiene* 60, 192–201.

Stretch, R. G. A., Leytham, G. W. H., and Kershaw, W. E. (1960a). The effect of acute schistosomiasis upon learning in rats under different levels of motivation. *Annals of Tropical Medicine and Parasitology* 54, 487–92.

Stretch, S. J. E., Stretch, R. G. A., Leytham, G. W. H., and Kershaw, W. E. (1960b). Discrimination learning and schistosomiasis in the rat: a low-grade infection of several months' duration. *Annals of Tropical Medicine and Parasitology* 54, 483–6.

Stretch, R. G. A., Stretch, S. J. E., Leytham, G. W. H., and Kershaw, W. E. (1960c). The effect of schistosomiasis upon discrimination learning and activity in mice. I. An acute infection. *Annals of Tropical Medicine and Parasitology* 54, 376–80.

Stuart, R. J. and Alloway, T. M. (1988). Aberrant yellow ants: North American *Leptothorax* species as intermediate hosts of cestodes. In *Advances in Myrmecology* (ed. J. C. Trager), pp. 537–45. E. J. Brill, London.

Sukhdeo, M. V. K. (1990). Habitat selection by helminths: a hypothesis. *Parasitology Today* 6, 234–35.

Sukhdeo, M. V. K. (1997). Earth's third environment: the worm's eye view. *BioScience* 47, 141–49.

Sukhdeo, M. V. K. and Bansemir, A. D. (1996). Critical resources that influence habitat selection decisions by gastrointestinal helminth parasites. *International Journal for Parasitology* 26, 483–98.

Sukhdeo, M. V. K. and Sukhdeo, S. C. (1994). Optimal habitat selection by helminths within the host environment. *Parasitology* 109, S41–55.

Sukhdeo, S. C., Sukhdeo, M. V. K., Black, M. B., and Vrijenhoek, R. C. (1997). The evolution of tissue migration in parasitic nematodes (Nematoda: Strongylida) inferred from a protein-coding mitochondrial gene. *Biological Journal of the Linnean Society 61*, 281–98.

Sutherst, R. W., Floyd, R. B., Bourne, A. S.,and Dallwitz, M. J. (1986). Cattle grazing behavior regulates tick populations. *Experientia* 42, 194–96.

Sweeting, R. A. (1976). Studies on *Ligula intestinalis* (L.) effects on a roach population in a gravel pit. *Journal of Fish Biology* 9, 515–22.

Sweeting, R. A. (1977). Studies on *Ligula intestinalis*: some aspects of the pathology in the second intermediate host. *Journal of Fish Biology* 10, 43–50.

Swennen, C. (1969). Crawling-tracks of trematode infected *Macoma balthica* (L.). *Netherlands Journal of Sea Research* 4, 376–79.

Swennen, C. and Ching, H. L. (1974). Observations on the trematode *Parvatrema affinis*, causative agent of crawling tracks of *Macoma balthica*. *Netherlands Journal of Sea Research* 8, 108–115.

Szidat, L. (1969). Structure, development, and behaviour of new strigeatoid metacercariae from subtropical fishes of South America. *Journal of the Fisheries Research Board of Canada* 26, 753–86.

Takaoka, H., Ochoa, J. O., Juarex, E. L., and Hansen, K. M. (1982). Effects of temperature on development of *Onchocerca volvulus* in *Simulium ochraceum*, and longevity of the simuliid vector. *Journal of Parasitology* 68, 478–83.

Takasaki, H. and Hunt, K. (1987). Further medicinal plant consumption in wild chimpanzees? *African Study Monographs* 8, 125–28.

Takken, W. (1991). The role of olfaction in host-seeking of mosquitoes: a review. *Insect Science and Its Application* 12, 287–95.

Tallmark, B. and Norrgren, G. (1976). The influence of parasitic trematodes on the ecology of *Nassarius reticulatus* (L.) in Gullmar Fjord (Sweden). *Zoon* 4, 149–54.

Tashiro, H. and Schwardt, H. H. (1953). Biological studies of horse flies in New York. *Journal of Economic Entomology* 46, 813–22.

Taylor, E. L. 1954. Grazing behaviour and helminthic disease. *British Journal of Animal Behaviour* 2, 61–62.

Théron, A. (1984). Early and late shedding patterns of *Schistosoma mansoni* cercariae: ecological significance in transmission to human and murine hosts. *Journal of Parasitology* 70, 652–55.

Théron, A. (1985). Polymorphisme du rythme d'émission des cercaires de *Schistosoma mansoni* et ses relations avec l'écologie de la transmission du parasite. *Vie et Milieu* 35, 23–31.

Théron, A. (1989). Hybrids between *Schistosoma mansoni* and *S. rodhaini*: characterization by cercarial emergence rhythms. *Parasitology* 99, 225–28.

Théron, A. and Combes, C. (1995). Asynchrony of infection timing, habitat preference, and sympatric speciation of schistosome parasites. *Evolution* 49, 372–375.

Théron, A., Bremond, P., and Imbert-Establet, D. (1989). Allelic frequency variations at the MDH-1 locus within *Schistosoma mansoni* strains from Guadeloupe (French West Indies): ecological interpretation. *Comparative Biochemistry and Physiology* 93B, 33–37.

Théron, A. and Combes, C. (1988). Genetic analysis of cercarial emergence rhythms of *Schistosoma mansoni*. *Behavior Genetics* 18, 201–9.

Thomas, F., Mete, K., Helluy, S., Santalla, F., Verneau, O., De Meeüs, T., Cezilly, F., and Renaud, F. (1997). Hitch-hiker parasites or how to benefit from the strategy of another parasite. *Evolution* 51, 1316–18.

Thomas, F. and Poulin, R. (1998). Manipulation of a mollusc by a trophically transmitted parasite: convergent evolution or phylogenetic inheritance? *Parasitology* 116, 431–36.

Thomas, F., Poulin, R., de Meeüs, T., Guegan, J-F. And F. Renaud. (1999). Parasites and ecosystem engineering: what roles could they play? *Oikos* 84, 167–171.

Thomas, F., Poulin, R., and Renaud, F. (1998a). Nonmanipulative parasites in manipulated hosts: 'hitch-hikers' or simply 'lucky passengers'? *Journal of Parasitology* 84, 1059–61.

Thomas, F., Renaud, F., and Cézilly, F. (1996a). Assortative pairing by parasitic prevalence in *Gammarus insensibilis* (Amphipoda): patterns and processes. *Animal Behaviour* 52, 683–90.

Thomas, F., Renaud, F., Derothe, J. M., Lambert, A., De Meeüs, T., and Cézilly, F. (1995). Assortative pairing in *Gammarus insensibilis* (Amphipoda) infected by a trematode parasite. *Oecologia* 104, 259–64.

Thomas, F., Renaud, F., and Poulin, R. (1998b). Exploitation of manipulators: 'hitch-hiking' as a parasite transmission strategy. *Animal Behaviour* 56, 199–206.

Thomas, F., Verneau, O., Santalla, F., Cézilly, F., and Renaud, F. (1996b). The influence of intensity of infection by a trematode parasite on the reproductive biology of *Gammarus insensibilis* (Amphipoda). *International Journal for Parasitology* 26, 1205–9.

Thomas, R. E. (1991). *Yersinia pestis* virulence characters and transmission events in the ecology of plague. Does *Yersinia pestis* moderate its own transmission? *Bulletin of the Society for Vector Ecology* 16, 43–9.

Thomas, R. E. (1996). Fleas and the agents they transmit. In *The biology of disease vectors* (eds. B. J. Beaty and W. C. Marquardt), pp. 146–59. University Press of Colorado, Niwot, Colorado.

Thomas, R. E., Karstens, R. H., and Schwan, T. G. (1993). Effect of *Yersinia pestis* infection on temperature preference and movement of the Oriental rat flea (*Xenopsylla cheopis*) (Siphonaptera: Pulicidae). *Journal of Medical Entomology* 30, 209–13.

Thompson, S. N. (1982a). Effects of parasitization by the insect parasite *Hyposoter exiguae* on the growth, development and physiology of its host *Trichoplusia ni*. *Parasitology* 84, 491–510.

Thompson, S. N. (1982b). Immediate effects of parasitization by the insect parasite, *Hyposoter exiguae* on the nutritional physiology of its host, *Trichoplusia ni*. *Journal of Parasitology* 68, 936–41.

Thompson, S. N. (1983). The nutritional physiology of *Trichoplusia ni* parasitized by the insect parasite, *Hyposoter exiguae*, and the effects of parallel feeding. *Parasitology* 87, 15–28.

Thompson, S. N. (1985). Metabolic integration during the host associations of multicellular animal parasites. *Comparative Biochemistry and Physiology* 81B, 21–42.

Thompson, S. N. and Kavaliers, M. (1994). Physiological bases for parasite-induced alterations of host behaviour. *Parasitology* 109, S119–38.

Thong, C. H. S. and Webster, J. M. (1975). Effects of the bark beetle nematode, *Contortylenchus reversus*, on gallery construction, fecundity, and egg viability of the douglas fir beetle, *Dendroctonus pseudotsugae* (Coleoptera: Scolytidae). *Journal of Invertebrate Pathology* 26, 235–38.

Tierney, J. F., Huntingford, F. A., and Crompton, D. W. T. (1993). The relationship between infectivity of *Schistocephalus solidus* (Cestoda) and anti-predator behaviour of its intermediate host, the three-spined stickleback, *Gasterosteus aculeatus*. *Animal Behaviour* 46, 603–5.

Tiner, J. D. 1954. The proportion of *Peromyscus leucopus* fatalities caused by raccoon ascarid larvae. *Journal of Mammalogy* 35, 589–92.

Tinley, K. L. (1964). Some observations on certain tabanid flies in North-Eastern Zululand (Diptera: Tabanidae). *Proceedings of the Royal Entomological Society of London A* 39, 73–75.

Tinsley, R. C. (1989). The effects of host sex on transmission success. *Parasitology Today* 5, 190–95.

Tinsley, R. C. (1990). Host behaviour and opportunism in parasite life cycles. In *Parasitism and host behaviour* (eds. C. J. Barnard and J. M. Behnke), pp. 158–92. Taylor & Francis, London.

Tinsley, R. C. and Jackson, H. C. (1988). Pulsed transmission of *Pseudodiplorchis americanus* (Monogenea) between desert hosts (*Scaphiopus couchii*). *Parasitology* 97, 437–52.

Titus, R. G. and Ribeiro, J. M. C. (1988). Salivary gland lysates from the sand fly *Lutzomyia longipalpis* enhance *Leishmania* infectivity. *Science* 239, 1306–8.

Titus, R. G. and Ribeiro, J. M. C.(1990). The role of vector saliva in transmission of arthropod-borne disease. *Parasitology Today* 6, 157–60.

Tocque, K. and Tinsley, R. C. (1991). The influence of desert temperature cycles on the reproductive biology of *Pseudodiplorchis americanus* (Monogenea). *Parasitology* 103, 111–20.

Tompkins, D. M. and Begon, M. (1999). Parasites can regulate wildlife populations. *Parasitology Today* 15, 311–13.

Tomlinson, I. (1987). Adaptive and non-adaptive suicide in aphids. *Nature* 330, 701.

Towers, G. H. N., Abramowski, Z., Finlayson, A. J., and Zucconi, A. (1985). Antibiotic properties of thiarubrine-A, a naturally occurring dithiacyclohedadiene polyine. *Planta Medica* 3, 225–29.

Townson, H. (1970). The effect of infection with *Brugia pahangi* on the flight of *Aedes aegypti*. *Annals of Tropical Medicine and Parasitology* 64, 411–20.

Trabalon, M., Plateaux, L., and Bagneres, A. G. (1994). Variation des hydrocarbures cuticulaires chez des ouvrières de Fourmi *Leptothorax nylanderi* parasitees. *Actes Colloques Insectes Sociaux* 9, 1–8.

Trabalon, M., Plateaux, L., Péru, L., Bagneres, A. G., and Hartmann, N. (2000). Modification of morphological characters and cuticular compounds in worker ants *Leptothorax nylanderi* induced by endoparasites *Anomotaenia brevis*. *Journal of Insect Physiology* 46, 169–178.

Tripet, F. and Richner, H. (1997). Host responses to ectoparasites: food compensation by parent blue tits. *Oikos* 78, 557–61.

Turell, M. J., Bailey, C. L., and Rossi, C. A. (1984). Increased mosquito feeding on Rift Valley Fever virus-infected lambs. *American Journal of Tropical Medicine and Hygiene* 33, 1232–38.

Turell, M. J., Gargan, T.P., II, and Bailey, C. L. (1985). *Culex pipiens* (Diptera: Culicidae) morbidity and mortality associated with Rift Valley Fever virus infection. *Journal of Medical Entomology* 22, 332–37.

Turell, M. J. and Lundstrom, J. O. (1990). Effect of environmental temperature on the vector competence of *Aedes aegypti* and *Ae. taeniorhynchus* for Ockelbo virus. *American Journal of Tropical Medicine and Hygiene* 43, 543–50.

Turner, P. E., Cooper, V. S., and Lenski, R. E. (1998). Tradeoff between horizontal and vertical modes of transmission in bacterial plasmids. *Evolution* 52, 315–29.

Tyrrell, D. (1990). Pathogenesis of *Entomophaga aulicae*. I. Disease symptoms and effect of infection on weight gain of infected *Choristoneura fumiferana* and *Malacosoma disstria* larvae. *Journal of Invertebrate Pathology* 56, 150–56.

Undeen, A. H. and Nolan, R. A. (1977). Ovarian infection and fungal spore oviposition in the blackfly *Prosimulium mixtum*. *Journal of Invertebrate Pathology* 30, 97–98.

Urdal, K., Tierney, J. F., and Jakobsen, P. J. (1995). The tapeworm *Schistocephalus solidus* alters the activity and response, but not the predation susceptibility of infected copepods. *Journal of Parasitology* 81, 330–33.

Utida, S. (1953). Interspecific competition between two species of bean weevil. *Ecology* 34, 301–7.

Vale, G. A. (1977). Feeding responses of tsetse flies (Diptera: Glossinidae) to stationary hosts. *Bulletin of Entomological Research* 67, 635–49.

van Alphen, J. J. M. and Vet, L. E. M. (1986). An evolutionary approach to host finding and selection. In *Insect Parasitoids* (eds. J. Waage and O. Greathead), pp. 23–61. Academic Press, London.

Vance, S. A. (1996). Morphological and behavioural sex reversal in mermithid-infected mayflies. *Proceedings of the Royal Society of London* B 263, 907–12.

Vance, S. A. and Peckarsky, B. L. (1996). The infection of nymphal *Baetis bicaudatus* by the mermithid nematode *Gasteromermis* sp. *Ecological Entomology* 21, 377–81.

Vanderberg, J. P. and Yoeli, M. (1966). Effects of temperature on sporogonic development of *Plasmodium berghei*. *Journal of Parasitology* 52, 559–64.

van Dobben, W. H. (1952). The food of the cormorant in the Netherlands. *Ardea* 40, 1–63.

van Riper, C., III, van Riper, S. G., Goff, M. L., and Laird, M. (1986). The epizootiology and ecological significance of malaria in Hawaiian land birds. *Ecological Monographs* 56, 327–44.

Vasconcelos, S. D., Cory, J. S., Wilson, K. R., Sait, S. M., and Hails, R. S. (1996). Modified behavior in baculovirus-infected lepidopteran larvae and its impact on the spatial distribution of inoculum. *Biological Control* 7, 299–306.

Vaughan, G. E. and Coble, D. W. (1975). Sublethal effects of three ectoparasites on fish. *Journal of Fish Biology* 7, 283–94.

Vaughn, L. K., Bernheim, H. A., and Kluger, M. J. (1974). Fever in the lizard *Dipsosaurus dorsalis*. *Nature* 252, 473–74.

Vernberg, W. B. (1969). Adaptations of host and symbionts in the intertidal zone. *American Zoologist* 9, 357–65.

Vernberg, W. B. and Vernberg, F. J. (1963). Influence of parasitism on thermal resistance of the mud-flat snail, *Nassarius obsoleta* Say. *Experimental Parasitology* 14, 330–32.

Vernberg, W. B. and Vernberg, F. J. (1967). Interrelationships between parasites and their hosts. III. Effect of larval trematodes on the thermal metabolic response of their molluscan host. *Experimental Parasitology* 20, 225–31.

Vernberg, W. B. and Vernberg, F. J. (1968). Interrelationships between parasites and their hosts. *Experimental Parasitology* 23, 347–54.

Völkl, W. (1992). Aphids or their parasitoids: who actually benefits from ant-attendance? *Journal of Animal Ecology* 61, 273–81.

von Brand, T., Weinstein, P. P., and Wright, W. H. (1954). The working ability of rats infected with *Trichinella spiralis*. *American Journal of Hygiene* 69, 26–31.

Vŏrišek, P., Votýpka, J. Zvára, K. and Svobodová, M. (1998). Heteroxenous coccidia increase the predation risk for parasitized rodents. *Parasitology* 117, 521–24.

Waage, J. K. (1979). The evolution of insect/vertebrate associations. *Biological Journal of the Linnean Society* 12, 187–224.

Waage, J. K. (1981). How the zebra got its stripes—biting flies as selective agents in the evolution of zebra coloration. *Journal of the Entomological Society of South Africa* 44, 351–58.

Waage, J. K. and Nondo, J. (1982). Host behaviour and mosquito feeding success: and experimental study. *Transactions of the Royal Society of Tropical Medicine and Hygiene* 76, 119–22.

Wakelin, D. (1993). *Trichinella spiralis*: immunity, ecology and evolution. *Journal of Parasitology* 79, 488–94.

Wakelin, D. (1997). Parasites and the immune system. *BioScience* 47, 32–40.

Wakelin, D. and Apanius, V. (1997). Immune defence: genetic control. In *Host–parasite evolution: general principles and avian models* (eds. D. H. Clayton and J. Moore), pp. 30–58. Oxford University Press, Oxford.

Walker, E. D. and Edman, J. D. (1985a). The influence of host defensive behavior on mosquito (Diptera: Culicidae) biting persistence. *Journal of Medical Entomology* 22, 370–72.

Walker, E. D. and Edman, J. D. (1985b). Feeding-site selection and blood-feeding behavior of *Aedes triseriatus* (Diptera: Culicidae) on rodent (Sciuridae) hosts. *Journal of Medical Entomology* 22, 287–94.

Walker, E. D. and Edman, J. D. (1986). Influence of defensive behavior of eastern chipmunks and gray squirrels (Rodentia: Sciuridae) on feeding success of *Aedes triseriatus* (Diptera: Culicidae). *Journal of Medical Entomology* 23, 1–10.

Walkey, M. and Meakins, R. H. (1970). An attempt to balance the energy budget of a host-parasite system. *Journal of Fish Biology* 2, 361–72.

Wallace, R. L., Ricci, C. and Malone, G. (1996). A cladistic analysis of pseudocoelomate (aschelhelminthes) morphology. *Invertebrate Biology* 115, 104–112.

Wang, Der-I and Moeller, F. E. (1970). The division of labor and queen attendance behavior of nosema-infected worker honey bees. *Journal of Economic Entomology* 63, 1539–41.

Ward, H. B. (1916). Notes on two free-living larval trematodes from North America. *Journal of Parasitology* 3, 10–19.

Wardle, W. J. (1988). A bucephalid larva, *Cercaria pleuromerae* n. sp. (Trematoda: Digenea), parasitizing a deepwater bivalve from the Gulf of Mexico. *Journal of Parasitology* 74, 692–94.

Warnes, M. L. and Finlayson, L. H. (1987). Effect of host behaviour on host preference in *Stomoxys calcitrans*. *Medical and Veterinary Entomology* 1, 53–57.

Washburn, J. O., Gross, M. E., Mercer, D. R., and Anderson, J. R. (1988). Predator-induced trophic shift of a free-living ciliate: parasitism of mosquito larvae and their prey. *Science* 240, 1193–95.

Watanabe, H. (1987). The host population. In *Epizootiology of insect diseases* (eds. J. R. Fuxa and Y. Tanada), pp. 71–112. John Wiley and Sons, New York.

Watson, D. W., Mullens, B. A., and Petersen, J. J. (1993). Behavioral fever response of *Musca domestica* (Diptera: Muscidae) to infection by *Entomophthora muscae* (Zygomycetes: Entomophthorales). *Journal of Invertebrate Pathology* 61, 10–16.

Webb, T. J. and Hurd, H. (1999). Direct manipulation of insect reproduction by agents of parasite origin. *Proceedings of the Royal Society of London B* 266, 1537–41.

Webber, L. A. and Edman, J. D. (1972). Anti-mosquito behaviour of ciconiiform birds. *Animal Behaviour* 20, 228–32.

Webber, R. A., Rau, M. E., and Lewis, D. J. (1987a). The effects of *Plagiorchis noblei* (Trematoda: Plagiorchiidae) metacercariae on the susceptibility of *Aedes aegypti* larvae to predation by guppies (*Peocilia reticulata*) and meadow voles (*Microtus pennsylvanicus*). *Canadian Journal of Zoology* 65, 2346–48.

Webber, R. A., Rau, M. E., and Lewis, D. J. (1987b). The effects of *Plagiorchis noblei* (Trematoda: Plagiorchiidae) metacercariae on the behavior of *Aedes aegypti* larvae. *Canadian Journal of Zoology* 65, 1340–42.

Webster, J. P. (1994). The effect of *Toxoplasma gondii* and other parasites on activity levels in wild and hybrid *Rattus norvegicus*. *Parasitology* 109, 583–89.

Webster, J. P., Brunton, C. F. A., and Macdonald, D. W. (1994). Effect of *Toxoplasma gondii* upon neophobic behaviour in wild brown rats, *Rattus norvegicus*. *Parasitology* 109, 37–43.

Webster, J. P., Gowtage-Sequeria, S., Berdoy, M., and Hurd, H. (2000). Predation of beetles (*Tenebrio molitor*) infected with tapeworms (*Hymenolepis diminuta*): a note of caution for the manipulation hypothesis. *Parasitology* 120, 313–18.

Wecker, S. C. (1962). The effects of bot fly parasitism on a local population of the white-footed mouse. *Ecology* 43, 561–3.

Wedekind, C. (1992). Detailed information about parasites revealed by sexual ornamentation. *Proceedings of the Royal Society of London B* 247, 169–74.

Wedekind, C. (1994). Mate choice and maternal selection for specific parasites resist-
ances before, during and after fertilization. *Philosophical Transactions of the Royal
Society of London B.* 346, 303–11.

Wedekind, C. (1997). The infectivity, growth, and virulence of the cestode *Schistocephalus solidus* in its first intermediate host, the copepod *Macrocyclops albidus*.
Parasitology 115, 317–24.

Wedekind, C. and Folstad, I. (1994). Adaptive or nonadaptive immunosuppression by
sex hormones? *The American Naturalist* 143, 936–38.

Wedekind, C. and Milinski, M. (1996). Do three-spined sticklebacks avoid consuming
copepods, the first intermediate host of *Schistocephalus solidus*?—an experimental analysis of behavioural resistance. *Parasitology* 112, 371–83.

Wedekind, C., Seebeck, T., Bettens, F. and Paepke, A. J. (1995). MHC-dependent mate
preferences in humans. *Proceedings of The Royal Society, London* 260, 245–249.

Weisel-Eichler, A., Haspel, G. and Libersat, F. (1999). Venom of a parasitoid wasp induces prolonged grooming in the cockroach. *The Journal of Experimental Biology*
202, 957–64.

Wekesa, J. W., Copeland, R. S., and Mwangi, R. W. (1992). Effect of *Plasmodium falciparum* on blood feeding behavior of naturally infected *Anopheles* mosquitoes in
western Kenya. *American Journal of Tropical Medicine and Hygiene* 47, 484–88.

Welch, H. E. (1960). *Hydromermis churchillensis* N. sp. (Nematoda: Mermithidae) a
parasite of *Aedes communis* (DeG.) from Churchill, Manitoba, with observations
of its incidence and bionomics. *Canadian Journal of Zoology* 38, 465–74.

Wesenberg-Lund, C. (1931). Contributions to the development of the trematoda digenea. Part I. The biology of *Leucochloridium paradoxum*. *Memoires de l'Academie
Royale des Sciences et des Lettres de Danemark, Copenhague, section des sciences*
4, 90–142.

Wheeler, W. M. (1907). The polymorphism of ants, with an account of some singular
abnormalities due to parasitism. *Bulletin of the American Museum of Natural History* 23, 1–93.

Wheeler, W. M. (1928). *Mermis* parasitism and intercastes among ants. *Journal of Experimental Zoology* 50, 165–237.

White, A., Ganter, P., McFarland, R., Stanton, N., and Lloyd, M. (1983). Spontaneous,
field tested and tethered flight in healthy and infected *Magicicada septendecim* L.
Oecologia 57, 281–86.

Whitlock, J. H., Crofton, H. D., and Georgi, J. R. (1972). Characteristics of parasite
populations in endemic trichostrongylidosis. *Parasitology* 64, 413–27.

Whitlock, V. H. (1974). Symptomatology of two viruses infecting *Heliothis armigera*.
Journal of Experimental Parasitology 23, 70–75.

Wickler, W. (1976). Evolution-oriented ethology, kin selection, and altruistic parasites.
Zeitschrift fur Tierpsychologie 42, 206–14.

Wikel, S. K. (1999). Modulation of the host immune system by ectoparasitic arthropods. *BioScience* 49, 311–20.

Wikelski, M. (1999). Influences of parasites and thermoregulation on grouping tendencies in marine iguanas. *Behavioral Ecology* 10, 22–29.

Williams, C. L. and Gilbertson, D. E. (1983). Altered feeding response as a cause for
the altered heartbeat rate and locomotor activity of *Schistosoma mansoni*-infected
Biomphalaria glabrata. *Journal of Parasitology* 69, 671–76.

Williams, E. E. (1964). The growth and distribution of *Littorina littorea* (L) on a rocky shore in Wales. *Journal of Animal Ecology* 33, 413–32.

Williams, G. C. and Nesse, R. M. (1991). The dawn of Darwinian medicine. *Quarterly Review of Biology* 66, 1–22.

Williams, H. H. (1960). The intestine in members of the genus *Raja* and host-specificity in the Tetraphyllidea. *Nature* 188, 514–16.

Williams, I. C. and Ellis, C. (1975). Movements of the common periwinkle, *Littorina littorea* (L.) On the Yorkshire coast in winter and the influence of infection with larval digenea. *Journal of Experimental Marine Biology and Ecology* 17, 47–58.

Willis, C. and Poulin, P. (1999). Effects of the tapeworm *Hymenolepis diminuta* on maternal investment in rats. *Canadian Journal of Zoology* 77, 1001–25.

Wilson, D. S. (1977). How nepotistic is the brain worm? *Behavioral Ecology and Sociobiology* 2, 421–25.

Wilson, E. O. (1975). *Sociobiology: the new synthesis*. Harvard University Press, Cambridge.

Wilson, K. and Edwards, J. (1986). The effects of parasitic infection on the behaviour of an intermediate host, the American cockroach, *Periplaneta americana*, infected with the acanthocephalan, *Moniliformis moniliformis*. *Animal Behaviour* 34, 942–44.

Wimberger, P. H. (1984). The use of green plant material in bird nests to avoid ectoparasitism. *Auk* 101, 615–18.

Winnepenninckx, B., Backeljau, T., Mackey, L. Y., Brooks, J. M., DeWachter, R., Kumar, S., and Garey, J. R. (1995). 18s rRNA data indicate that aschelminthes are polyphyletic in origin and consist of at least three distinct clades. *Molecular Biology and Evolution* 12, 1132–37.

Woodhead, A. E. (1935). The mother sporocysts of *Leucochloridium*. *Journal of Parasitology* 21, 337–46.

Woodrow, A. W. and Holst, E. C. (1942). The mechanism of colony resistance to American foulbrood. *Journal of Economic Entomology* 35, 327–30.

Worden, B. D., Parker, P. G. and Pappas, P. W. (2000). Parasites reduce attractiveness and reproductive success in male grain beetles. *Animal Behaviour* 59, 543–50.

Wrangham, R. W. (1995). Relationship of chimpanzee leaf-swallowing to a tapeworm infection. *American Journal of Primatology* 37, 297–303.

Wrangham, R. W. and Goodall, J. (1989). Chimpanzee use of medicinal leaves. In *Understanding chimpanzees* (ed. P. G. Heltne and L. A. Marquardt), pp. 22–37. Harvard University Press, Cambridge, MA.

Wrangham, R. W. and T. Nishida. (1983). *Aspilia* spp. leaves: a puzzle in the feeding behavior of wild primates. *Primates* 24, 276–282.

Wülker, W. (1964). Parasite-induced changes of internal and external sex characters in insects. *Experimental Parasitology* 15, 561–97.

Wülker, W. (1985). Changes in behaviour, flight tone and wing shape in nematode-infested *Chironomus* (Insecta, Diptera). *Zeitschrift fur Parasitenkunde* 71, 409–18.

Würtz, J., Taraschewski, H. and Pelster, B. (1996). Changes in gas composition in the swimbladder of the European eel (*Anguilla anguilla*) infected with *Anguillicola crassus* (Nematoda). *Parasitology* 112, 233–38.

Wygant, N. D. (1941). An infestation of the Pandora moth, *Coloradia pandora* Blake, on lodgepole pine in Colorado. *Journal of Economic Entomology* 34, 697–702.

Yan, G. and Phillips, T. W. (1996). Influence of tapeworm infection on the production of aggregation pheromone and defensive compounds in *Tribolium castaneum*. *Journal of Parasitology* 82, 1037–39.

Yan, G., Severson, D. W., and Christensen, B. M. (1997). Costs and benefits of mosquito refractoriness to malaria parasites: implications for genetic variability of mosquitoes and genetic control of malaria. *Evolution* 51, 441–50.

Yan, G., Stevens, L., and Schall, J. J. (1994). Behavioral changes in *Tribolium* beetles infected with a tapeworm: variation in effects between beetle species and among genetic strains. *American Naturalist* 143, 830–47.

Yan, G., Stevens, L., Goodnight, C. J., and Schall, J. J. (1998). Effects of a tapeworm parasite on the competition of *Tribolium* beetles. *Ecology* 79, 1093–1103.

Yeboah, D. O., Undeen, A. H., and Colbo, M. H. (1984). Phycomycetes parasitizing the ovaries of blackflies (Simuliidae). *Journal of Invertebrate Pathology* 43, 363–73.

Yen, D. F. (1962). An *Entomophthora* infection in the larva of the tiger moth, *Creatonotus gangis* (Linnaeus). *Journal of Insect Pathology* 4, 88–94.

Yendol, W. G. and Paschke, J. D. (1967). Infection of a looper complex by *Entomophthora sphaerosperma*. *Journal of Invertebrate Pathology* 9, 274–76.

Yezerinac, S. M. and Weatherhead, P. J. (1995). Plumage coloration, differential attraction of vectors and haematozoa infections in birds. *Journal of Animal Ecology* 64, 528–37.

Yoshinaga, T., Nagakura, T., Ogawa, K. and Wakabayashi, H. (2000). Attachment-inducing capacities of fish tissue extracts on oncomiracidia of *Neobenedenia girellae* (Monogenea, Capsalidae). *Journal of Parasitology* 86, 214–19.

Young, S. Y. and Yearian, W. C. (1989). Nuclear polyhedrosis virus-infected and healthy *Anticarsia gemmatalis* larvae as prey for *Nabis roseipennis* adults in the laboratory. *Journal of Invertebrate Pathology* 54, 139–43.

Yuhl, D. E., Burright, R. G., Donovick, P. J., and Cypess, R.H. (1985). Behavioral effects of early lead exposure and subsequent toxocariasis in mice. *Journal of Toxicology and Environmental Health* 16, 315–21.

Yuill, T. M. (1987). Diseases as components of mammalian ecosystems: mayhem and subtlety. *Canadian Journal of Zoology* 65, 1061–66.

Zakikhani, M. and Rau, M. E. (1999). *Plagiorchis elegans* (Digenea: Plagiorchiidae) infections in *Stagnicola elodes* (Pulmonata: Lymnaeidae): host susceptibility, growth, reproduction, mortality, and cercarial production. *Journal of Parasitology* 85, 454–63.

Zelmer, D. A. and Esch, G. W. (1998). Interactions between *Halipegus occidualis* and its ostracod second intermediate host: evidence for castration? *Journal of Parasitology* 84, 773–82.

Zielke, E. (1976). Studies on quantitative aspects of the transmission of *Wuchereria bancrofti*. *Tropenmedizin und Parasitologie* 27, 160–64.

Zinsser, H. (1963). *Rats, lice and history.* Little, Brown & Co., Boston.

Zohar, A. S. (1993). The effects of two acanthocephalans, *Polymorphus paradoxus* and *P. marilis,* on the reproduction of their intermediate host, *Gammarus lacustris* (Crustacea). Ph.D. thesis, University of Alberta, Edmonton.

Zohar, A. S. and Holmes, J. C. (1998). Pairing success of male *Gammarus lacustris* infected by two acanthocephalans: a comparative study. *Behavioral Ecology* 9, 206–11.

Zohar, A. S. and Rau, M. E. (1986). The role of muscle larvae of *Trichinella spiralis* in the behavioral alterations of the mouse host. *Journal of Parasitology* 72, 464–66.

Zuk, M. (1992). The role of parasites in sexual selection: current evidence and future directions. *Advances in the Study of Behavior* 21, 39–68.

Zuk, M., Kim, T., Kristan, D. M., and Luong, L. T. (1997). Sex, pain and parasites. *Parasitology Today* 13, 332–33.

Zuk, M., Kim, T., Robinson, S. I., and Johnsen, T. S. (1998b). Parasites influence social rank and morphology, but not mate choice, in female red junglefowl, *Gallus gallus. Animal Behaviour* 56, 493–99.

Zuk, M., Rotenberry, J. T., and Simmons, L. W. (1998a). Calling songs of field crickets (*Teleogryllus oceanicus*) with and without phonotactic parasitoid infection. *Evolution* 52,166–71.

Zuk, M., Simmons, L. W., and Cupp, L. (1993). Calling characteristics of parasitized and unparasitized populations of the field cricket *Teleogryllus oceanicus. Behavioral Ecology and Sociobiology* 33, 339–43.

Zuk, M., Simmons, L. W., and Rotenberry, J. T. (1995). Acoustically-orienting parasitoids in calling and silent males of the field cricket *Teleogryllus oceanicus. Ecological Entomology* 20, 380–83.

Species Index

Subject Index

Printed in the United States
18550LVS00002B/31-40